EDEXCEL A LEVEL

CHEMISTRY

2

Graham Curtis
Andrew Hunt
Graham Hill

HODDER
EDUCATION
AN HACHETTE UK COMPANY

Photo credits: p. 1 yodiyim/Fotolia; **p. 8** FOOD-micro/Fotolia; **p. 12** Maximilian Stock Ltd/Science Photo Library; **p. 13** Andrew Lambert Photography/Science Photo Library; **p. 16** *tl* Charles D. Winters/ Science Photo Library, *c* Manfred Kage/Science Photo Library; **p. 20** *tl* toa555/Fotolia, *c* Science Photo Library; **p. 25** Charles D. Winters/Science Photo Library; **p. 29** nenetus/Fotolia; **p. 34** Andrew Lambert Photography/Science Photo Library; **p. 38** Andrew Lambert Photography/Science Photo Library (*all*); **p. 40** gabriffaldi/Fotolia; **p. 42** JPC-PROD/Fotolia; **p. 44** Sebastian Kaulitzki/Fotolia; **p. 50** Andrew Lambert Photography/Science Photo Library; **p. 61** Vit Kovalcik/Fotolia; **p. 67** *cl* maros_bauer/ Fotolia, *cr* ju_skz/Fotolia, *bl* ova/Fotolia; **p. 68** *t* Anna Khomulo/Fotolia, *c* Ghen/Fotolia, *b* Science Photo Library; **p. 69** Charles D. Winters/Science Photo Library (*both*); **p. 71** WavebreakmediaMicro/ Fotolia; **p. 74** kalpis/Fotolia; **p. 81** *c* Fuse/Thinkstock, *r* Mark Williamson/Science Photo Library; **p. 82** marcel/Fotolia; **p. 86** Stocktrek Images/Getty images; **p. 91** Andrew Lambert Photography/Science Photo Library (*both*); **p. 95** Martyn F. Chillmaid/Science Photo Library; **p. 101** design56/Fotolia; **p. 103** *tr* Andrew Lambert Photography/Science Photo Library, *br* Peticolas/Megna/Fundamental Photos/Science Photo Library; **p. 109** mikanaka/Thinkstock; **p. 116** Andrew Lambert Photography/ Science Photo Library; **p. 121** sumnersgraphicsinc/Fotolia; **p. 123** Andrew Lambert Photography/ Science Photo Library; **p. 124** Andrew Lambert Photography/Science Photo Library; **p. 128** Andrew Lambert Photography/Science Photo Library; **p. 133** Interfoto/Alamy; **p. 134** Science Photo Library; **p. 141** Biosym Technologies, Inc./Science Photo Library; **p. 147** Kadmy/Fotolia; **p. 169** *l* mosinmax/ Fotolia, *r* atoss/Fotolia; **p. 170** full image/Fotolia; **p. 172** *c* indigolotos/Fotolia, *b* James Watson; **p. 180** Steve Gschmeissner/Science Photo Library; **p. 182** Vesna Cvorovic/Fotolia; **p. 188** Andrew Lambert Photography/Science Photo Library; **p. 189** Andrew Lambert Photography/Science Photo Library (*both*); **p. 194** *l* skynet/Fotolia, *r* Debu55y/Fotolia; **p. 195** Susan Wilkinson; **p. 197** xeni4ka/ Thinkstock; **p. 201** Andrew Lambert Photography/Science Photo Library; **p. 207** *tr* Philippe Hallé/ Thinkstock, *br* Sally and Richard Greenhill/Alamy; **p. 212** Corbis Super RF/Alamy; **p. 219** Andrew Lambert Photography/Science Photo Library; **p. 221** sashagrunge/Fotolia; **p. 226** Martyn F. Chillmaid/ Science Photo Library; **p. 233** WavebreakmediaMicro/Fotolia; **p. 235** *tr* Mediablitzimages/Alamy, *br* Nomadsoul1/Thinkstock; **p. 237** Martyn F. Chillmaid/Science Photo Library; **p. 250** *tl* Charles D. Winters/Science Photo Library, *cl* Jeff Morgan 09/Alamy; **p. 252** Ashley Cooper/Corbis; **p. 254** Eye of Science/Science Photo Library; **p. 259** James Bell/Science Photo Library; **p. 260** Steffen Hauser/ botanikfoto/Alamy; **p. 273** Peggy Greb/US Department of Agriculture/Science Photo Library; **p. 274** Phototake Inc./Alamy; **p. 283** Geoff Tompkinson/Science Photo Library; **p. 292** Food Collection/ Alamy; **p. 297** Vince Bevan/Alamy; **p. 298** Roger Hutchings/Alamy; **p. 304** Jerry Mason/Science Photo Library; **p. 306** Michael Donne/Science Photo Library; **p. 308** l ESA/ATG medialab, *r* STFC

b = bottom, *c* = centre, *l* = left, *r* = right

Acknowledgement

Data used for the proton NMR spectrum in Figure 19.21 and the two IR spectra on page 314 come from the SDBS, National Institute of Advanced Industrial Science and Technology, Japan.

Although every effort has been made to ensure that website addresses are correct at time of going to press, Hodder Education cannot be held responsible for the content of any website mentioned in this book. It is sometimes possible to find a relocated web page by typing in the address of the home page for a website in the URL window of your browser.

Hachette UK's policy is to use papers that are natural, renewable and recyclable products and made from wood grown in well-managed forests and other controlled sources. The logging and manufacturing processes are expected to conform to the environmental regulations of the country of origin.

Orders: please contact Hachette UK Distribution, Hely Hutchinson Centre, Milton Road, Didcot, Oxfordshire, OX11 7HH. Telephone: +44 (0)1235 827827. Email education@hachette.co.uk Lines are open from 9 a.m. to 5 p.m., Monday to Friday. You can also order through our website: www.hoddereducation.co.uk

Cover photo © Gina Sanders – Fotolia

Illustrations by Aptara, Inc.

Typeset in 11/13 pt Bembo by Aptara, Inc.

Printed by CPI Group (UK) Ltd, Croydon CR0 4YY

A catalogue record for this title is available from the British Library

ISBN 9781471807497

Contents

Get the most from this book

Welcome to the **Edexcel A level Chemistry 2 Student's Book**! This book covers Year 2 of the Edexcel A level Chemistry specification.

The following features have been included to help you get the most from this book.

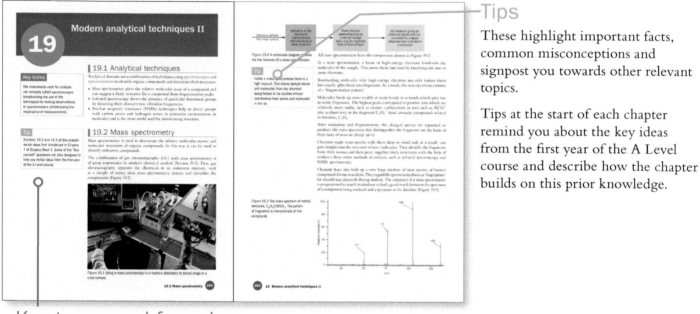

Tips

These highlight important facts, common misconceptions and signpost you towards other relevant topics.

Tips at the start of each chapter remind you about the key ideas from the first year of the A Level course and describe how the chapter builds on this prior knowledge.

Key terms and formulae

These are highlighted in the text and definitions are given in the margin to help you pick out and learn these important concepts.

Test yourself questions

These short questions, found throughout each chapter, are useful for checking your understanding as you progress through a topic.

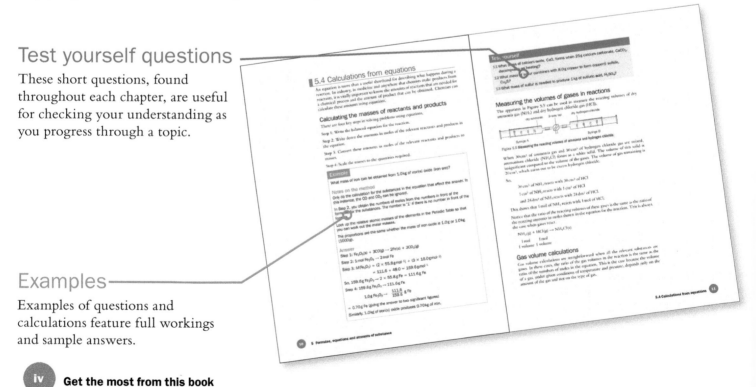

Examples

Examples of questions and calculations feature full workings and sample answers.

Activities and Core practicals

These practical-based activities will help consolidate your learning and test your practical skills. Edexcel's Core practicals are clearly highlighted.

In this edition the authors describe many important experimental procedures to conform to recent changes in the A Level curriculum. Teachers should be aware that, although there is enough information to inform students of techniques and many observations for exam purposes, there is not enough information for teachers to replicate the experiments themselves, or with students, without recourse to CLEAPSS Hazcards or Laboratory worksheets which have undergone a risk assessment procedure.

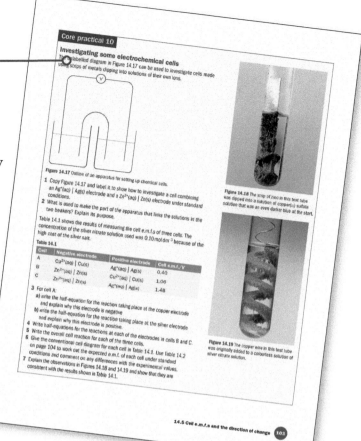

Exam practice questions

You will find Exam practice questions at the end of every chapter. These follow the style of the different types of questions you might see in your examination and are colour coded to highlight the level of difficulty.

In general, a question with ● is straightforward and based directly on the information, ideas and methods described in the chapter. Each problem-solving part of the question typically only involves one step in the argument or calculation. A question with ● is a more demanding, but still structured, question involving the application of ideas and methods to solve a problem with the help of data or information from this chapter or elsewhere. Arguments and calculations typically involve more than one step. The questions marked by ● are hard and they may well expect you to bring together ideas from different areas of the subject. In these harder questions you may have to structure an argument or work out the steps required to solve a problem. In the earlier chapters, you may well decide not attempt the questions with ● until you have gained wider experience and knowledge of the subject.

Introduction

This book is an extensively revised, restructured and updated version of *Edexcel Chemistry for A2* by Graham Hill and Andrew Hunt. We have relied heavily on the contributions that Graham Hill made to the original book and are most grateful that he has encouraged us to build on his work. We also thank these teachers who contributed authentic student data as a basis for activities and Core Practicals: Christopher Buckley (The Manchester Grammar School) and Michael Yates (Bolton School). The team at Hodder Education, led initially by Hanneke Remsing and then by Emma Braithwaite, has made an extremely valuable contribution to the development of this book. In particular, we would like to thank Abigail Woodman, the project manager, for her expert advice and encouragement. We are also grateful for the skilful work on the print and electronic resources by Anne Trevillion.

Practical work is of particular importance in A Level chemistry. Each of the Core Practicals in the specification features in the main chapters of this book with an outline of the procedure and data for you to analyse and interpret. Throughout the text there are references to Practical skills sheets which can be accessed via www.hoddereducation.co.uk/EdexcelAChemistry2. Sheets 1 to 5 provide general guidance and the remainder provide more detailed guidance for the Core Practicals:

1 Assessment of practical work

2 Overview of practical skills

3 Assessing hazards and risks

4 Researching and referencing

5 Identifying errors and estimating uncertainties

6 Finding the K_a value for a weak acid

7 Measuring chemical amounts by titration

8 Investigating reaction orders and activation energies

9 Analysing inorganic unknowns

10 Analysing organic unknowns

11 Synthesis of an organic solid

You will need to refer to the Edexcel Data booklet when answering some of the questions in this book. It is important that you become familiar with the booklet because you will need to use it in the examinations. You can download the Data booklet from the Edexcel website. It is part of the specification. The booklet includes the version of the periodic table that you use in the examinations.

Andrew Hunt and Graham Curtis

May 2015

Equilibrium II

11

11.1 Reversible reactions and dynamic equilibrium

All chemical reactions tend towards a state of dynamic equilibrium. An understanding of equilibrium ideas helps to explain changes in the natural environment, the biochemistry of living things and the conditions used in the chemical industry to manufacture new products (Figure 11.1).

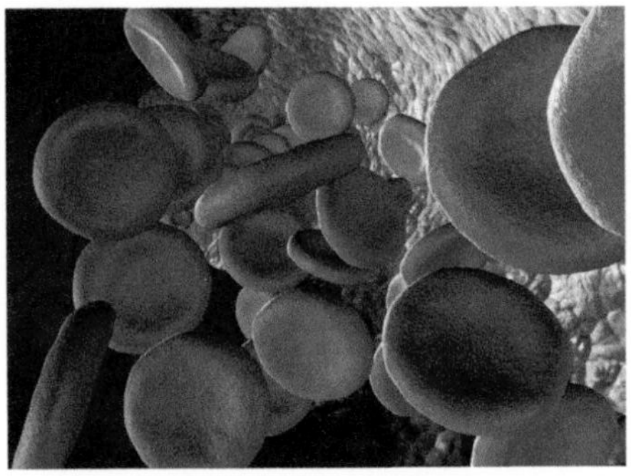

> **Tip**
>
> The first two sections of this chapter, and parts of Section 11.5, revisit ideas first introduced in Chapter 10 of Student Book 1. In Chapter 10 the treatment of equilibrium was qualitative. Chapter 11 builds on what you already know and shows you how to apply the equilibrium law quantitatively. Some of the 'Test yourself' questions are also designed to help revise ideas from the first year of the A Level course.

Reversible reactions reach equilibrium when neither the forward change nor the backward change is complete, but both changes are still going on at equal rates. They cancel each other out and there is no overall change. This is **dynamic equilibrium** (see Chapter 10 in Student Book 1).

Under given conditions the same equilibrium state can be reached either by starting with the chemicals on one side of the equation for a reaction or by starting with the chemicals on the other side. Figures 11.2 and 11.3 illustrate this for the reversible reaction between hydrogen and iodine:

$$H_2(g) + I_2(g) \rightleftharpoons 2HI(g)$$

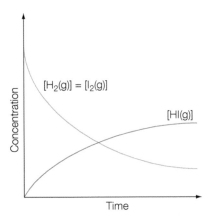

Figure 11.1 Red blood cells flowing through a blood vessel magnified ×3000. The protein haemoglobin has just the right properties to take up oxygen in the lungs and release it to cells throughout the body. The position of equilibrium of this reversible process varies with the concentration of oxygen.

> **Key term**
>
> In **dynamic equilibrium** the forward and backward changes continue at equal rates so that overall there is no change. At the molecular level there is continuous activity. At the macroscopic level nothing appears to be happening.

Figure 11.2 Reaching an equilibrium state by the reaction of equal amounts of hydrogen gas and iodine gas.

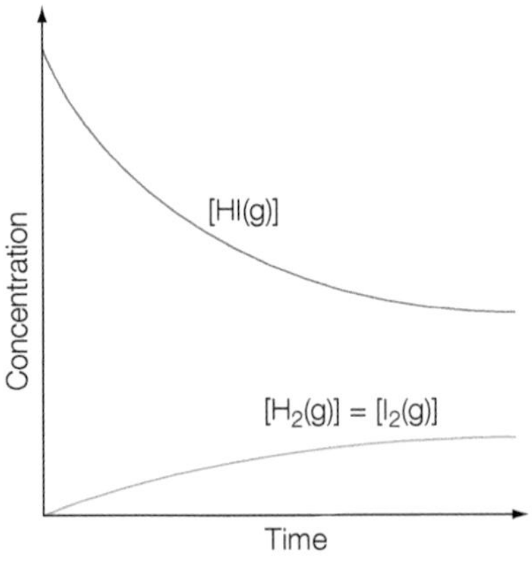

Figure 11.3 Reaching the same equilibrium state by decomposing HI(g) under the same conditions as for Figure 11.2.

Activity

Testing the equilibrium law

The reversible reaction involving hydrogen, iodine and hydrogen iodide has been used to test the equilibrium law experimentally. In a series of six experiments, samples of the chemicals were sealed in reaction tubes and then heated at 731 K until the mixtures reached equilibrium. Four of the tubes started with different mixtures of hydrogen and iodine. Two of the tubes started with just hydrogen iodide.

Once the tubes had reached equilibrium they were rapidly cooled to stop the reactions. Then the contents of the tubes were analysed to find the compositions of the equilibrium mixture. The results for the six tubes are shown in Table 11.1.

Table 11.1

Tube	Initial concentrations/10^{-2} mol dm^{-3}			Equilibrium concentrations/10^{-2} mol dm^{-3}		
	[$H_2(g)$]	[$I_2(g)$]	[HI(g)]	[$H_2(g)$]$_{eqm}$	[$I_2(g)$]$_{eqm}$	[HI(g)]$_{eqm}$
1	2.40	1.38	0	1.14	0.12	2.52
2	2.40	1.68	0	0.92	0.20	2.96
3	2.44	1.98	0	0.77	0.31	3.34
4	2.46	1.76	0	0.92	0.22	3.08
5	0	0	3.04	0.345	0.345	2.35
6	0	0	7.58	0.86	0.86	5.86

1 Write the equation for the reversible reaction to form hydrogen iodide from hydrogen and iodine.

2 Show that the equilibrium concentration of:

   a) hydrogen in tube 1 is as expected, given the value of [$I_2(g)$]$_{eqm}$

   b) hydrogen iodide in tube 2 is as expected, given the value of [$I_2(g)$]$_{eqm}$.

3 Explain why [$H_2(g)$]$_{eqm}$ = [$I_2(g)$]$_{eqm}$ for tubes 5 and 6.

4 For each of the tubes, work out the value of:

   a) $\dfrac{[HI(g)]_{eqm}}{[H_2(g)]_{eqm}[I_2(g)]_{eqm}}$

   b) $\dfrac{[HI(g)]^2_{eqm}}{[H_2(g)]_{eqm}[I_2(g)]_{eqm}}$.

   Enter your values in a table and comment on the results.

5 What is the value of K_c for the reaction of hydrogen with iodine at 731 K?

11.2 The equilibrium law

The equilibrium law has been established by experiment. It is a quantitative law for predicting the amounts of reactants and products when a reversible reaction reaches a state of dynamic equilibrium.

In general, for a reversible reaction at equilibrium:

$$aA + bB \rightleftharpoons cC + dD$$

$$K_c = \frac{[C]^c[D]^d}{[A]^a[B]^b}$$

This is the form for the equilibrium constant, K_c, when the concentrations of the reactants and products are measured in moles per cubic decimetre. [A], [B] and so on are the equilibrium concentrations, sometimes written as $[A]_{eqm}$ and $[B]_{eqm}$ to make this clear.

The concentrations of the chemicals on the right-hand side of the equation appear on the top line of the expression. The concentrations of reactants on the left appear on the bottom line. Each concentration term is raised to the power of the number in front of its formula in the equation.

Equilibrium constants and balanced equations

An equilibrium constant always applies to a particular chemical equation and can be deduced directly from the equation.

There can be different ways of writing the equation for a reversible reaction at equilibrium. As a result, there are different forms for the equilibrium constant, each with a different value. So long as the matching equation and equilibrium constant are used in any calculation, the predictions based on the equilibrium law are the same.

For example, for this equilibrium involving dinitrogen oxide, oxygen and nitrogen monoxide:

$$2N_2O(g) + O_2(g) \rightleftharpoons 4NO(g)$$

$$K_c = \frac{[NO(g)]^4}{[N_2O(g)]^2[O_2]}$$

But for this equilibrium:

$$N_2O(g) + \tfrac{1}{2}O_2(g) \rightleftharpoons 2NO(g)$$

$$K_c = \frac{[NO(g)]^2}{[N_2O(g)][O_2]^{\frac{1}{2}}}$$

Reversing the equation also changes the form of the equilibrium constant because the concentration terms for the chemicals on the right-hand side of the equation always appear on the top of the expression for K_c.

So, for this equilibrium:

$$4NO(g) \rightleftharpoons 2N_2O(g) + O_2(g)$$

$$K_c = \frac{[N_2O(g)]^2[O_2]}{[NO(g)]^4}$$

> **Tip**
>
> Equilibrium constants are only constant at a particular temperature.
>
> The form of the expression for an equilibrium constant can be deduced from the balanced chemical equation (unlike rate equations – see Section 16.3).

> **Tip**
>
> It is important to write the balanced equation and the equilibrium constant together.

In an equilibrium mixture of dinitrogen oxide, oxygen and nitrogen monoxide, all three substances are gases. They are all in the same gaseous phase. This is an example of a homogeneous equilibrium.

Test yourself

4 Write the expression for K_c for each equation and state the units of the equilibrium constant.

a) $N_2(g) + O_2(g) \rightleftharpoons 2NO(g)$

b) $\frac{1}{2}N_2(g) + \frac{1}{2}O_2(g) \rightleftharpoons NO(g)$

c) $N_2(g) + 3H_2(g) \rightleftharpoons 2NH_3(g)$

d) $2NH_3(g) \rightleftharpoons N_2(g) + 3H_2(g)$

5 Explain what is meant by the term 'homogeneous'.

Finding equilibrium constants by experiment

The strategy for determining the value of an equilibrium constant involves three main steps:

Step 1: Mix measured quantities of reactants and/or products. Then allow the mixture to reach equilibrium under steady conditions.

Step 2: Analyse the mixture to find the equilibrium concentration of one of the chemicals at equilibrium.

Step 3: Use the equation for the reaction and the information from steps 1 and 2 to work out the values for the equilibrium concentrations of all the atoms, molecules or ions. Then substitute these values into the expression for K_c.

The challenge when investigating reactions at equilibrium is to find ways to measure equilibrium concentrations without upsetting the equilibrium. Many methods of analysis use up the chemical being analysed. Analytical methods of this kind are generally unsuitable because, as Le Chatelier's principle shows (Section 10.4 in Student Book 1), the position of equilibrium shifts whenever one of the reactants or products is removed from the equilibrium mixture.

There are two main ways to measure equilibrium concentrations. One way is to find a method for 'freezing' the reaction and thus slowing down the rate so much that it is possible to measure one of the equilibrium concentrations by titration. The most obvious way to slow down the rate and 'freeze' the equilibrium is by cooling. Other possibilities are to dilute the equilibrium mixture or to remove a catalyst.

The second way of measuring equilibrium concentrations is to use an instrument that responds to a property of the mixture that varies with concentration. Well-established methods for doing this include measuring the pH with a pH meter or measuring the intensity of a colour with a colorimeter.

Measuring K_c for a reaction used to make an ester

The equilibrium between ethanoic acid, ethanol, ethyl ethanoate and water (Section 17.3.3) is one of the few reaction systems (other than acid–base equilibria) that lends itself to study in an advanced chemistry course:

$$CH_3COOH(l) + C_2H_5OH(l) \rightleftharpoons CH_3COOC_2H_5(l) + H_2O(l)$$

This esterification reaction is very, very slow at room temperature in the absence of a catalyst. In the presence of an acid catalyst the reaction mixture reaches equilibrium in about 48 hours.

Diluting the equilibrium mixture means that the reaction is slow enough to find the equilibrium concentration of ethanoic acid by titration without the position of equilibrium shifting perceptibly in the time taken for the titration.

The procedure follows these three steps.

Step 1: Mix measured quantities of chemicals and allow the mixture to reach equilibrium.

Precisely measured quantities of the chemicals are added to sample tubes. The masses of the components of the mixture can be found by weighing. The sample tubes are tightly stoppered to avoid loss by evaporation and set aside at constant temperature for 48 hours.

Some of the tubes at first contain just ethanol, ethanoic acid and hydrochloric acid. Others start with only ethyl ethanoate, water and hydrochloric acid. Working in this way shows that it is possible to reach equilibrium from either side of the equation.

Step 2: Analyse the mixture to find the equilibrium concentration of the acid.

Each equilibrium mixture is transferred quantitatively to a flask and diluted with water. Titration with a standard solution of sodium hydroxide determines the total amount of acid in the sample at equilibrium: both hydrochloric acid and ethanoic acid.

Step 3: Use the equation for the reaction and the information from steps 1 and 2 to work out the values for all the equilibrium concentrations.

Some of the sodium hydroxide used in the titration reacts with the hydrochloric acid. Since the amount of HCl(aq) does not change as the reactants reach equilibrium, it is possible to work out how much of the titre was used to neutralise it, knowing how much HCl(aq) was added at the start. The remainder of the alkali added during the titration reacts with ethanoic acid. Hence the amount of ethanoic acid at equilibrium can be calculated. The other equilibrium concentrations can be found given the starting amounts of chemicals present and the equation for the reaction.

> **Tip**
>
> The equilibrium mixtures for the reaction of ethanoic acid with ethanol includes water from the dilute hydrochloric acid as well as any added water.

> **Tip**
>
> In this equilibrium system water is present in relatively small amounts as a reactant and not just as a solvent. The concentration of water is a variable and appears in the expression for K_c.

Analysing the results of an experiment to measure K_c

This activity is based on two sets of results from an experiment carried out by students to measure the value of K_c for the equilibrium between ethanoic acid, ethanol, ethyl ethanoate and water.

The approach to the calculation can be illustrated from the results for one of the samples which initially contained only the ester but no added ethanoic acid or ethanol. The sample was initially made up of the following:

- ethyl ethanoate (ester), 3.64 g (0.0413 mol)
- water, 0.99 g
- 5.0 cm^3 of 2.00 mol dm^{-3} HCl(aq) containing 0.010 mol HCl.

Mass of added hydrochloric acid = 5.17 g

This contained 4.81 g water

So, the total mass of water at the start = 0.99 g + 4.81 g = 5.80 g
= 0.322 mol H_2O

Titration of the equilibrium mixture found that 39.20 cm^3 of 1.00 mol dm^{-3} sodium hydroxide neutralised the total acid present. Both HCl and CH_3COOH react 1:1 with NaOH, so this shows that the total amount of acid at equilibrium = 0.0392 mol

So, taking away the amount of hydrochloric acid added at the start, this shows that the amount of ethanoic acid at equilibrium = 0.0292 mol. The results are summarised in Table 11.2.

1 Explain why the amount of hydrochloric acid in the mixture did not change as the mixture of reactants reached equilibrium.
2 Confirm, by calculation, that the mass of water in the hydrochloric acid added to the reaction mixture was 4.81 g.
3 Show that initially 0.322 mol of water was present.
4 Use the titration result to confirm that the total acid (as H^+ ions) in the equilibrium mixture was 0.0392 mol. Hence show that the amount of ethanoic acid at equilibrium was 0.0292 mol.
5 Explain why the answer to Question 3 shows that the equilibrium amounts of ethanol, ethyl ethanoate and water were as shown in Table 11.2.
6 Write the expression for K_c for the equilibrium reaction; then use the values in the table to calculate the value of K_c.
7 Explain why it is not necessary to know the volume of the reaction mixture to calculate K_c and why the equilibrium constant has no units.
8 Titration of the equilibrium mixture in another sample tube from the experiment required 41.30 cm^3 of 1.00 mol dm^{-3} NaOH(aq). Calculate a value for K_c given that, at the start, the tube contained 4.51 g ethyl ethanoate with 5.0 cm^3 of 2.00 mol dm^{-3} HCl(aq) but no added water other than the water in the dilute acid.
9 Compare and comment on the two values for K_c.

Table 11.2

Reaction	CH$_3$COOH(l)	+	C$_2$H$_5$OH(l)	⇌	CH$_3$COOC$_2$H$_5$(l)	+	H$_2$O(l)
Initial amount/mol	0.000		0.000		0.0413		0.322
Measured amount at equilibrium/mol	0.0292						
Other equilibrium amounts calculated from the starting amounts and the equation/mol			0.0292		0.0121		0.293

Calculating equilibrium constants

Experimental results can be used to calculate the values for equilibrium constants along similar lines to the determination of the value of K_c for the formation of ethyl ethanoate. Once a value of K_c is known, it can be used to calculate the concentrations of the reactants and products in specific equilibrium mixture.

Example

1.00 mol of NOCl gas was enclosed in a 0.50 dm³ flask at 298 K. The amount of NO gas in the flask at equilibrium was found to be 0.33 mol. Calculate the value of K_c for this reaction:

$$2NOCl(g) \rightleftharpoons 2NO(g) + Cl_2(g)$$

Note on the method

Write down the equation. Underneath write first the initial amounts, then write the given amount at equilibrium. Next calculate the equilibrium amounts not given, taking into account the numbers of moles of each substance shown in the equation for the reaction.

Calculate the equilibrium concentrations given the volume of the solution. Substitute the values and units in the expression for K_c.

Answer

Equation:	$2NOCl(g)$	$\rightleftharpoons$	$2NO(g)$	$+$	$Cl_2(g)$
Initial amounts/mol:	1.00		0		0
Equilibrium amount given/mol:			0.33		
Equilibrium amounts calculated/mol:	(1.00 − 0.33)				(0.33 ÷ 2)
Equilibrium concentrations/mol dm⁻³:	0.67 ÷ 0.5 = 1.34		0.33 ÷ 0.5 = 0.66		(0.33 ÷ 2) ÷ 0.5 = 0.33

$$K_c = \frac{[NO(g)]^2[Cl_2(g)]}{[NOCl(g)]^2} = \frac{(0.66 \, mol \, dm^{-3})^2(0.33 \, mol \, dm^{-3})}{(1.34 \, mol \, dm^{-3})^2}$$

Hence $K_c = 0.080 \, mol \, dm^{-3}$

Test yourself

6 On mixing 1.68 mol $PCl_5(g)$ with 0.36 mol $PCl_3(g)$ in a 2.0 dm³ container, and allowing the mixture to reach equilibrium, the amount of PCl_5 in the equilibrium mixture was 1.44 mol. Calculate K_c for the reaction:

$$PCl_5(g) \rightleftharpoons PCl_3(g) + Cl_2(g)$$

7 Consider the equilibrium between sulfur dioxide, oxygen and sulfur trioxide:

$$2SO_2(g) + O_2(g) \rightleftharpoons 2SO_3(g) \quad K_c = 1.6 \times 10^6 \, dm^3 \, mol^{-1}$$

a) Show that the units for the equilibrium constant, K_c, for the equation are $dm^3 \, mol^{-1}$.

b) What is the value of K_c for this equation at the same temperature?

$$SO_2(g) + \frac{1}{2}O_2(g) \rightleftharpoons SO_3(g)$$

c) What is the value of K_c for this equation at the same temperature?

$$2SO_3(g) \rightleftharpoons 2SO_2(g) + O_2(g)$$

8 $K_c = 170 \, dm^3 \, mol^{-1}$ at 298 K for the equilibrium system: $2NO_2(g) \rightleftharpoons N_2O_4(g)$. If a 5 dm³ flask contains 1.0×10^{-3} mol of NO_2 and 7.5×10^{-4} mol N_2O_4, is the system at equilibrium? Is there any tendency for the concentration of NO_2 to change and, if so, does it tend to increase or decrease?

Heterogeneous equilibria

In many equilibrium systems the substances involved are not all in the same phase. An example is the equilibrium state formed when steam is heated with coke (carbon) in a closed container. This is an example of a heterogeneous equilibrium.

$$H_2O(g) + C(s) \rightleftharpoons H_2(g) + CO(g)$$

The concentrations of solids do not appear in the expression for the equilibrium constant. Pure solids have, in effect, a constant 'concentration' so their values are incorporated into the value for the equilibrium constant. Hence for the reaction of steam with carbon:

$$K_c = \frac{[H_2(g)][CO(g)]}{[H_2O(g)]}$$

The same applies to heterogeneous systems which have a separate liquid phase as one of the reactants or products.

Another example is the equilibrium state between solid calcium carbonate and a dilute solution containing dissolved carbon dioxide and calcium hydrogencarbonate.

$$CaCO_3(s) + CO_2(aq) + H_2O(l) \rightleftharpoons Ca^{2+}(aq) + 2HCO_3^-(aq)$$

This example illustrates another general rule. The K_c expression for dilute solutions does not include a concentration term for water. There is so much water present that its concentration is effectively constant.

So the expression for the equilibrium constant becomes:

$$K_c = \frac{[Ca^{2+}(aq)][HCO_3^-(aq)]^2}{[CO_2(aq)]}$$

Tip

Remember that in dilute solution, the water is in such large excess that the value of $[H_2O(l)]$ is effectively constant. As a result, it does not appear in the equilibrium law expression.

Figure 11.4 Pouring fizzy water from a glass bottle.

Test yourself

9 Explain what is meant by the term 'heterogeneous'.

10 Explain why the bottle shown in Figure 11.4 contained a heterogeneous equilibrium before the top was unscrewed.

11 Write the expression for K_c for each equation and state the units of the equilibrium constant.

   a) $NH_4HS(s) \rightleftharpoons H_2S(g) + NH_3(g)$

   b) $Pb^{2+}(aq) + Sn(s) \rightleftharpoons Pb(s) + Sn^{2+}(aq)$

   c) $BiCl_3(aq) + H_2O(l) \rightleftharpoons BiOCl(s) + 2HCl(aq)$

12 In the natural world, where is it possible to find solid calcium carbonate and a dilute solution containing dissolved carbon dioxide and calcium hydrogencarbonate close to a state of dynamic equilibrium?

13 Calculate the concentration of water in water (in $mol\,dm^{-3}$) to show that it is reasonable to regard the concentration of water as a constant when writing the expression for K_c for equilibria in dilute aqueous solution.

The extent and direction of change

Chemists can use equilibrium constants to predict quantitatively the direction and extent of chemical change.

Table 11.3 shows that if the value of an equilibrium constant is large, then the position of equilibrium is over to the right-hand side of the equation. Conversely, if the value of an equilibrium constant is small, then the position of equilibrium is over to the left-hand side of the equation. If the value of K_c is close to 1, then there are significant quantities of both reactants and products present at equilibrium.

Table 11.3 Relating the value of K_c to the direction and extent of change.

Direction and extent of change	Value of K_c
Reaction does not go	$K_c < 1 \times 10^{-10}$
Reaction reaches an equilibrium in which the reactants predominate	$K_c \approx 0.01$
Roughly equal amounts of reactants and products at equilibrium	$K_c = 1$
Reaction reaches an equilibrium in which the products predominate	$K_c \approx 100$
Reaction goes to completion	$K_c > 1 \times 10^{10}$

It is very important to keep in mind that the equilibrium constant gives no information about the time it takes for a reaction mixture to reach equilibrium. The system may reach equilibrium rapidly or slowly. The value of K_c for the reaction of hydrogen with chlorine to make hydrogen chloride, for example, is about 1×10^{31} at room temperature, but in the absence of a catalyst, ultraviolet light or a flame there is no reaction.

Test yourself

14 What can you conclude about the direction and extent of change in each of these examples?

   a) $Zn(s) + Cu^{2+}(aq) \rightleftharpoons Zn^{2+}(aq) + Cu(s)$ $K_c = 1 \times 10^{37}$ at 298 K

   b) $2HBr(g) \rightleftharpoons H_2(g) + Br_2(g)$ $K_c = 1 \times 10^{-10}$ at 298 K

   c) $N_2(g) + 3H_2(g) \rightleftharpoons 2NH_3(g)$ $K_c = 2.2$ at 623 K

15 In general, if the equilibrium constant for a forward reaction is large, what is the size of the equilibrium constant for the reverse of the same reaction?

▌11.3 Gaseous equilibria

Many important industrial processes involve reversible reactions between gases. Applying the equilibrium law to these reactions helps to determine the optimal conditions for manufacturing chemicals. When it comes to gas reactions it is often easier to measure the pressure rather than the concentration and to use a modified form of the equilibrium law.

Gas mixtures and partial pressures

In any mixture of gases the total pressure of the mixture can be 'shared out' between the gases. The contribution each gas makes to the total pressure is its **partial pressure**. It is possible to calculate a partial pressure for each gas in the mixture. In a mixture of gases A, B and C, the sum of the three partial pressures equals the total pressure.

$$p_A + p_B + p_C = p_{total}$$

Partial pressures are a useful alternative to concentrations when studying mixtures of gases and gas reactions.

In gas mixtures it is the amounts (in moles) of gas molecules that matter and not the chemical nature of the molecules. As a result the molar volume of a gas is the same for all gases under the same conditions of temperature and pressure. The gas laws show that this means that the total pressure is shared between the gases simply according to their mole fractions in the mixture.

In a mixture of n_A moles of A with n_B moles of B and n_C moles of C, the total amount in moles is $(n_A + n_B + n_C)$. The mole fractions (symbol X) are given by the following:

$$X_A = \frac{n_A}{n_A + n_B + n_C} \qquad X_B = \frac{n_B}{n_A + n_B + n_C} \qquad X_C = \frac{n_C}{n_A + n_B + n_C}$$

So, the mole fraction of A is the fraction of the total number of molecules which are molecules of A.

The sum of all the mole fractions is 1, so $X_A + X_B + X_C = 1$.

On this basis the partial pressures of three gases A, B and C in a gas mixture with total pressure p are:

$$p_A = X_A p, \qquad p_B = X_B p \quad \text{and} \quad p_C = X_C p$$

The partial pressure for each gas is the pressure it would exert if it was the only gas in the container under the same conditions. The partial pressure of a gas is proportional to the concentration of the gas in the mixture. This makes it possible to work in partial pressures when applying the equilibrium law to gas reactions.

Test yourself

16 A 20 mol sample of a gas mixture contains 15.6 mol nitrogen and 4.4 mol oxygen.

   a) Calculate the mole fractions of the two gases in the mixture.

   b) Calculate the partial pressures of each of the two gases if the total pressure is 1 atm.

17 A mixture of 22 g propane gas and 11 g 2-methylpropane gas is compressed into an aerosol can to give a total pressure of 1.5 atm.

   a) What are the mole fractions of the two gases?

   b) Calculate the partial pressures of each of the two gases.

11.4 K_p

K_p is the symbol for the equilibrium constant for an equilibrium involving gases when the concentrations are measured by partial pressures. The rules for writing equilibrium expressions are the same for K_p as for K_c, with partial pressures replacing concentrations. This is shown in Table 11.4.

Tip

Do not use square brackets when writing K_p expressions. In the context of the equilibrium law, square brackets signify concentrations in $mol\,dm^{-3}$.

Table 11.4 Examples of equilibrium expressions for K_p. Note that when writing an expression for K_p for a heterogeneous reaction the same rules apply as for K_c. The expression does not include terms for any separate pure solid phases.

Equilibrium	K_p	Units of K_p
$H_2(g) + I_2(g) \rightleftharpoons 2HI(g)$	$K_p = \dfrac{(p_{HI})^2}{p_{H_2} \times p_{I_2}}$	no units
$N_2(g) + 3H_2(g) \rightleftharpoons 2NH_3(g)$	$K_p = \dfrac{(p_{NH_3})^2}{p_{N_2} \times (p_{H_2})^3}$	atm^{-2}
$N_2O_4(g) \rightleftharpoons 2NO_2(g)$	$K_p = \dfrac{(p_{NO_2})^2}{p_{N_2O_4}}$	atm
$HCl(g) + LiH(s) \rightleftharpoons H_2(g) + LiCl(s)$	$K_p = \dfrac{p_{H_2}}{p_{HCl}}$	no units

Example

An experimental study of the equilibrium between $N_2(g)$, $H_2(g)$ and $NH_3(g)$ found that one equilibrium mixture contained 2.15 mol of $N_2(g)$, 6.75 mol of $H_2(g)$ and 1.41 mol of $NH_3(g)$ at a total pressure 10.0 atm. Calculate the value for K_p under the conditions that the measurements were taken.

Notes on the method

First work out the mole fractions of the gases.

Multiply the total pressure by the mole fractions to get the partial pressures.

Check that the sum of the partial pressures equals the total pressure.

Finally substitute in the expression for K_p and give the units.

Tip

Changing the total pressure or the composition of the gas mixture has no effect on the value of K_p as long as the temperature stays constant.

Answer

Total number of moles	$= 2.15\,mol + 6.75\,mol + 1.41\,mol = 10.31\,mol$

Mole fraction of $N_2(g)$ $= \dfrac{2.15\,mol}{10.31\,mol} = 0.208$

Mole fraction of $H_2(g)$ $= \dfrac{6.75\,mol}{10.31\,mol} = 0.655$

Mole fraction of $NH_3(g)$ $= \dfrac{1.41\,mol}{10.31\,mol} = 0.137$

Partial pressure of $N_2(g)$ $= 0.208 \times 10\,atm = 2.08\,atm$

Partial pressure of $H_2(g)$ $= 0.655 \times 10\,atm = 6.55\,atm$

Partial pressure of $NH_3(g)$ $= 0.137 \times 10\,atm = 1.37\,atm$

Check: the total pressure $= 2.08\,atm + 6.55\,atm + 1.37\,atm = 10.0\,atm$

For the equilibrium: $N_2(g) + 3H_2(g) \rightleftharpoons 2NH_3(g)$

$$K_p = \frac{(p_{NH_3})^2}{p_{N_2} \times (p_{H_2})^3}$$

$$= \frac{(1.37)^2}{2.08 \times (6.55)^3} = 3.21 \times 10^{-3}\,atm^{-2}$$

Test yourself

18 Write the expression for K_p for each equilibrium. Give the units with pressures measured in atmospheres, atm.

 a) $2SO_2(g) + O_2(g) \rightleftharpoons 2SO_3(g)$

 b) $4NH_3(g) + 3O_2(g) \rightleftharpoons 2N_2(g) + 6H_2O(g)$

 c) $CaCO_3(s) \rightleftharpoons CaO(s) + CO_2(g)$

19 Calculate K_p at 330 K for this equilibrium mixture:

 $N_2O_4(g) \rightleftharpoons 2NO_2(g)$

 At this temperature, a sample of the gas mixture at 1.20 atm pressure consists of 8.1 mol $N_2O_4(g)$ and 3.8 mol $NO_2(g)$.

20 Calculate K_p for this reversible reaction at 1000 K:

 $C_2H_6(g) \rightleftharpoons C_2H_4(g) + H_2(g)$

 At this temperature, starting with just 5 mol ethane yields an equilibrium mixture containing 1.8 mol ethene at 1.80 atm pressure.

11.5 Factors affecting systems at equilibrium

The chemical industry is being reinvented to make it more sustainable. It is no longer acceptable to operate processes that make inefficient use of valuable resources. Chemists and chemical engineers are devising new methods by applying the theoretical ideas and models that explain how fast reactions go, in which direction and how far (Figure 11.5).

Figure 11.5 This vast catalytic cracker in Germany is used to make ethene from natural gas or oil. Controlling chemical reactions carried out on such a large scale requires precise application of chemical principles.

The effect of changing concentrations on systems at equilibrium

The equilibrium law makes it possible to explain the effect of changing the concentration of one of the chemicals in an equilibrium mixture.

An example is the equilibrium in solution involving chromate(VI) and dichromate(VI) ions in water (Figure 11.6):

$$2CrO_4{}^{2-}(aq) + 2H^+(aq) \rightleftharpoons Cr_2O_7{}^{2-}(aq) + H_2O(l)$$

yellow   orange

At equilibrium: $K_c = \dfrac{[Cr_2O_7{}^{2-}(aq)]}{[CrO_4{}^{2-}(aq)]^2 \, [H^+(aq)]^2}$

where these are equilibrium concentrations.

Adding a few drops of concentrated acid increases the concentration of $H^+(aq)$ on the left-hand side of the equation.

This briefly upsets the equilibrium. For an instant after adding acid:

$$\dfrac{[Cr_2O_7{}^{2-}(aq)]}{[CrO_4{}^{2-}(aq)]^2 \, [H^+(aq)]^2} < K_c$$

The system restores equilibrium as chromate(VI) ions react with hydrogen ions to produce more of the products. There is very soon a new equilibrium. Once again:

$$\dfrac{[Cr_2O_7{}^{2-}(aq)]}{[CrO_4{}^{2-}(aq)]^2 \, [H^+(aq)]^2} = K_c$$

but now with new values for the various equilibrium concentrations.

Chemists sometimes say that adding acid makes the 'position of equilibrium shift to the right'. The effect is visible because the yellow colour of the chromate(VI) ions turns to the orange colour of dichromate(VI) ions. This is as Le Chatelier's principle predicts. The advantage of using K_c is that it makes quantitative predictions possible.

Figure 11.6 On the left, a yellow solution of chromate(VI) ions in water. On the right, the solution has turned orange as more dichromate(VI) ions form after adding a few drops of strong acid.

> ### Tip
>
> Changing the concentrations does not alter the value of the equilibrium constant so long as the temperature stays constant. Remember that in dilute solutions $[H_2O(l)]$ is constant, so it does not appear in the equilibrium law expression.

Test yourself

21 Describe and explain the effect of adding alkali to a solution of dichromate(VI) ions.

22 a) Use the equilibrium law to predict and explain the effect of adding pure ethanol to an equilibrium mixture of ethanoic acid, ethanol, ethyl ethanoate and water:

$$CH_3COOH(l) + C_2H_5OH(l) \rightleftharpoons CH_3COOC_2H_5(l) + H_2O(l)$$

  b) Show that your prediction is consistent with Le Chatelier's principle.

The effects of pressure changes on systems at equilibrium

Changes of pressure are not generally significant for equilibria that involve only solids or liquids, but they have a marked effect on gaseous equilibria. Lowering the pressure on a system at equilibrium favours the direction of change that produces more molecules, while increasing the pressure favours the change that produces fewer molecules. This is what Le Chatelier's principle predicts.

$$\text{low pressure} \longrightarrow$$
$$\text{fewer molecules} \rightleftharpoons \text{more molecules}$$
$$\longleftarrow \text{high pressure}$$

The effects of increasing or decreasing the total pressure of a gas mixture at equilibrium can be predicted quantitatively with the help of the equilibrium law. Take the example of the reaction used to make ammonia in the Haber process:

$$N_2(g) + 3H_2(g) \rightleftharpoons 2NH_3(g)$$

The equilibrium law expression in terms of partial pressures is:

$$K_p = \frac{(p_{NH_3})^2}{p_{N_2} \times (p_{H_2})^3}$$

Suppose that the equilibrium partial pressures of nitrogen, hydrogen and ammonia are a atm, b atm and c atm, respectively. Substituting in the expression for K_p gives:

$$K_p = \frac{c^2}{ab^3} \text{ atm}^{-2}$$

Now suppose that the total pressure is suddenly doubled. At that instant all the partial pressures double so that $p_{N_2} = 2a$ atm, $p_{H_2} = 2b$ atm and $p_{NH_3} = 2c$ atm. Substituting these values in the 'equilibrium constant ratio' gives:

$$\frac{(p_{NH_3})^2}{p_{N_2} \times (p_{H_2})^3} = \frac{(2c)^2}{2a \times (2b)^3} = \frac{1}{4} \times \frac{c^2}{ab^3} \text{ atm}^{-2}$$

So at that moment the 'equilibrium constant ratio' is one-quarter of the value of K_p. The system is not at equilibrium. In order to restore equilibrium, some of the nitrogen and hydrogen must react to decrease their partial pressures and form more ammonia to increase its partial pressure. This happens until the values are such that the 'equilibrium constant ratio' again equals K_p. In other words, increasing the pressure causes the equilibrium to shift to the right. In this way the equilibrium law makes it possible to predict not only the direction, but also the extent of the shift (Table 11.5).

> **Tip**
>
> Changing pressure does not alter the value of the equilibrium constant K_p so long as the temperature stays constant.

Table 11.5

Total pressure/atm	10	50	100	200
Percentage by volume of ammonia at equilibrium at 773 K	1.2	5.6	10.6	18.3

23 Predict the effect of increasing the pressure on these systems at equilibrium:

   a) $2SO_2(g) + O_2(g) \rightleftharpoons 2SO_3(g)$

   b) $CH_4(g) + H_2O(g) \rightleftharpoons CO(g) + 3H_2(g)$

   c) $N_2(g) + O_2(g) \rightleftharpoons 2NO(g)$

24 For the reaction $N_2O_4(g) \rightleftharpoons 2NO_2(g)$, the value of K_p is 0.11 atm at 298 K. Is a mixture containing $N_2O_4(g)$ with a partial pressure of 2.4 atm and $NO_2(g)$ with a partial pressure of 1.2 atm at equilibrium at 298 K? If not, which gas tends to increase its partial pressure?

25 Use the expression for K_p to predict and explain the effect of the following changes on an equilibrium mixture of hydrogen, carbon monoxide and methanol:

   $2H_2(g) + CO(g) \rightleftharpoons CH_3OH(g)$

   a) adding more hydrogen to the gas mixture at constant total pressure

   b) compressing the mixture to increase the total pressure

   c) adding an inert gas such as argon while keeping the total pressure constant.

The effects of temperature changes on systems at equilibrium

Le Chatelier's principle predicts that raising the temperature makes the equilibrium shift in the direction which is endothermic. For example, for the reaction which produces sulfur trioxide during the manufacture of sulfuric acid, raising the temperature lowers the percentage of sulfur trioxide at equilibrium.

$$2SO_2(g) + O_2(g) \rightleftharpoons 2SO_3(g) \qquad \Delta H = -98\,kJ\,mol^{-1}$$

The equilibrium shifts to the left as the temperature rises because this is the direction in which the reaction is endothermic (Figure 11.7).

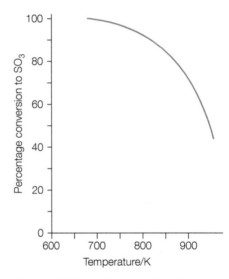

Figure 11.7 The effect of raising the temperature on the equilibrium between SO_2, O_2 and SO_3.

Tip

If ΔH for the forward reaction is negative, then ΔH for the reverse reaction has the same magnitude but the opposite sign. So if the forward reaction is exothermic, then the reverse reaction is endothermic.

The reason that temperature changes cause a shift in the position of equilibrium is that the value of the equilibrium constant changes. This is illustrated by the values in Table 11.6.

Table 11.6 Values of K_p at four temperatures for the equilibrium $2SO_2(g) + O_2(g) \rightleftharpoons 2SO_3(g)$.

Temperature/K	K_p/atm^{-1}
298	4.0×10^{24}
500	2.5×10^{10}
1100	1.3×10^{-1}
700	3.0×10^4

Figure 11.8 Sealed tubes containing equilibrium mixtures of $NO_2(g)$ which is orange-brown and $N_2O_4(g)$ which is colourless. The tube on the left is in hot water and the tube on the right in ice.

Test yourself

26 Show that graph in Figure 11.7 and the values in Table 11.6 are consistent with predictions for the equilibrium based on Le Chatelier's principle.

27 The value of K_p for the equilibrium $N_2O_4(g) \rightleftharpoons 2NO_2(g)$ is 4.79 atm at 400 K and 347 atm at 500 K.

   a) What is the effect of raising the temperature on the position of equilibrium?

   b) How can your answer to (a) account for the appearance of the gas mixtures in Figure 11.8?

   c) What is the sign of ΔH for the reaction?

28 For the reaction between hydrogen and iodine to form hydrogen iodide, the value of K_p is 794 at 298 K but 54 at 700 K. What can you deduce from this information?

The effects of catalysts on systems at equilibrium

Catalysts speed up reactions but are not used up as they do so (Figure 11.9). It is important to note that while a catalyst speeds up the rate at which a reaction gets to an equilibrium state, it has no effect on the final position of equilibrium. In other words, a catalyst provides a faster route to the same equilibrium state. The alternative route with a catalyst has a lower activation energy but speeds up the forward and back reactions to the same extent, so that the dynamic equilibrium is unchanged.

Figure 11.9 Crystals of palladium seen under an electron microscope. Palladium is a rare and precious metal which is used to catalyse hydrogenation reactions.

Exam practice questions

1 At 298 K the value of K_c for the following equilibrium is 1×10^{10}:

$$Sn^{2+}(aq) + 2Fe^{3+}(aq) \rightleftharpoons Sn^{4+}(aq) + 2Fe^{2+}(aq)$$

a) i) Write the expression for K_c. *(2)*
ii) What are the units of K_c for this reaction? Explain your answer. *(2)*
b) What is the value of K_c for:

   i) $Sn^{4+}(aq) + 2Fe^{2+}(aq)$
$$\rightleftharpoons Sn^{2+}(aq) + 2Fe^{3+}(aq) \quad (2)$$

   ii) $\frac{1}{2}Sn^{2+}(aq) + Fe^{3+}(aq)$
$$\rightleftharpoons \frac{1}{2}Sn^{4+}(aq) + Fe^{2+}(aq)? \quad (2)$$

2 a) A flask contains an equilibrium mixture of hydrogen gas ($0.010 \, mol \, dm^{-3}$), iodine gas ($0.010 \, mol \, dm^{-3}$) and hydrogen iodide gas ($0.070 \, mol \, dm^{-3}$) at a constant temperature. Calculate K_c for the reaction of hydrogen with iodine to form hydrogen iodide. *(3)*
b) Enough hydrogen is added to the mixture in (a) to suddenly double the hydrogen concentration in the flask to $0.020 \, mol \, dm^{-3}$. After a while the mixture settles down with a new iodine concentration of $0.0070 \, mol \, dm^{-3}$ at the same temperature as before.

   i) What are the new concentrations of hydrogen and hydrogen iodide? *(2)*
   ii) Show that the new mixture is at equilibrium. *(2)*

c) How does a sudden doubling of the hydrogen concentration affect the position of equilibrium? *(2)*

3 Nitrogen and hydrogen react to form ammonia when heated under pressure in the presence of a catalyst.

$$N_2(g) + 3H_2(g) \rightleftharpoons 2NH_3(g)$$
$$\Delta H = -92 \, kJ \, mol^{-1}$$

Analysis of an equilibrium mixture of the gases at 10 atmospheres and 650 K found that it contained $1.41 \, mol \, NH_3$, $6.75 \, mol \, H_2$ and $2.15 \, mol \, N_2$.

a) Explain, in this context, the term 'dynamic equilibrium'. *(2)*

b) i) Calculate the mole fraction of each of the three gases in the mixture at 10 atm and 650 K. *(3)*
ii) Calculate the partial pressures of the three gases. *(2)*
iii) Calculate a value for K_p and give the units. *(3)*

4 Explain what is wrong with each of the following statements. To what extent, if at all, is there any truth in each of the statements?

a) Once a reaction mixture reaches equilibrium, there is no further reaction. *(3)*
b) Adding more of one of the reactants to an equilibrium mixture increases the yield of products because the value of the equilibrium constant increases. *(3)*
c) Adding a catalyst to make a reaction go faster can increase the amount of product at equilibrium. *(3)*
d) Raising the temperature to make a reaction go faster can increase the amount of product at equilibrium. *(3)*
e) Adding a catalyst can mean that a reaction that is only feasible at a high temperature becomes feasible at a much lower temperature. *(3)*

5 A solution of ammonia in water was shaken with an equal volume of an organic solvent until the system reached equilibrium with the ammonia distributed between the two solvents. In a series of titrations, $10 \, cm^3$ of the aqueous layer was neutralised by an average of $17.0 \, cm^3$ of $0.50 \, mol \, dm^{-3}$ hydrochloric acid. In a second series of titrations, $10 \, cm^3$ of the organic layer was neutralised by $6.0 \, cm^3$ of $0.010 \, mol \, dm^{-3}$ hydrochloric acid.

a) What was the concentration of the ammonia in the aqueous layer at equilibrium? *(2)*
b) What was the concentration of the ammonia in the organic solvent at equilibrium? *(2)*
c) What is the value of K_c for the equilibrium:

$$NH_3(org) \rightleftharpoons NH_3(aq)? \quad (2)$$

6 At $473\,K$, the value of K_c for the decomposition of PCl_5 is $8 \times 10^{-3}\,mol\,dm^{-3}$.

$$PCl_5(g) \rightleftharpoons PCl_3(g) + Cl_2(g)$$
$$\Delta H = +124\,kJ\,mol^{-1}$$

a) Write the expression for K_c for the reaction. *(1)*

b) What is the value of K_c for the reverse reaction at $473\,K$ and what are its units? *(2)*

c) A sample of pure PCl_5 is heated to $473\,K$ in a vessel containing no other chemicals. At equilibrium the concentration of PCl_5 is $5 \times 10^{-2}\,mol\,dm^{-3}$. What are the equilibrium concentrations of PCl_3 and Cl_2? *(3)*

d) Explain how the concentrations of PCl_5, PCl_3 and Cl_2 change in the equilibrium mixture if:
   i) more PCl_5 is added *(2)*
   ii) the pressure is increased *(2)*
   iii) the temperature is increased. *(2)*

e) Explain the effect on the value of K_c if:
   i) more PCl_5 is added *(1)*
   ii) the pressure is increased *(1)*
   iii) the temperature is increased. *(2)*

7 Hydrogen is made from natural gas by partial oxidation with steam. This involves the following reaction:

$$CH_4(g) + H_2O(g) \rightleftharpoons CO(g) + 3H_2(g)$$
$$\Delta H = +210\,kJ\,mol^{-1}$$

a) Write an expression for K_p for this reaction. *(2)*

b) How is the value of K_p affected by:
   i) increasing the pressure *(1)*
   ii) increasing the temperature *(1)*
   iii) using a catalyst? *(1)*

c) How does the composition of an equilibrium mixture of the gases change when:
   i) the pressure rises *(1)*
   ii) the temperature rises *(1)*
   iii) a catalyst is added? *(1)*

8 Ammonia is converted to nitric acid on a large scale. In the first step, ammonia is mixed with air and compressed before passing through a reactor containing catalyst gauzes. The catalyst is an alloy of platinum and rhodium. The reversible, exothermic reaction produces nitrogen monoxide (NO) and steam.

a) Write an equation for the reaction of ammonia with oxygen to form nitrogen monoxide. *(2)*

b) Predict, qualitatively, the conditions which favour a high yield of NO in the equilibrium mixture. *(3)*

c) The industrial process typically runs at $1175\,K$ and a pressure of about $7\,atm$ with a mixture of 10% ammonia and 90% air. How and why are these conditions similar to, or different from, those you predicted in (b)? *(3)*

d) Explain the advantage of using a heterogeneous catalyst in this process. *(1)*

e) Why does the gas mixture leaving the reactor not contain as high a percentage of NO as predicted by the equilibrium law? *(2)*

f) The hot gas mixture leaving the reactor has to be cooled before the next step. Suggest how this might be done in a way that improves the overall energy efficiency of the process. *(2)*

9 At $488\,K$, for this equilibrium:

$$COCl_2(g) \rightleftharpoons CO(g) + Cl_2(g)$$

$K_p = 0.2\,Pa$. The fraction of $COCl_2$ that splits up in this way can be represented as the degree of dissociation, α.

a) Derive a relationship between K_p, α, and the total pressure P. *(6)*

b) Assuming that the temperature remains constant, calculate the degree of dissociation at these two pressures:
   i) $10^5\,Pa$ *(2)*
   ii) $2 \times 10^5\,Pa$. *(2)*
   Make the assumption that α is small so that $(1 + \alpha) \approx 1$ and $(1 - \alpha) \approx 1$.

10 During the manufacture of sulfuric acid, sulfur dioxide and air pass through reactors at about $700\,K$, which convert the sulfur dioxide to sulfur trioxide. The volume ratio of oxygen : sulfur dioxide is $1:1$. The gas pressure is $1–2$ atmospheres.

$$2SO_2(g) + O_2(g) \rightleftharpoons 2SO_3(g)$$
$$\Delta H = -192\,kJ\,mol^{-1}$$

The gases pass through a series of four beds of catalyst. The gas mixture is cooled in heat exchangers as it flows from one catalyst bed to the next.

The catalyst is vanadium(v) oxide. The catalyst is not effective if the temperature is lower than $700\,K$. Between the third and fourth beds of catalyst, the gases pass through an absorption tower to remove the sulfur trioxide produced in the first three stages. At the end of the process over 99.5% of the sulfur dioxide is converted to sulfur trioxide.

a) i) Show that the oxygen is in excess in this process. (2)

  ii) Why is excess oxygen used? (2)

  iii) Suggest a reason why the process does not operate with an even larger excess of oxygen. (2)

b) Explain the choice of 700 K as the temperature for the process. (3)

c) Explain why the process operates at a pressure close to atmospheric pressure. (3)

d) Explain why the gas mixture is cooled between the catalyst beds. (2)

e) Explain why the sulfur trioxide is removed from the gas stream before the gases pass into the fourth catalyst bed. (2)

f) Suggest reasons why it is important to convert nearly 100% of the sulfur dioxide to sulfur trioxide. (2)

11 Note that each part of this question requires you to use algebra and solve a quadratic equation. This is not expected for the Edexcel specification. The general form of a quadratic equation is:

$ax^2 + bx + c = 0$, where a, b and c are constants.

The general solution to the equation is

$$x = \frac{-b \pm \sqrt{b^2 - 4ac}}{2a}$$

Solving for x with a calculator always gives two possible values. In equilibrium calculations, only one of the values turns out to be a possible solution to the problem.

In each question work out the amounts at equilibrium in terms of the unknown quantity x. Where necessary then convert to concentrations (or partial pressures) and substitute in the expression for the equilibrium constant. Rearrange the expression to arrive at a quadratic equation that you can solve using the formula given.

a) 2.0 mol hydrogen and 1.0 mol iodine are mixed in a 1.0 dm^3 container at 710 K. The value of K_c for the decomposition of hydrogen iodide into hydrogen and iodine is 0.020 at this temperature. What amounts of hydrogen, iodine and hydrogen iodide are present when the system reaches equilibrium? (6)

b) 1.0 mol ethanoic acid is mixed with 3.0 mol ethanol and 3.0 mol water in the presence of an acid catalyst. The value of K_c for the reaction to form the ester is 4.0 at the temperature of the mixture. What amount of ethyl ethanoate forms at equilibrium? (6)

c) 0.01 mol iodine is mixed with 0.01 mol of iodide ions in 1 dm^3 of an aqueous solution at 298 K. What are the equilibrium concentrations of I_3^- ions and iodide ions in the solution? (6)

$I_3^-(aq) \rightleftharpoons I_2(aq) + I^-(aq)$

$K_c = 1.5 \times 10^{-3}$ mol dm^{-3} at 298 K.

d) 0.20 mol of carbon monoxide and 0.10 mol of chlorine are mixed in a 3.0 dm^3 container and allowed to reach equilibrium. What is the equilibrium concentration of $COCl_2$?

$CO(g) + Cl_2(g) \rightleftharpoons COCl_2(g)$

$K_c = 0.41$ dm^3 mol^{-1} at the temperature of the mixture. (6)

e) $K_p = 7.1$ atm^{-1} for this equilibrium at 298 K:

$2NO_2(g) \rightleftharpoons N_2O_4(g)$

Calculate the partial pressures of the two gases at equilibrium when starting with pure $N_2O_4(g)$ at 1 atmosphere pressure. (6)

12 Acid–base equilibria

Acids and bases are very common, not only in laboratories but also in living things, in the home and in the natural environment. Acid–base reactions are reversible and governed by the equilibrium law. This means that chemists are able to predict reliably and quantitatively how acids and bases behave. This is important for the supply of safe drinking water, the care of patients in hospital, the formulation of shampoos and cosmetics, as well as the processing of food and many other aspects of life.

12.1 Theories to explain reactions of acids and bases

Jabir ibn-Hayyan and his discoveries

The mineral acids that are now taken for granted were discovered by Jabir ibn-Hayyan (*c.* 722–*c.* 815), who worked in the alchemical tradition but pioneered experimental chemistry (Figure 12.2). He developed the techniques of crystallisation and distillation and used them to discover sulfuric, hydrochloric and nitric acids. He also studied ethanoic acid in vinegar and tartaric acid in wine.

Acids were first recognised by their chemical properties. Acidic solutions have a sour taste; they tend to corrode metals and they change the colour of indicators.

Figure 12.1 Red ants attacking a insect on leaf. Methanoic acid is an ingredient of the sting of red ants. The traditional name for the acid was formic acid based on the Latin name for ants: *formica*.

Tip
Section 12.1 traces the development of ideas about acids and bases. This section reminds you of Arrhenius's theory, which you have used until now to explain the reactions of acids and bases. Section 12.2 introduces a new theory which can be applied to reactions that you studied in the first year of the course. The chapter goes on to show how this theory uses the equilibrium law to explain quantitatively the behaviour of acids and bases.

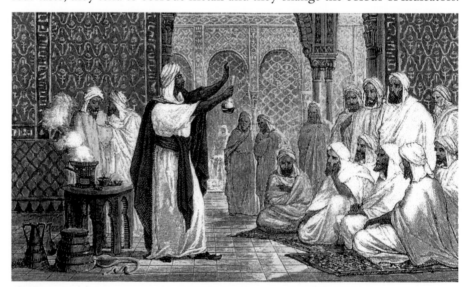

Figure 12.2 Jabir ibn-Hayyan in a coloured engraving, published in 1883, which shows him teaching at the school at Edessa in Mesopotamia (now Sanliurfa in Turkey). He played a key part in turning chemistry from a mystical practice (alchemy) into a science. He pioneered experimental techniques and invented much of the equipment that is still commonly used in laboratories.

The theories of Antoine Lavoisier and Humphry Davy

The scientist who laid the foundations of modern chemical theory was the French nobleman Antoine Lavoisier (1743–1794). His experiments and insights led to the oxygen-theory of burning and a systematic approach to quantitative chemistry.

The name 'oxygen' means 'acid former'. Lavoisier gave the gas this name because he thought that all acids were compounds containing this element.

The generalisation that all acids contain oxygen was disproved by a series of experiments carried out by the English scientist Humphry Davy between 1809 and 1810. He heated hydrogen chloride (then called muriatic acid) to high temperatures with a range of metals and non-metals. He could find no trace of oxygen in the compound. After further work, Davy proposed, in 1816, that what all acids have in common is that they contain hydrogen.

Arrhenius's theory

As a young man in his mid-20s, the Swedish chemist Svante Arrhenius wrote a doctoral thesis which proposed that some compounds are ionised in solution all the time. This was the start of the ionic theory of solutions that we now take for granted. In 1884 it was highly controversial. At the time, Arrhenius was bitterly disappointed to be awarded the bottom grade for his paper. Later he was vindicated and awarded the Nobel prize for chemistry in 1903.

Arrhenius used his ionic theory to come up with an explanation of why it is that all acids have similar properties when dissolved in water. His theory could also account for what happens when an acid is neutralised by an alkali and explain the difference between strong and weak acids. He realised that acids dissociate when they dissolve in water to form ions in the solution. The extent of dissociation into ions distinguishes strong and weak acids.

In 1887, Arrhenius defined an acid as a compound that could produce hydrogen ions when dissolved in water, and an alkali as a compound that could produce hydroxide ions in water. According to Arrhenius's theory, hydrochloric acid is a **strong acid** which is fully ionised when dissolved in water.

$$HCl(aq) \rightarrow H^+(aq) + Cl^-(aq)$$

Ethanoic acid is a **weak acid** which is only slightly ionised.

$$CH_3COOH(aq) \rightleftharpoons CH_3COO^-(aq) + H^+(aq)$$

Arrhenius's theory was a big advance in its time. It could account for the similarities between acids. In this theory, the typical reactions of dilute acids in water are the reactions of aqueous hydrogen ions.

With metals:	$Mg(s) + 2H^+(aq) \rightarrow Mg^{2+}(aq) + H_2(g)$
With carbonates:	$CO_3{}^{2-}(s) + 2H^+(aq) \rightarrow CO_2(g) + H_2O(l)$
With bases:	$O^{2-}(s) + 2H^+(aq) \rightarrow H_2O(l)$

The Arrhenius theory is still useful today and equations for the reactions of acids and alkalis are often written in a form based on the theory. However, the theory has a number of weaknesses, one of which is that it is limited to aqueous solutions.

Test yourself

5 What, according to Arrhenius's theory, happens when an acid neutralises an alkali?

6 Suggest a simple practical demonstration of the difference between equimolar solutions of a strong acid and of a weak acid.

7 Write ionic equations to show how Arrhenius's theory describes the reactions of nitric acid with:

a) zinc   b) potassium carbonate

c) calcium oxide   d) lithium hydroxide.

The theory of Johannes Brønsted and Thomas Lowry

The preferred theory for discussing acid–base equilibria today was put forward independently in 1923 by the Danish chemist Johannes Brønsted and the English chemist Thomas Lowry. This theory describes acids as proton donors, and bases as proton acceptors, as explained in the next section.

12.2 Acids, bases and proton transfer

Acids as proton donors

According to the Brønsted–Lowry theory, hydrogen chloride molecules give hydrogen ions (**protons**) to water molecules when they dissolve in water, producing hydrated hydrogen ions called oxonium ions. The water acts as a base.

$$HCl(aq) + H_2O(l) \rightleftharpoons H_3O^+(aq) + Cl^-(aq)$$

The proton transfer between hydrogen chloride molecules and water molecules is reversible. A proton from the oxonium ion can transfer back to the chloride ion to give hydrogen chloride and water.

Hydrochloric acid is a strong acid. What this means is that it readily gives up its protons to water molecules and the equilibrium in solution lies well over to the right. Hydrochloric acid is effectively completely ionised in solution. Other examples of strong acids are sulfuric acid and nitric acid.

Some **acids** can give away (donate) one proton per molecule. Examples are hydrochloric acid, HCl; nitric acid, HNO_3; and ethanoic acid, CH_3COOH. These acids are sometimes described as monobasic acids because one mole of the acid neutralises one mole of hydroxide ions (a **base**). They are also called monoprotic acids because one proton per molecule can ionise.

There are other acids that can give away (donate) two protons per molecule. Examples are sulfuric acid, H_2SO_4; and ethanedioic acid, HOOC–COOH. These acids are sometimes described as dibasic acids because one mole of the acid neutralises two moles of a base such as sodium hydroxide ions. They are also called diprotic acids.

Tip

It is more correct to represent hydrogen ions in aqueous solution as $H_3O^+(aq)$; however, chemists commonly use a shorter symbol for hydrated protons, $H^+(aq)$.

Key terms

According to the Brønsted-Lowry theory, **acids** are proton donors.

According to the Brønsted-Lowry theory, a **base** is a proton acceptor.

Test yourself

8 a) What type of bond links the water molecule to a proton in an oxonium ion?

b) Draw a dot-and-cross diagram to show the bonding in an oxonium ion. Use your diagram to explain why the ion has a positive charge.

c) Predict the shape of an oxonium ion.

9 Write a balanced ionic equation for the reaction of 1 mol ethanedioic acid with 2 mol NaOH, showing the structural formulae for the acid and for the ethanedioate ion formed.

Bases as proton acceptors

According to the Brønsted–Lowry theory, a base is a molecule or ion which can accept a hydrogen ion (proton) from an acid. A base has a lone pair of electrons which can form a dative covalent bond with a proton (Figures 12.3 and 12.4).

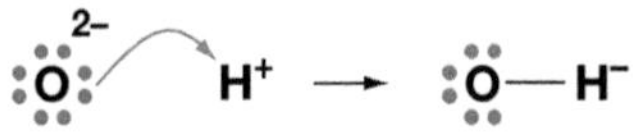

Figure 12.3 Oxide ions have lone pairs of electrons which can form dative covalent bonds with hydrogen ions.

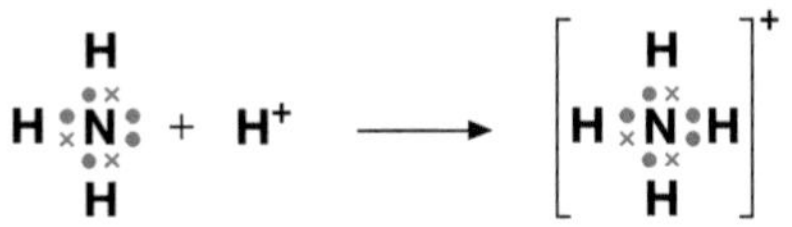

Figure 12.4 The lone pair on the nitrogen atom of ammonia allows it to act as a base.

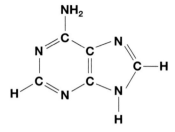

Figure 12.5 The displayed formula of adenine. This is one of the bases in DNA.

An ionic oxide, such as calcium oxide, reacts completely with water to form calcium hydroxide. The calcium ions do not change; but the oxide ions, which are powerful proton acceptors, all take protons from water molecules. An oxide ion is a strong base. Common bases include the oxide and hydroxide ions, ammonia, amines, as well as the carbonate and hydrogencarbonate ions.

In biochemistry the term 'base' often refers to one of the five nitrogenous bases which make up nucleotides and the nucleic acids DNA and RNA. These compounds (adenine, guanine, cytosine, uracil and thymine) are bases in the chemical sense because they have lone pairs on nitrogen atoms which can accept hydrogen ions (Figure 12.5).

> **Tip**
>
> Many compounds of the Group 1 and Group 2 metals form alkaline solutions. This is because metals such as sodium, potassium, magnesium and calcium (unlike other metals) form oxides, hydroxides and carbonates which are soluble (to a greater or lesser extent) in water. It is important to realise that it is the oxide, hydroxide or carbonate ions in these compounds that are bases, and not the metal ions.

Test yourself

10 a) Identify the products of the reaction when concentrated sulfuric acid reacts with sodium chloride.

   b) Show that this is a proton transfer reaction and identify the base.

   c) Account for the fact that this reaction can give a good yield of hydrogen chloride gas, despite the fact that concentrated sulfuric acid and hydrogen chloride are strong acids.

11 Show that the reactions between these pairs of compounds are acid–base reactions and identify as precisely as possible the molecules or ions which are the acid and the base in each example.

   a) $MgO + HCl$

   b) $H_2SO_4 + NH_3$

   c) $NH_4NO_3 + NaOH$

   d) $HCl + Na_2CO_3$

12 a) What type of bond links the ammonia molecule to a proton in an ammonium ion?

   b) Draw a dot-and-cross diagram to show the bonding in an ammonium ion.

   c) Predict the shape of an ammonium ion.

Conjugate acid–base pairs

Any acid–base reaction involves competition for protons. This is illustrated by a solution of an ammonium salt, such as ammonium chloride, in water.

$$NH_4^+(aq) + H_2O(l) \rightleftharpoons NH_3(aq) + H_3O^+(aq)$$

acid 1 base 2 base 1 acid 2

In this example there is competition for protons between ammonia molecules and water molecules. On the left-hand side of the equation the protons are held by lone pairs on the ammonia molecules. On the right-hand side they are held by lone pairs on water molecules. The position of equilibrium shows which of the two bases has the stronger hold on the protons.

Chemists use the term **conjugate acid–base pair** to describe a pair of molecules or ions which can be converted from one to the other by the gain or loss of a proton. The equilibrium in a solution of the ammonium salt above involves two examples of conjugate acid–base pairs:

- NH_4^+ and NH_3
- H_3O^+ and H_2O.

Key terms

An acid turns into its conjugate base when it loses a proton. A base turns into its conjugate acid when it gains a proton. Any pair of compounds made up of an acid and a base that can be converted from one to the other by proton transfer is a **conjugate acid–base pair.**

The definition of **pH** is:

$pH = -\log_{10}[H^+(aq)]$

Test yourself

13 Identify and name the conjugate bases of these acids: HNO_3, CH_3COOH, H_2SO_4, HCO_3^-.

14 Identify and name the conjugate acids of these bases: O^{2-}, OH^-, NH_3, CO_3^{2-}, HCO_3^-, SO_4^{2-}.

15 Explain and illustrate these two statements:

a) The stronger the acid, the weaker its conjugate base.

b) The stronger the base, the weaker its conjugate acid.

12.3 The pH scale

The concentration of hydrogen ions in aqueous solutions commonly ranges from about $2\,mol\,dm^{-3}$ to about $1 \times 10^{-14}\,mol\,dm^{-3}$. The concentration of aqueous hydrogen ions in dilute hydrochloric acid is about 100 000 000 000 000 times greater than the concentration of hydrogen ions in dilute sodium hydroxide solution.

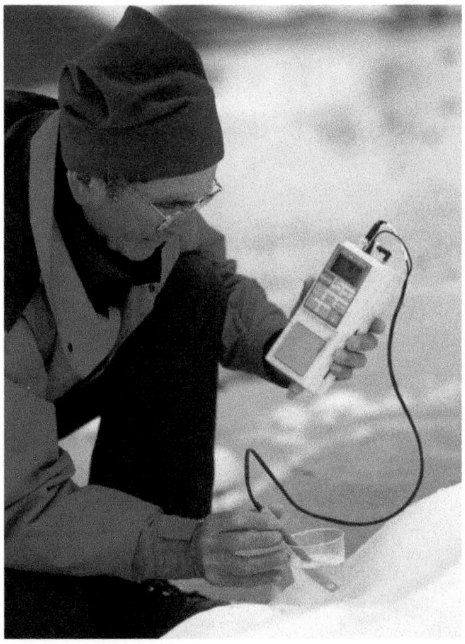

Figure 12.6 A scientist measuring the pH of glacier melt water during research into air and water pollution.

Given such a wide range of concentrations, scientists find it convenient to use a logarithmic scale to measure the concentration of aqueous hydrogen ions in acidic or alkaline solutions (Figures 12.6 and 12.7). This is the **pH** scale, where:

$$pH = -\log_{10}[H^+(aq)]$$

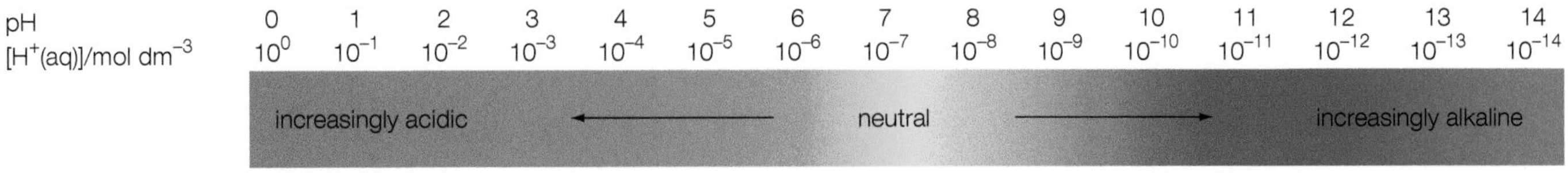
pH
$[H^+(aq)]/mol\,dm^{-3}$
0 1 2 3 4 5 6 7 8 9 10 11 12 13 14
10^0 10^{-1} 10^{-2} 10^{-3} 10^{-4} 10^{-5} 10^{-6} 10^{-7} 10^{-8} 10^{-9} 10^{-10} 10^{-11} 10^{-12} 10^{-13} 10^{-14}
increasingly acidic
neutral
increasingly alkaline

Figure 12.7 The pH scale showing the colours of a full-range indicator at the different pH values.

Examples

Example 1

What is the pH of $0.020 \, mol \, dm^{-3}$ hydrochloric acid?

Notes on the method

Hydrochloric acid is a strong acid so it is fully ionised. Note that 1 mol HCl gives 1 mol $H^+(aq)$. Use the log button on your calculator. Do not forget the minus sign in the definition of pH. Give your answer to 2 decimal places.

Answer

$[H^+(aq)] = [HCl(aq)] = 0.020 \, mol \, dm^{-3}$

$pH = -\log(0.020) = 1.70$

Example 2

The pH of human blood is 7.40. What is the aqueous hydrogen ion concentration in blood?

Notes on the method

$pH = -\log[H^+(aq)]$

From the definition of logarithms this rearranges to $[H^+(aq)] = 10^{-pH}$

Use the inverse log button (10^x) on your calculator. Do not forget the minus sign in the definition of pH.

Give your answer to 2 decimal places.

Answer

$pH = 7.40$

$[H^+(aq)] = 10^{-7.40} = 3.98 \times 10^{-8} \, mol \, dm^{-3}$

Test yourself

16 What is the pH of solutions of hydrochloric acid with these concentrations?

   a) $0.10 \, mol \, dm^{-3}$

   b) $0.010 \, mol \, dm^{-3}$

   c) $0.0010 \, mol \, dm^{-3}$

17 Calculate the pH of a $0.080 \, mol \, dm^{-3}$ solution of nitric acid. Give your answer to 2 decimal places.

18 Calculate the concentration of hydrogen ions in each solution:

   a) orange juice with a pH of 3.30

   b) coffee with a pH of 5.40

   c) saliva with a pH of 6.70

   d) a suspension of an antacid in water with a pH of 10.50.

The ionic product of water

There are hydrogen and hydroxide ions even in pure water because of a transfer of hydrogen ions between water molecules. This only happens to a very slight extent.

$$H_2O(l) + H_2O(l) \rightleftharpoons H_3O^+(aq) + OH^-(aq)$$

This equilibrium can be written more simply as:

$$H_2O(l) \rightleftharpoons H^+(aq) + OH^-(aq)$$

The equilibrium constant is:

$$K_c = \frac{[H^+(aq)][OH^-(aq)]}{[H_2O(l)]}$$

There is such a large excess of water that $[H_2O(l)]$ is a constant, so the relationship simplifies to:

$$K_w = [H^+(aq)][OH^-(aq)]$$

where K_w is the **ionic product** of water.

The pH of pure water at 298 K is 7.0. So the hydrogen ion concentration at equilibrium is:

$$[H^+(aq)] = 1.0 \times 10^{-7} \, mol \, dm^{-3}$$

Also, in pure water $[H^+(aq)] = [OH^-(aq)]$, so:

$$[OH^-(aq)] = 1.0 \times 10^{-7} \, mol \, dm^{-3}$$

Hence:

$$K_w = 1.0 \times 10^{-14} \, mol^2 \, dm^{-6}$$

K_w is a constant in all aqueous solutions at 298 K. This makes it possible to calculate the pH of alkaline solutions.

> ## Key term
>
> K_w is the **ionic product** of water. It is the equilibrium constant for the ionisation of water. It is defined by the expression:
>
> $$K_w = [H^+(aq)][OH^-(aq)]$$

> ## Tip
>
> Refer to Section 2 in 'Mathematics in A Level chemistry', which you can access via the QR code for Chapter 12 on page 320, for help with multiplying together numbers in standard form.

> ## Example
>
> What is the pH of a $0.050 \, mol \, dm^{-3}$ solution of sodium hydroxide at 298 K?
>
> ### Notes on method
> Sodium hydroxide is fully ionised in solution. So in this solution:
>
> $$[OH^-(aq)] = 0.050 \, mol \, dm^{-3}$$
>
> $$pH = -\log[H^+(aq)]$$
>
> ### Answer
> For this solution:
>
> $$K_w = [H^+(aq)] \times 0.050 \, mol \, dm^{-3} = 1.0 \times 10^{-14} \, mol^2 \, dm^{-6}$$
>
> So $\quad [H^+(aq)] = \dfrac{1.0 \times 10^{-14} \, mol^2 \, dm^{-6}}{0.050 \, mol \, dm^{-3}} = 2.0 \times 10^{-13} \, mol \, dm^{-3}$
>
> Hence pH $= -\log(2.0 \times 10^{-13}) = 12.7$

Test yourself

19 The value of K_w varies with temperature. At 273 K its value is $1.10 \times 10^{-15}\,mol^2\,dm^{-6}$, while at 303 K it is $1.50 \times 10^{-14}\,mol^2\,dm^{-6}$.

a) Is the ionisation of water an exothermic or an endothermic process?

b) What happens to the hydrogen ion concentration in pure water, and hence the pH, as the temperature rises?

c) Does pure water stop being neutral if its temperature is above or below 298 K?

20 Calculate the pH of these solutions at 298 K:

a) $1.0\,mol\,dm^{-3}$ NaOH

b) $0.020\,mol\,dm^{-3}$ KOH

c) $0.0010\,mol\,dm^{-3}$ $Ba(OH)_2$.

Working in logarithms

The logarithmic form of equilibrium constants is particularly useful for pH calculations. Taking logarithms produces a conveniently small range of values.

$$K_w = [H^+(aq)][OH^-(aq)] = 1.0 \times 10^{-14} \text{ at } 298\,K$$

Taking logarithms, and applying the rule that $\log xy = \log x + \log y$, gives:

$$\log K_w = \log[H^+(aq)] + \log[OH^-(aq)] = \log 10^{-14} = -14$$

Multiplying through by −1 reverses the signs:

$$-\log K_w = -\log[H^+(aq)] - \log[OH^-(aq)] = 14$$

The term $-\log K_w$ is given the symbol pK_w.

The term $-\log[OH^-(aq)]$ is represented as pOH.

Hence: $pK_w = pH + pOH = 14$

So: $pH = 14 - pOH$

This makes it easy to calculate the pH of alkaline solutions at 298 K.

Key term

pK_w is defined as $-\log K_w$.

Example

What is the pH of a $0.050\,mol\,dm^{-3}$ solution of sodium hydroxide?

Note on method

Sodium hydroxide, NaOH, is a strong base so it is fully ionised.

Use the log button on your calculator to find the values of the logarithms.

Answer

$[OH^-(aq)] = 0.050\,mol\,dm^{-3}$

$pOH = -\log 0.050 = 1.3$

$pH = 14 - pOH = 14 - 1.3 = 12.7$

Tip

See Section 3 in 'Mathematics in A Level chemistry', which you can access via the QR code for Chapter 12 on page 320, for help with logarithms. You do not have to be able to work in logarithms, but some people find it easier. Do not try to remember the formula pH = 14 − pOH. Only use it if you can work it out quickly for yourself from the definition of K_w.

12.4 Weak acids and bases

Most organic acids and bases ionise to only a slight extent in aqueous solution. Carboxylic acids (see Section 17.3.3), such as ethanoic acid in vinegar, citric acid in fruit juices and lactic acid in sour milk, are all weak acids (Figure 12.8). Ammonia and amines (see Section 18.2.3) are weak bases.

Weak acids

In a $0.10\,mol\,dm^{-3}$ solution of ethanoic acid, only about 1 in 100 molecules ionise to produce hydrogen ions. In other words, they only dissociate into ions to a very slight extent.

$$CH_3COOH(aq) \rightleftharpoons CH_3COO^-(aq) + H^+(aq)$$

This means that the pH of a $0.10\,mol\,dm^{-3}$ solution of ethanoic acid is 2.9 and not 1.0, as it would be if it were a strong acid.

There is a very important distinction between acid strength and concentration. Strength is the extent of ionisation. Concentration is the amount in moles of acid in a cubic decimetre. It takes just as much sodium hydroxide to neutralise $25\,cm^3$ of a $0.10\,mol\,dm^{-3}$ solution of a weak acid such as ethanoic acid as it does to neutralise $25\,cm^3$ of a $0.10\,mol\,dm^{-3}$ solution of a strong acid such as hydrochloric acid.

Figure 12.8 Bacteria added to milk ferment the lactose sugar and turn it into lactic acid. Lactic acid is a weak acid that turns the milk into yogurt and also restricts the growth of food poisoning bacteria.

> **Test yourself**
>
> 21 Explain why measuring the pH of a solution of an acid does not provide enough evidence to show whether or not the acid is strong or weak.
>
> 22 Explain why it takes the same amount of sodium hydroxide to neutralise $25.0\,cm^3$ of $0.10\,mol\,dm^{-3}$ ethanoic acid as it does to neutralise $25\,cm^3$ of $0.10\,mol\,dm^{-3}$ hydrochloric acid.

Tip

In everyday life people use 'weak' to mean dilute, as in, "I'll have a cup of weak tea, please." In chemistry the word has a technical meaning and refers to the degree of ionisation and not to the concentration. In chemistry: weak ≠ dilute.

Weak bases

Weak bases only react to form ions to a slight extent when they dissolve in water. In a $0.1\,mol\,dm^{-3}$ solution of ammonia, for example, 99 in every 100 molecules do not react but remain as dissolved molecules. Only 1 molecule in 100 reacts to form ammonium ions.

$$NH_3(aq) + H_2O(l) \rightleftharpoons NH_4^+(aq) + OH^-(aq)$$

As with weak acids, it is important to distinguish between strength and concentration.

Acid dissociation constants

Chemists use the equilibrium constant for the reversible ionisation of a weak acid as a measure of its strength. The equilibrium constant shows the extent to which acids dissociate into ions in solution.

A weak acid can be represented by the general formula HA, where A^- is the ion produced when the acid ionises.

$$HA(aq) \rightleftharpoons H^+(aq) + A^-(aq)$$

Key term

Acid dissociation constant is the name given to the equilibrium constant K_c for the ionisation of a weak acid. This is such an important type of equilibrium constant that it is given its own symbol: K_a.

According to the equilibrium law, the equilibrium constant K_c, takes this form with the subscript 'c' for concentration replaced by subscript 'a' for acid:

$$K_a = \frac{[H^+(aq)][A^-(aq)]}{[HA(aq)]}$$

In this context the equilibrium constant K_a is called the **acid dissociation constant**. Given the value for K_a and the concentration, it is possible to calculate the pH of a solution of a weak acid.

Test yourself

23 If a weak acid is shown as HA, what is A^- in the particular case of:

  a) hydrogen fluoride

  b) methanoic acid

  c) chloric(I) acid?

Example

Calculate the hydrogen ion concentration and the pH of a $0.010\,mol\,dm^{-3}$ solution of propanoic acid. K_a for the acid is $1.3 \times 10^{-5}\,mol\,dm^{-3}$.

Notes on the method
Two approximations simplify the calculation.

1 The first assumption is that at equilibrium $[H^+(aq)] = [A^-(aq)]$. In this example A^- is the propanoate ion $CH_3CH_2COO^-$. This assumption seems obvious from the equation for the ionisation of a weak acid, but it ignores the hydrogen ions from the ionisation of water. Water produces far fewer hydrogen ions than most weak acids, so its ionisation can usually be ignored. This assumption is acceptable so long as the pH of the acid is below 6.

2 The second assumption is that so little of the propanoic acid ionises in water that at equilibrium $[HA(aq)] \approx 0.01\,mol\,dm^{-3}$. Here HA represents propanoic acid. This is a riskier assumption which has to be checked, because in very dilute solutions the degree of ionisation may become quite large relative to the amount of acid in the solution. Chemists generally agree that this assumption is acceptable so long as less than 5% of the acid ionises.

Answer

$$CH_3CH_2COOH(aq) \rightleftharpoons H^+(aq) + CH_3CH_2COO^-(aq)$$

$$K_a = \frac{[H^+(aq)][CH_3CH_2COO^-(aq)]}{[CH_3CH_2COOH(aq)]} = \frac{[H^+(aq)]^2}{0.010 \text{ mol dm}^{-3}}$$

$$K_a = 1.3 \times 10^{-5}\,mol\,dm^{-3}$$

Therefore

$$[H^+(aq)]^2 = 0.010\,mol\,dm^{-3} \times 1.3 \times 10^{-5}\,mol\,dm^{-3} = 1.3 \times 10^{-7}\,mol^2\,dm^{-6}$$

So $[H^+(aq)] = 3.61 \times 10^{-4}\,mol\,dm^{-3}$

$$pH = -\log[H^+(aq)]$$

$$= -\log(3.61 \times 10^{-4})$$

$$= 3.4$$

Check the second assumption: in this case less than $0.0004\,mol\,dm^{-3}$ of the $0.0100\,mol\,dm^{-3}$ of acid (4%) has ionised. In this instance the degree of dissociation is small enough to justify the assumption that $[HA(aq)] \approx$ the concentration of un-ionised acid.

One method which can, in principle, be used to measure K_a for a weak acid is to measure the pH of a solution when the concentration of the acid is accurately known. This is not a good method for determining the size of K_a because the pH values of dilute solutions are very susceptible to contamination – for example by dissolved carbon dioxide from the air.

Example

Calculate the K_a of lactic acid given that pH = 2.43 for a $0.10\,mol\,dm^{-3}$ solution of the acid.

Notes on the method

The same two approximations simplify the calculation.

1 Assume that $[H^+(aq)] = [A^-(aq)]$, where $A^-(aq)$ here represents the aqueous lactate ion. Since the pH is well below 6 this is certainly justified.

2 Also assume that so little of the lactic acid ionises in water that at equilibrium $[HA(aq)] \approx 0.1\,mol\,dm^{-3}$. Here HA represents lactic acid. This is a riskier assumption, which again can be checked during the calculation.

Answer

$$pH = 2.43$$

$$[H^+(aq)] = 10^{-2.43} = 3.72 \times 10^{-3}\,mol\,dm^{-3}$$

$$[H^+(aq)] = [A^-(aq)] = 3.72 \times 10^{-3}\,mol\,dm^{-3}$$

In this example less than 5% of the acid is ionised (less than 0.004 out of 0.100 mol in each litre).

So $[HA(aq)] \approx 0.1\,mol\,dm^{-3}$

Substituting in the expression for K_a:

$$K_a = \frac{[H^+(aq)]\,[A^-(aq)]}{[HA(aq)]} = \frac{(3.72 \times 10^{-3}\,mol\,dm^{-3})^2}{0.1\,mol\,dm^{-3}}$$

$$K_a = 1.4 \times 10^{-4}\,mol\,dm^{-3}$$

24 Calculate the pH of a $0.010\,mol\,dm^{-3}$ solution of hydrogen cyanide given that $K_a = 4.9 \times 10^{-10}\,mol\,dm^{-3}$.

25 Calculate the pH of a $0.050\,mol\,dm^{-3}$ solution of ethanoic acid given that $K_a = 1.7 \times 10^{-5}\,mol\,dm^{-3}$.

26 Calculate K_a for methanoic acid given that pH = 2.55 for a $0.050\,mol\,dm^{-3}$ solution of the acid.

27 Calculate K_a for butanoic acid, C_3H_7COOH, given that pH = 3.42 for a $0.010\,mol\,dm^{-3}$ solution of the acid.

Working in logarithms

Chemists find it convenient to define a quantity $pK_a = -\log K_a$ when working with weak acids. This definition means that hydrocyanic acid, HCN, with a pK_a value of 9.3, is a much weaker acid than nitrous acid, HNO_2, with a pK_a value of 3.3.

Data tables show pK_a values. The relationship between acid strength and pH can be expressed simply because both are logarithmic quantities.

$$K_a = \frac{[H^+(aq)][A^-(aq)]}{[HA(aq)]}$$

The two common assumptions when using this expression in calculations are that:

$$[H^+(aq)] = [A^-(aq)]$$

$[HA(aq)] = c_A$, where c_A = the concentration of the un-ionised acid.

Substituting in the expression for K_a gives:

$$K_a = \frac{[H^+(aq)]^2}{c_A}$$

Hence: $\qquad K_a \times c_A = [H^+(aq)]^2$

Taking logarithms:

$$\log(K_a \times c_A) = \log[H^+(aq)]^2$$

Applying the rules that $\log xy = \log x + \log y$ and that $\log x^n = n\log x$, gives:

$$\log K_a + \log c_A = 2\log[H^+(aq)]$$

which on multiplying by -1 becomes:

$$-\log K_a - \log c_A = -2\log[H^+(aq)]$$

Hence: $\qquad pK_a - \log c_A = 2 \times pH$

This shows that, for a solution of a weak acid which is less than 5% ionised:

$$pH = \tfrac{1}{2}(pK_a - \log c_A) \quad \text{which rearranges to} \quad pK_a = 2\,pH + \log c_A.$$

Tip

See Section 3 in 'Mathematics in A Level chemistry', which you can access via the QR code for Chapter 12 on page 320, for help with logarithms. You do not have to be able to work in logarithms, but some people find it easier. Do not try to remember this logarithmic form of the equilibrium law. The relationship is easy to use, but only apply it if you can derive it quickly from first principles as shown here. Do not forget that this form of the law has two built-in assumptions, so it only applies when these assumptions are acceptable.

28 What is the value of pK_a for methanoic acid, given that $K_a = 1.6 \times 10^{-4} \, mol \, dm^{-3}$?

29 What is the value of K_a for benzoic acid, given that $pK_a = 4.2$?

30 Show that the logarithmic relationship $pK_a = 2pH + \log c_A$ gives the same answers from the data as the methods used in the worked examples on pages 30 and 31.

Activity

The effect of dilution on the degree of dissociation of a weak acid

Two students used a pH meter to investigate the effect of dilution on the dissociation of ethanoic acid. They started by preparing the solutions shown in Table 12.1 by diluting a $0.10 \, mol \, dm^{-3}$ solution of the acid.

Next they calibrated a pH meter by dipping the probe into a solution of known pH (a buffer solution, see Section 12.8). After rinsing the probe with distilled water, they dipped it into the least concentrated of the solutions to measure the pH. They continued to rinse the probe and then measure the pH of the next solution, until they had recorded pH values for all four solutions.

Table 12.1

Concentration of ethanoic acid/$mol \, dm^{-3}$	Measured pH of ethanoic acid	Calculated pH of solutions of hydrochloric acid with the same concentration
0.00010	4.2	
0.0010	3.5	
0.010	3.0	
0.10	2.7	1.0

1 Describe how the students could prepare a $0.010 \, mol \, dm^{-3}$ solution of ethanoic acid from the $0.10 \, mol \, dm^{-3}$ solution.

2 The water that the students used for the dilutions had been boiled and then allowed to cool to room temperature. Explain why this improved the accuracy of the measurements.

3 Explain why it was necessary to calibrate the pH meter.

4 Why did the students measure the pH of the most dilute solution first and then work up to the more concentrated solutions in the order they are listed in Table 12.1?

5 What are the calculated pH values for hydrochloric acid that are missing from the table?

6 a) What does the table tell you about the degree of dissociation of ethanoic acid compared to hydrochloric acid at any concentration?

   b) Study the difference in the pH values for the two acids at each dilution. What do the differences show about the effect of dilution on the degree of dissociation of ethanoic acid?

   c) Use the equilibrium law to explain the effect of dilution on the degree of dissociation of ethanoic acid.

12.5 Acid–base titrations

The equilibrium law helps to explain what happens during acid–base titrations and it provides a rationale for the selection of the right indicator for a titration.

During a titration, the pH changes as a solution of an alkali runs from a burette and mixes with an acid in a flask (Figure 12.9). Plotting pH against volume of alkali added gives a graph whose shape is determined by the nature of the acid and the base. Usually there is a marked change in pH near the equivalence point, and it is this which makes it possible to detect the end-point of the titration with an indicator.

At the end-point, the colour change of the indicator shows that enough of the solution in the burette has been added to react with the amount of the chemical in the flask. In a well-planned titration, the colour change observed at the end-point corresponds exactly with the equivalence point.

The equivalence point is the point during any titration when the amount in moles of one reactant added from a burette is just enough to react exactly with all of the measured amount of chemical in the flask as shown by the balanced equation.

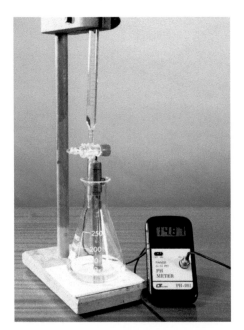

Figure 12.9 A pH meter can be used to measure the pH of the solution in the flask during an acid-base titration, and to detect the end-point.

Key terms

The **equivalence point** during a titration is reached when the amount of reactant added from a burette is just enough to react exactly with all the measured amount of chemical in the flask according to the balanced equation.

The **end-point** in a titration the point at which a colour change or pH change indicates that just enough of the solution in the burette has been added to react with the chemical in the flask.

Activity

Titration of a strong acid with a strong base

Strong acids and bases are fully ionised in solution. Figure 12.10 shows the shape of the pH curve for the titration of a strong acid, hydrochloric acid, with a strong base, sodium hydroxide.

1 Show that pH = 1.0 for a solution of $0.10 \, \text{mol dm}^{-3}$ HCl(aq).
2 Why does pH = 7 at the equivalence point of a titration of a strong acid with a strong base?
3 Calculate the pH of $25 \, \text{cm}^3$ of a solution of sodium chloride after adding:
   a) $0.05 \, \text{cm}^3$ (1 drop) of $0.10 \, \text{mol dm}^{-3}$ HCl(aq)
   b) $0.05 \, \text{cm}^3$ (1 drop) of $0.10 \, \text{mol dm}^{-3}$ NaOH(aq).
   (In both instances assume that the volume change on adding 1 drop is insignificant.)
4 Calculate the pH of the solution produced by adding $5.0 \, \text{cm}^3$ of $0.10 \, \text{mol dm}^{-3}$ NaOH(aq) to $25.0 \, \text{cm}^3$ of a solution of sodium chloride.
5 Show that your answers to Questions 1, 2, 3 and 4 are consistent with Figure 12.10.
6 What features of the curve plotted in Figure 12.10 are important for the accuracy of acid–base titrations of this kind?

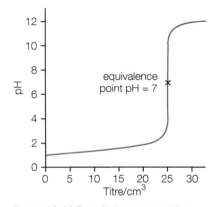

Figure 12.10 The pH change on adding $0.10 \, \text{mol dm}^{-3}$ NaOH(aq) from a burette to $25.0 \, \text{cm}^3$ of a $0.10 \, \text{mol dm}^{-3}$ solution of HCl(aq).

Titration of a weak acid with a strong base

If the acid in the titration flask is weak, then the equilibrium law applies and the pH curve up to the equivalence point has to be calculated with the help of the expression for K_a.

Consider, for example, the reaction of ethanoic acid with sodium hydroxide during a titration (Figure 12.11). At the start the flask contains the pure acid.

$$CH_3COOH(aq) \rightleftharpoons H^+(aq) + CH_3COO^-(aq)$$

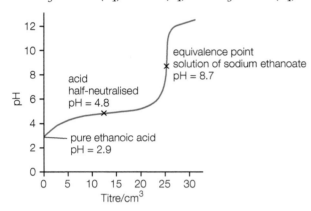

Figure 12.11 The pH change on adding $0.10\,mol\,dm^{-3}$ NaOH(aq) from a burette to $25.0\,cm^3$ of a $0.10\,mol\,dm^{-3}$ solution of $CH_3COOH(aq)$.

The pH of the pure acid can be calculated from K_a, as shown in Section 12.4. However, when some strong alkali runs in from the burette, some of the ethanoic acid reacts to produce sodium ethanoate. Once this has happened, $[H^+(aq)] \neq [CH_3COO^-(aq)]$, and the method of calculating the pH has to change to account for this. The following worked example shows how this is done. (The reason why this method is necessary is explained in Section 12.8.)

Example

What is the pH of a mixture formed during a titration after adding $20.0\,cm^3$ of $0.10\,mol\,dm^{-3}$ NaOH(aq) to $25.0\,cm^3$ of a $0.10\,mol\,dm^{-3}$ solution of $CH_3COOH(aq)$ if $K_a = 1.7 \times 10^{-5}\,mol\,dm^{-3}$?

Notes on the method

The pH of the mixture can be estimated quite accurately using the equilibrium law by assuming that:

- the concentration of ethanoic acid molecules at equilibrium is determined by the amount of acid which has yet to be neutralised
- the concentration of ethanoate ions is determined by the amount of acid converted to sodium ethanoate.

Answer

$$K_a = \frac{[H^+(aq)][CH_3COO^-(aq)]}{[CH_3COOH(aq)]}$$

This rearranges to give:

$$[H^+(aq)] = \frac{K_a \times [CH_3COOH(aq)]}{[CH_3COO^-(aq)]}$$

The total volume of the solution = $45.0\,cm^3$.

$5.0\,cm^3$ of the $0.10\,mol\,dm^{-3}$ ethanoic acid remains not neutralised. This is now diluted to a total volume of $45\,cm^3$ solution.

Concentration of ethanoic acid molecules

$$= \frac{5.0\,cm^3}{45.0\,cm^3} \times 0.10\,mol\,dm^{-3}$$

Also the concentration of ethanoate ions

$$= \frac{20.0\,cm^3}{45.0\,cm^3} \times 0.10\,mol\,dm^{-3}$$

So the ratio $\dfrac{[CH_3COOH(aq)]}{[CH_3COO^-(aq)]} = \dfrac{5.0}{20.0}$

Substituting in the rearranged expression for the equilibrium law gives:

$$[H^+(aq)] = K_a \times \frac{5.0}{20.0} = 1.7 \times 10^{-5}\,mol\,dm^{-3} \times \frac{5.0}{20.0}$$

$$[H^+(aq)] = 4.25 \times 10^{-6}\,mol\,dm^{-3}$$

$$pH = -\log[H^+(aq)] = -\log(4.25 \times 10^{-6}) = 5.4$$

The pH change for this titration is shown in Figure 12.11. Note that halfway to the equivalence point, the added alkali converts half of the weak acid to its salt. In this example, at this point:

$$[CH_3COOH(aq)] = [CH_3COO^-(aq)]$$

So: $[H^+(aq)] = \dfrac{K_a \times [CH_3COOH(aq)]}{[CH_3COO^-(aq)]} = K_a$

Hence, at this point: $[H^+(aq)] = K_a$

It follows that $pH = pK_a$ halfway to the equivalence point (see Core practical 9 in Section 12.8).

At the equivalence point, the solution contains sodium ethanoate. As Figure 12.11 shows, the solution at this point is not neutral. A solution of a salt of a weak acid and a strong base is alkaline (see Section 12.7). Sodium ions have no effect on the pH of a solution, but ethanoate ions are basic. The ethanoate ion is the conjugate base of a weak acid.

Beyond the equivalence point, the curve is determined by the excess of strong base, and so the shape is very close to the shape of the curve after the end-point in Figure 12.10.

Test yourself

31 Calculate the pH of a 0.10 mol dm^{-3} solution of CH$_3$COOH(aq).

32 Calculate the pH of a mixture formed during a titration after adding 10.0 cm^3 of 0.10 mol dm^{-3} NaOH(aq) to 25.0 cm^3 of a 0.10 mol dm^{-3} solution of CH$_3$COOH(aq).

33 Explain why a solution of sodium ethanoate is alkaline.

34 Why is the equivalence point reached at 25.0 cm^3 in both the titrations illustrated in Figures 12.10 and 12.11?

Titration of a strong acid with a weak base

During the titration of a strong acid with a weak base, the flask contains a strong acid at the start and the titration curve follows the same line as in Figure 12.10. In a titration of hydrochloric acid with ammonia solution, for example, the salt formed at the equivalence point is ammonium chloride.

Since ammonia is a weak base, the ammonium ion is an acid. So a solution of ammonium chloride is acidic and the pH is below 7 at the equivalence point. As shown in Figure 12.12, after the equivalence point the curve rises less far than in Figure 12.10 because the excess alkali is a weak base and is not fully ionised.

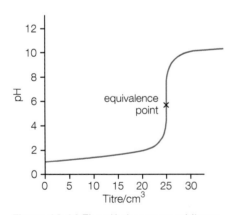

Figure 12.12 The pH change on adding a 0.10 mol dm^{-3} solution of the weak base NH$_3$(aq) from a burette to 25.0 cm^3 of a 0.10 mol dm^{-3} solution of the strong acid HCl(aq).

Test yourself

35 Explain why a solution of ammonium chloride is acidic.

36 Write a balanced equation for the neutralisation of ethanoic acid by ammonia solution.

Titration of a weak acid with a weak base

In practice it is not usual to titrate a weak acid with a weak base. As shown in Figure 12.13, the change of pH around the equivalence point is gradual and not very marked if both the acid and base are weak. This means that it is hard to fix the end-point precisely. If the dissociation constants for the weak acid and for the weak base are approximately equal (as is the case for ethanoic acid and ammonia) then the salt formed at the equivalence point is neutral and pH = 7 at this point.

Working with logarithms

There can be an advantage to working with a logarithmic form of the equilibrium law when calculating the pH of a mixture of a weak acid and one of its salts during titrations.

In general, for a weak acid HA:

$$HA(aq) \rightleftharpoons H^+(aq) + A^-(aq)$$

$$K_a = \frac{[H^+(aq)][A^-(aq)]}{[HA(aq)]}$$

This rearranges to give:

$$[H^+(aq)] = K_a \times \frac{[HA(aq)]}{[A^-(aq)]}$$

Taking logs and substituting pH for $-\log[H^+(aq)]$ and pK_a for $-\log K_a$ gives:

$$pH = pK_a + \log\left(\frac{[A^-(aq)]}{[HA(aq)]}\right)$$

Note the change of sign and the inversion of the log ratio. This follows because:

$$-\log\left(\frac{[A^-(aq)]}{[HA(aq)]}\right) = +\log\left(\frac{[HA(aq)]}{[A^-(aq)]}\right)$$

In a mixture of a weak acid and one of its salts, the weak acid is only slightly ionised, while the salt is fully ionised, so it is often accurate enough to make the assumption that all the negative ions come from the salt present and all the un-ionised molecules from the acid.

Hence: $$pH = pK_a + \log\frac{[salt]}{[acid]}$$

This form of the equilibrium law cannot be used to calculate the pH of a solution of a weak acid on its own. However, it can help to explain the properties of acid–base indicators (Section 12.6) and to account for the behaviour of buffer solutions (Section 12.8).

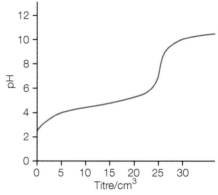

Figure 12.13 The pH change on adding $0.10\,mol\,dm^{-3}$ $NH_3(aq)$ from a burette to $25.0\,cm^3$ of a $0.10\,mol\,dm^{-3}$ solution of $CH_3COOH(aq)$. Before the end-point the curve is essentially the same as in Figure 12.11, while after the end-point it is as in Figure 12.12. Note the resulting small change of pH around the equivalence point.

Tip

You are not required to use the logarithmic form of the equilibrium law. If you choose to do so, make sure that you can derive it for yourself. Also check that you understand the assumptions made when deriving this form of the law, so that you know when it applies.

12.6 Indicators

Acid–base indicators change colour when the pH changes. They signal the end-point of a titration. No one indicator is right for all titrations and the equilibrium law can help chemists to choose the indicator that gives accurate results for a particular combination of acid and base.

The indicator chosen for a titration must change colour completely in the pH range of the near vertical part of the pH curve (see Figures 12.10, 12.11 and 12.12). This is essential if the visible end-point is to correspond to the equivalence point when exactly equal amounts of acid and base are mixed.

Table 12.2 gives some data for four common indicators. Note that each indicator changes colour over a range of pH values, which differs from one indicator to the next (Figures 12.14, 12.15 and 12.16). These indicators are themselves weak acids or bases which change colour when they lose or gain protons.

Table 12.2 Properties of selected indicators (the Edexcel Data booklet includes information about more indicators).

Indicator	pK_a	Colour change HIn/In⁻	pH range over which the colour change occurs
Methyl orange	3.7	Red/yellow	3.2–4.4
Methyl red	5.1	Yellow/red	4.2–6.3
Bromothymol blue	7.0	Yellow/blue	6.0–7.6
Phenolphthalein	9.3	Colourless/pink	8.2–10.0

> **Tip**
>
> Methyl orange can be a difficult indicator to use because it is hard to spot the point at which an orange colour marks the end-point. Sometimes a dye is mixed with the indicator to produce 'screened methyl orange'. This changes from purple to grey at the end-point and then goes green with excess alkali. Some people find it much easier to detect the end-point with the screened indicator.

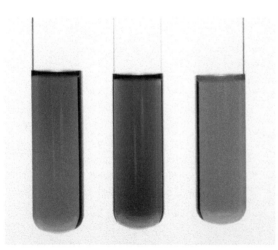

Figure 12.14 The colours of screened methyl orange indicator at pH 6 (left), pH 4 (middle) and pH 2 (right). This indicator includes a green dye to make the colour change easier to see.

Figure 12.15 The colours of phenolphthalein indicator at pH 7 (left) and pH 11 (right).

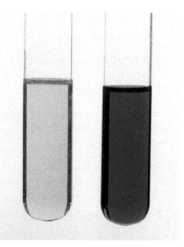

Figure 12.16 The colours of bromothymol blue indicator at pH 5 (left) and pH 8 (right).

When added to a solution, an indicator gains or loses protons depending on the pH of the solution. It is conventional to represent a weak acid indicator as HIn, where In is a shorthand for the rest of molecule other than the ionisable hydrogen atom. In water:

$$HIn(aq) \quad \rightleftharpoons \quad H^+(aq) \quad + \quad In^-(aq)$$

un-ionised indicator   indicator after losing a proton
colour 1   colour 2

Note that an analyst only adds a drop or two of indicator during a titration. This means that there is so little indicator that it cannot affect the pH of the mixture. The pH is determined by the titration (as shown in Figures 12.10, 12.11 and 12.12). The position of the equilibrium for the ionisation of the indicator shifts one way or the other as dictated by the changing pH of the solution in the titration flask.

The pH range over which an indicator, HIn, changes colour is determined by its strength as the acid (Figure 12.17). Typically the range is given roughly by $pK_a \pm 1$. The logarithmic form of the equilibrium law derived at the end of Section 12.5 shows why this is so. For an indicator it takes this form:

$$pH = pK_a + \log\left(\frac{[In^-(aq)]}{[HIn(aq)]}\right)$$

When $pH = pK_a$, $[HIn(aq)] = [In^-(aq)]$ and the two different colours of the indicator are present in equal amounts. The indicator is mid-way through its colour change.

Add a few drops of acid and the pH falls. If the two colours of the indicator are equally intense, it turns out that the human eye sees the characteristic acid colour of the indicator clearly when $[HIn] = 10 \times [In^-(aq)]$.

At this point $pH = pK_a + \log 0.1 = pK_a - 1$ (since $\log 0.1 = -1$).

Add a few drops of alkali and the pH rises. Similarly, the human eye sees the characteristic alkaline colour of the indicator clearly when $[In^-(aq)] = 10 \times [HIn(aq)]$.

At this point: $pH = pK_a + \log 10 = pK_a + 1$ (since $\log 10 = +1$).

Figure 12.17 shows structures of methyl orange in acid and alkaline solutions. In acid solution the added hydrogen ion (proton) localises two electrons to form a covalent bond. In alkaline solution the removal of the hydrogen ion allows the two electrons to join the other delocalised electrons (see Section 18.1.3). The change in the number of delocalised electrons causes a shift in the peak of the wavelengths of light absorbed, so the colour changes and the molecule acts as an indicator.

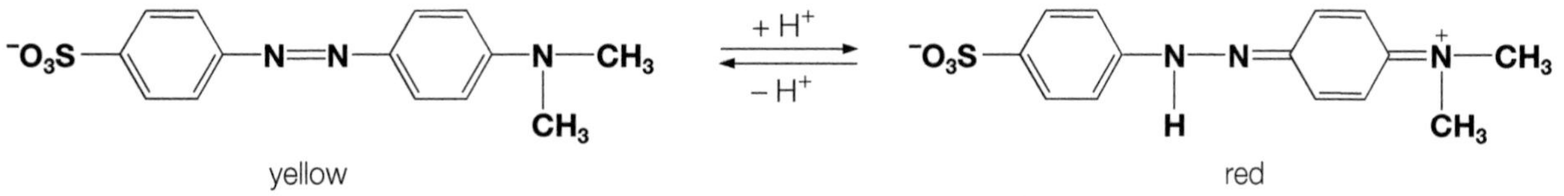

Figure 12.17 The structures of methyl orange in acid (right) and alkaline solutions (left).

12.7 Neutralisation reactions

Chemists use the term 'neutralisation' to describe any reaction in which an acid reacts with a base to form a salt, even when the pH does not equal 7 on mixing equivalent amounts of the acid and the alkali.

The pH of salts

Mixing equal amounts (in moles) of hydrochloric acid with sodium hydroxide produces a neutral solution of sodium chloride. Strong acids, such as hydrochloric acid, and strong bases, such as sodium hydroxide, are fully ionised in solution. The salt formed from the reaction of hydrochloric acid and sodium hydroxide, sodium chloride, is also fully ionised. Writing ionic equations for these examples shows that neutralisation is essentially a reaction between aqueous hydrogen ions and hydroxide ions. This is supported by the values for enthalpies of neutralisation – see the next part of this section.

$$H^+(aq) + OH^-(aq) \rightleftharpoons H_2O(l)$$

The surprise is that 'neutralisation reactions' do not always produce neutral solutions. 'Neutralising' a weak acid, such as ethanoic acid, with an equal amount, in moles, of a strong base, such as sodium hydroxide, produces a solution of sodium ethanoate, which is alkaline (see Figure 12.11).

'Neutralising' a weak base, such as ammonia, with an equal amount of the strong acid hydrochloric acid produces a solution of ammonium chloride, which is acidic (see Figure 12.12).

Where either the 'parent acid' or 'parent base' of a salt is weak, the salt dissolves to give a solution which is not neutral (Figure 12.18). The 'strong parent' in the partnership 'wins':

- weak acid/strong base – the salt is alkaline in solution
- strong acid/weak base – the salt is acidic in solution

Figure 12.18 Raponzolo di roccia grows in the moist and shady crevices of limestone cliffs of the Italian Alps. Weathering of the limestone keeps the pH high so that the soil water is alkaline.

Enthalpy change of neutralisation

Strong acids and bases

The standard enthalpy change of neutralisation is the enthalpy change for the reaction when an acid neutralises an alkali to form one mole of water.

$$HCl(aq) + NaOH(aq) \rightarrow NaCl(aq) + H_2O(l)$$

$$\Delta_n H^{\ominus} = -57.5\,kJ\,mol^{-1}$$

The standard enthalpy change of neutralisation for dilute solutions of strong acid with strong base is always close to $-57.5\,kJ\,mol^{-1}$. The reason is that these acids and alkalis are fully ionised. So, in every instance, the reaction is the same:

$$H^+(aq) + OH^-(aq) \rightarrow H_2O(l) \quad \Delta H^{\ominus} = -57.5\,kJ\,mol^{-1}$$

Enthalpy changes of neutralisation can be measured approximately by mixing solutions of acids and alkalis in a calorimeter (Figure 12.19).

Weak acids and bases

The standard enthalpy changes of neutralisation reactions involving weak acids and weak bases are less negative than those for neutralisation reactions between strong acids and bases. The standard enthalpy changes for the neutralisation of ethanoic acid by sodium hydroxide, for example, is $-56.1\,kJ\,mol^{-1}$.

This is partly because the weak acids and bases are not fully ionised at the start, so that the neutralisations cannot be described simply as reactions between aqueous hydrogen ions and aqueous hydroxide ions. Also, the solutions are not neutral at the equivalence point.

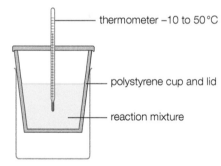

Figure 12.19 Apparatus for measuring the enthalpy change of the neutralisation of an acid by a base.

Example

50 cm^3 of 1.0 mol dm^{-3} dilute nitric acid were mixed with 50 cm^3 of 1.0 mol dm^{-3} dilute potassium hydroxide solution in an expanded polystyrene cup. The temperature rise was 6.7 °C. Calculate the enthalpy change of neutralisation for the reaction.

Notes on the method

Note that the total volume of solution on mixing is 100 cm^3.

Assume that the density and specific heat capacity of the solutions is the same as for pure water.

The density of water is $1.0\,g\,cm^{-3}$, so the mass of $100\,cm^3$ water is $100\,g$.

The specific heat capacity of water is $4.18\,J\,g^{-1}\,K^{-1}$.

A temperature change of $6.7\,°C$ is the same as a change of $6.7\,K$ on the Kelvin scale.

The energy from the exothermic reaction is trapped in the system by the expanded polystyrene, so it heats up the mixture.

Answer

The energy change $= 4.18\,J\,g^{-1}\,K^{-1} \times 100\,g \times 6.7\,K = 2800\,J$

Amount of acid neutralised $= \dfrac{50}{1000}\,dm^3 \times 1.0\,mol\,dm^{-3} = 0.050\,mol$

$$HNO_3(aq) + KOH(aq) \rightarrow KNO_3(aq) + H_2O(l)$$

$$\Delta_nH = \dfrac{-2800\,J}{0.050\,mol} = -56\,000\,J\,mol^{-1} = -56\,kJ\,mol^{-1}$$

Test yourself

41 Account for the discrepancy between the value calculated in the worked example from experimental results and the expected value of about $-57.5\,kJ\,mol^{-1}$.

42 Suggest an explanation for the difference in the values of $\Delta_nH^{\ominus}$ for HCl/NaOH and CH_3COOH/NaOH.

43 Here are three pairs of acids and bases which can react to form salts: HBr/NaOH, HCl/NH_3, CH_3COOH/NH_3.

Here are three values for the standard enthalpy change of neutralisation:

- $-50.4\,kJ\,mol^{-1}$
- $-53.4\,kJ\,mol^{-1}$
- $-57.6\,kJ\,mol^{-1}$.

Write the equations for the three neutralisation reactions and match them with the corresponding value of $\Delta_nH^{\ominus}$.

12.8 Buffer solutions

Buffer solutions are mixtures of molecules and ions in solution which help to keep the pH more or less constant when small quantities of acid or alkali are added to a solution. Buffer solutions help to stabilise the pH of blood, medicines, shampoos, water in swimming pools, and of many other solutions in living things, domestic products and in the environment (Figure 12.20).

Buffer mixtures are important in the food industry because many properties of drinks and foods depend upon their being formulated to the correct pH. The quality of food can be affected significantly if the pH shifts too far from the required value. Maintaining a low pH can help to prevent the growth of bacteria or fungi that spoil food or cause food poisoning.

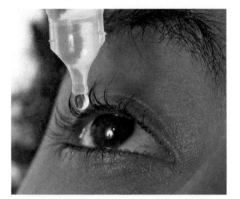

Figure 12.20 Eye drops contain a buffer solution to make sure that they do not irritate the sensitive surface of the eye.

Buffers are also important in living organisms. The pH of blood, for example, is closely controlled by buffers within the narrow range 7.35 to 7.45. Chemists use buffers when they want to investigate chemical reactions at a fixed pH.

Buffers are equilibrium systems which illustrate the practical importance of the equilibrium law. A typical buffer mixture consists of a solution of a weak acid and one of its salts; for example, a mixture of ethanoic acid and sodium ethanoate (Figure 12.21). There must be plenty of both the acid and its salt.

Key term

A **buffer solution** is a mixture of molecules and ions in solution which help to keep the pH more or less constant. A buffer solution cannot prevent pH changes, but it evens out the large changes in pH which can happen without a buffer when small amounts of acid or alkali are added to an aqueous solution. A typical buffer mixture consists of a solution of a weak acid and one of its salts.

$CH_3COOH(aq)$ + $H_2O(l)$ $\rightleftharpoons$ $CH_3COO^-(aq)$ + $H_3O^+(aq)$

acid molecules are a reservoir of H^+ ions

base ions – with the capacity to accept H^+ ions

stays roughly constant so the pH hardly changes

plenty of weak acid to supply more H^+ ions if alkali is added

plenty of the ions from the salt able to combine with H^+ ions if acid is added

Figure 12.21 The action of a buffer solution.

Le Chatelier's principle provides a qualitative interpretation of the buffering action. Adding a little strong acid temporarily increases the concentration of $H^+(aq)$ so at that instant the system is not equilibrium. So the reaction mixture shifts towards the left-hand side of the equation to reduce the hydrogen ion concentration, thus counteracting the change and establishing a new equilibrium. Conversely, the effect of adding a little strong alkali temporarily decreases the concentration of $H^+(aq)$ so the reaction mixture shifts to the right of the equation to replace some, though not all, of the hydrogen ions that have been neutralised.

The pH of buffer solutions

By choosing the right weak acid, it is possible to prepare buffers at any pH value throughout the pH scale. If the concentrations of the weak acid and its salt are the same, then the pH of the buffer is equal to pK_a for the acid. The pH of a buffer mixture can be calculated with the help of the equilibrium law.

$$K_a = \frac{[H^+(aq)][A^-(aq)]}{[HA(aq)]}$$

This rearranges to give: $[HA(aq)] = K_a \times \dfrac{[A^-(aq)]}{[H^+(aq)]}$

So the equilibrium law makes it possible to calculate the pH of a buffer solution made from a mixture of a weak acid and its conjugate base.

In a mixture of a weak acid and its salt, the weak acid is only slightly ionised while the salt is fully ionised. This means that it is often accurate enough to assume that:

- all the molecules HA come from the added acid
- all the negative ions, $A^-(aq)$, come from the added salt.

So the calculation of the hydrogen ion concentration of a buffer solution can be based on the formula:

$$[H^+(aq)] = K_a \times \frac{[acid]}{[salt]}$$

Alternatively, you can use the logarithmic form of this relationship (see Section 12.5).

Test yourself

44 Explain why a weak acid on its own cannot make a buffer solution, but a mixture of a weak acid and one of its salts can.

Tip

The theory of acid–base indicators and the theory of buffer solutions is essentially the same. The difference is that a large amount of buffer mixture is added to dictate the pH of a solution, whereas the drop or two of an indicator in a titration flask is too little to affect the pH. An indicator follows the pH changes dictated by the mixture of acid and alkali during the titration.

Blood buffers

In a healthy person the pH of blood lies within a narrow range (7.35–7.45). Chemical reactions in cells tend to upset the normal pH. Respiration, for example, continuously produces carbon dioxide. The carbon dioxide diffuses into the blood where it is mainly in the form of carbonic acid. However, the blood pH stays constant because it is stabilised by buffer solutions, in particular by the buffer system based on the equilibrium between carbon dioxide, water and hydrogencarbonate ions. This is the carbonic acid–hydrogencarbonate buffer.

Proteins in blood, including haemoglobin, can also contribute to the buffering action of blood pH. This is because the molecules contain both acidic and basic functional groups (see Section 18.2.7).

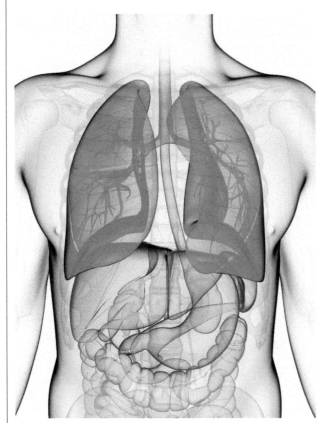

Figure 12.22 The lungs have a vital part to play in maintaining the pH of the blood.

Two major organs help to control the total amounts of carbonic acid and hydrogencarbonate ions in the blood. The lungs remove excess carbon dioxide from the blood (Figure 12.22) and the kidneys remove excess hydrogencarbonate ions.

The brain responds to the level of carbon dioxide in the blood. During exercise, for example, the brain speeds up the rate of breathing.

The consequences can be fatal if the blood pH moves outside the normal range. Patients who have been badly burned or suffered other serious injuries are treated quickly with a drip into a vein. One of the purposes of an intravenous drip is to help maintain the pH of the blood close to its normal value.

1　Write an equation to show aqueous carbon dioxide reacting with water to form hydrogen ions and hydrogencarbonate ions.

2　Explain why breathing faster and more deeply tends to raise the blood pH.

3　a) Suggest two reasons why the blood pH tends to fall during strenuous exercise.

　　b) Why do people breathe faster and more deeply when running?

4　a) Write the expression for K_a for the equilibrium between carbon dioxide, water, hydrogen ions and hydrogencarbonate ions

　　b) In a sample of blood, the concentration of hydrogencarbonate ions = $2.50 \times 10^{-2}\,mol\,dm^{-3}$. The concentration of aqueous carbon dioxide = $1.25 \times 10^{-3}\,mol\,dm^{-3}$. The value of $K_a = 4.5 \times 10^{-7}\,mol\,dm^{-3}$. Use this information to calculate:

　　　i) the hydrogen ion concentration in the blood

　　　ii) the pH of the blood.

　　c) What can you conclude about the person who gave the blood sample?

5　Explain why a mixture of carbon dioxide, water and hydrogencarbonate ions can act as a buffer solution.

6　Why are blood buffers on their own unable to maintain the correct blood pH for any length of time?

7　Suggest reasons why people may need treatment to adjust their blood pH if they have been rescued after breathing thick smoke during a fire.

Diluting a buffer solution with water does not change the ratio of the concentrations of the salt and acid, so the pH does not change, unless the dilution is so great that the assumptions used to arrive at this formula break down.

Buffer solutions form during the titration of a weak acid with a strong base. This is illustrated by Figure 12.23. Once some of the base has been added the pH does not change by much until the titration begins to approach the end-point.

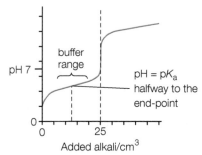

Figure 12.23 In the buffering range, the pH changes little on adding substantial volumes of strong alkali. Over this range the flask contains significant amounts of both the acid and the salt formed from the acid.

Test yourself

45 Calculate the pH of these buffer mixtures.

   a) A solution containing equal amounts in moles of $H_2PO_4^-(aq)$ and $HPO_4^{2-}(aq)$. K_a for the dihydrogenphosphate(v) ion is 6.3×10^{-7} mol dm^{-3}.

   b) A solution containing 12.2 g benzoic acid (C_6H_5COOH) and 7.2 g of sodium benzoate in 250 cm^3 solution. K_a for benzoic acid is 6.3×10^{-5} mol dm^{-3}.

   c) A solution containing 12.2 g benzoic acid (C_6H_5COOH) and 7.2 g of sodium benzoate in 1000 cm^3 solution.

46 What must be the ratio of the concentrations of the ethanoic acid molecules and ethanoate ions in a buffer solution with pH = 5.4 if $K_a = 1.7 \times 10^{-5}$ mol dm^{-3} for ethanoic acid?

Finding the K_a value for a weak acid

A good method to determine K_a for a weak acid is to measure the pH of the solution in the flask during a titration of the acid with a dilute solution of a strong alkali such as sodium hydroxide. The procedure can be related to the steps for determining equilibrium constants described in Section 11.2.

Procedure

Step 1: Mix measured quantities of chemicals and allow the mixture to come to equilibrium.

Sodium hydroxide solution is added from a burette to a measured volume of the weak acid solution in a flask. The added alkali neutralises some of the acid and turns it into its sodium salt. After each addition of alkali there is a new equilibrium mixture in the flask. The reaction is fast, so after each addition the mixture reaches an equilibrium state instantly.

Step 2: Analyse the mixture to find the equilibrium concentration of the one of the reactants or products.

In this case the hydrogen ion concentration in the solution can be determined easily using a pH meter.

Step 3: Use the equation for the reaction and the equilibrium law to find the value of the equilibrium constant.

In this example the value of K_a can be determined by taking readings from the graph.

This procedure was used to determine the acid dissociation constant for chloroethanoic acid, $CH_2ClCOOH$.

Results

Figure 12.24 shows the results of plotting pH against titre for a titration of $25.0\,cm^3$ of a roughly $0.1\,mol\,dm^{-3}$ solution of chloroethanoic acid, $CH_2ClCOOH$, with $0.10\,mol\,dm^{-3}$ sodium hydroxide solution.

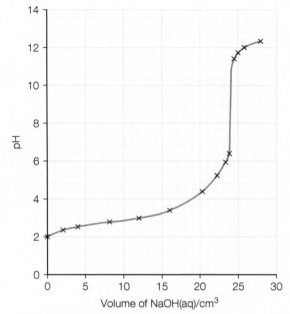

Figure 12.24 Plot of pH against titre for a titration of chloroethanoic acid with sodium hydroxide.

1 Why is it possible to determine the equilibrium concentrations in acid–base equilibria without upsetting the position of equilibrium?

2 a) Show, with the help of values read from the graph in Figure 12.24, that the flask contained a series of buffer solutions during the titration.

b) Write the equation for the reversible reaction in the buffer solutions.

3 a) Take and note down the readings from the graph that you need to work out the value of K_a for chloroethanoic acid.

b) Calculate the value for K_a, showing your working. State any assumptions that you make in the calculation.

4 Why is it not necessary to know the concentration of the acid or the alkali precisely when this method is used to measure K_a?

Tip

Refer to Practical skills sheet 6, 'Finding the K_a value for a weak acid', which you can access via the QR code for Chapter 12 on page 320.

Exam practice questions

1 a) Write an equation for the reaction of ammonia with water and identify the two conjugate acid–base pairs in the solution. *(3)*

b) Explain why a solution of sodium chloride in water is neutral but a solution of sodium ethanoate is alkaline. *(4)*

c) Explain why acid dissociation constants are used to compare the strengths of weak acids and not the pH of the acids in aqueous solution. *(3)*

2 a) Write an equation for the reaction which occurs when a weak acid HX is added to water. *(1)*

b) Write an expression for the acid dissociation constant of the weak acid HX. *(1)*

c) The ionisation of HX in aqueous solution is endothermic. Predict the effect, if any, of:

i) an increase in temperature on the value of the acid dissociation constant for HX *(1)*

ii) an increase in temperature on the pH of an aqueous solution of the weak acid HX *(1)*

iii) a decrease in concentration of the acid HX on the value of its acid dissociation constant. *(1)*

3 a) Explain what is meant by the term 'ionic product of water'. *(2)*

b) i) Show that the value of the ionic product of water is 1×10^{-14} based on the fact that for pure water pH = 7 at 298 K. *(2)*

ii) At 303 K the value of the ionic product of water is 1.47×10^{-14}. What can be deduced from the difference in values between 298 K and 303 K? *(2)*

c) i) Calculate the pH of a $0.300\,mol\,dm^{-3}$ solution of NaOH at 298 K. *(2)*

ii) Calculate the pH of a solution formed by mixing $25.0\,cm^3$ of a $0.300\,mol\,dm^{-3}$ solution of NaOH with $225\,cm^3$ of water at 298 K. *(3)*

iii) Calculate the pH of a solution formed by mixing $25.0\,cm^3$ of $0.300\,mol\,dm^{-3}$ NaOH with $75.0\,cm^3$ of $0.200\,mol\,dm^{-3}$ hydrochloric acid at 298 K. *(4)*

4 a) Give examples to explain the difference between a strong acid and a weak acid. *(4)*

b) At 298 K, what is the pH of:

i) $0.010\,mol\,dm^{-3}$ $HNO_3(aq)$ *(1)*

ii) $0.010\,mol\,dm^{-3}$ KOH(aq)? *(2)*

c) Butanoic acid, C_3H_7COOH, has an acid dissociation constant, K_a, of $1.5 \times 10^{-5}\,mol\,dm^{-3}$ at 298 K.

i) Calculate the pH of a $0.020\,mol\,dm^{-3}$ solution of butanoic acid at this temperature. *(2)*

ii) Draw a sketch graph to show the change of pH when $50\,cm^3$ of $0.020\,mol\,dm^{-3}$ NaOH(aq) is added to $25\,cm^3$ of $0.020\,mol\,dm^{-3}$ butanoic acid. *(4)*

iii) Choose from the table the indicator that would be most suitable for detecting the end-point of a titration between butanoic acid and sodium hydroxide. Give your reasons. *(2)*

Indicator	Colour change acid/alkaline	pH range over which colour change occurs
Thymol blue	Red/yellow	1.2–2.8
Congo red	Violet/red	3.0–5.0
Thymolphthalein	Colourless/blue	8.3–10.6

5 a) Describe and explain the use of buffer solutions with the help of examples. *(6)*

b) i) What is the pH of a buffer solution in which the concentration of ethanoic acid is $0.080\,mol\,dm^{-3}$ and the concentration of sodium ethanoate is $0.040\,mol\,dm^{-3}$?
K_a for ethanoic acid is $1.7 \times 10^{-5}\,mol\,dm^{-3}$. *(2)*

ii) Calculate the new pH value if 0.020 mol of NaOH is dissolved in $1\,dm^3$ of the buffer solution in (i). *(2)*

iii) Calculate the pH of a solution of 0.020 mol of NaOH in $1\,dm^3$ of water. *(2)*

iv) Comment on the effectiveness of the buffer solution. *(1)*

6 For a solution containing $0.050\,mol\,dm^{-3}$ chloric(I) acid (HClO) and $0.050\,mol\,dm^{-3}$ sodium chlorate(I), the pH = 7.43 at 298 K.

   a) i) Write an equation for the ionisation of chloric(I) acid. *(1)*

      ii) Write an expression for the acid dissociation constant of chloric(I) acid. *(1)*

   b) Work out the value of K_a for chloric(I) acid, showing your working. *(4)*

7 The graph shows the changes in pH during the titration of $10\,cm^3$ of a monobasic acid with a $0.010\,mol\,dm^{-3}$ solution of sodium hydroxide.

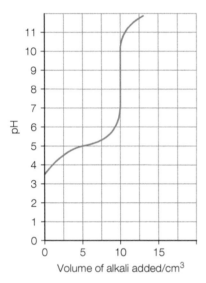

Volume of alkali added/cm³

   a) Answer these questions, giving your reasons.

      i) What was the concentration of the acid at the start? *(2)*

      ii) What was the pH of the acid before any alkali was added? *(1)*

      iii) Was the acid strong or weak? *(1)*

   b) Use your answers to part (a) to calculate a value of K_a for the acid. *(2)*

   c) i) Over what range of titration readings was there an effective buffer solution in the flask? Explain your answer. *(3)*

      ii) Use a value read from the buffer region to determine a value for the K_a of the acid. *(2)*

   d) i) What was the pH of the mixture in the flask at the equivalence point? *(1)*

      ii) How do you account for your answer to (i)? *(2)*

8 Introducing halogen atoms into the structure of carboxylic acids can have a marked effect on their acid strength. This is illustrated by the values in the table.

Acid	$K_a/mol\,dm^{-3}$	pK_a
Ethanoic acid	1.7×10^{-5}	4.8
Fluoroethanoic acid	2.2×10^{-3}	2.7
Chloroethanoic acid	1.3×10^{-3}	2.9
Iodoethanoic acid	7.6×10^{-4}	3.1
Dichloroethanoic acid	5.0×10^{-2}	1.3
Trichloroethanoic acid	2.3×10^{-1}	0.7
Butanoic acid	1.5×10^{-5}	4.8
2-Chlorobutanoic acid	1.4×10^{-3}	2.8
3-Chlorobutanoic acid	8.7×10^{-5}	4.0
4-Bromobutanoic acid	3.0×10^{-5}	4.5

   a) Calculate the pH of a $0.10\,mol\,dm^{-3}$ solution of:

      i) butanoic acid *(2)*

      ii) trichloroethanoic acid. *(2)*

   b) i) What is the pattern in the acid strength of fluoro-, chloro- and iodo-ethanoic acids when compared with the value for ethanoic acid? *(1)*

      ii) Suggest an explanation for the pattern. *(4)*

   c) i) What is the pattern in the acid strength of chloro-, dichloro- and trichloro-ethanoic acids when compared with the value for ethanoic acid? *(1)*

      ii) Is the pattern consistent with your suggested explanation in (b)(ii)? *(2)*

   d) i) What is the pattern in the acid strength of the chlorinated butanoic acids when compared with the value for butanoic acid? *(1)*

      ii) Suggest an explanation for the pattern. *(3)*

9 The graph opposite shows the results from an experiment in which measured volumes of a $2.0\,mol\,dm^{-3}$ solution of an acid, H_nX, were mixed with measured volumes of $2.0\,mol\,dm^{-3}$ NaOH(aq). The temperatures of the two solutions were the same before mixing. The temperature rise after mixing was measured and recorded.

12 Acid–base equilibria

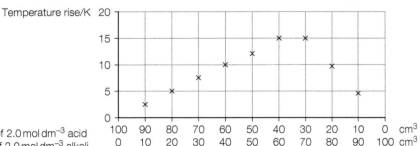

Temperature rise/K

Volume of $2.0\,mol\,dm^{-3}$ acid
Volume of $2.0\,mol\,dm^{-3}$ alkali

a) Suggest a suitable container for mixing the two solutions. *(1)*

b) Account for the shape of the plot on the graph. *(6)*

c) i) Estimate the volumes of the acid and the alkali which would react to exactly neutralise each other. *(1)*

   ii) What does your answer to (i) tell you about the value of n in the formula of the acid? *(2)*

d) Consider the mixture of acid and alkali which would react to exactly neutralise each other.

   i) Use the graph to estimate the temperature rise on making this mixture. *(1)*

   ii) Assuming that the mixed solution has a specific heat capacity of $4.18\,J\,g^{-1}\,K^{-1}$ and a density of $1.0\,g\,cm^{-3}$, calculate the energy change from the reaction in this mixture. *(2)*

   iii) Calculate the enthalpy change of neutralisation per mole of the acid. *(2)*

   iv) Does your answer to (iii) suggest that H_nX is a strong or a weak acid? *(2)*

13.1.1 Ionic bonding and structures

Compounds of metals with non-metals, such as sodium chloride and magnesium oxide, are composed of ions. When such compounds form, the metal atoms lose electrons and form positive ions. At the same time, the non-metal atoms gain electrons and form negative ions. For example, when sodium reacts with chlorine (Figure 13.1.1), each sodium atom loses its one outer electron forming a sodium ion, Na^+. Chlorine atoms gain these electrons and form chloride ions, Cl^- (Figure 13.1.2).

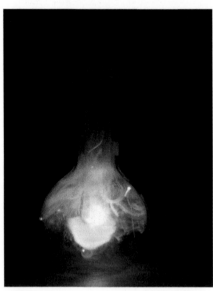

Figure 13.1.1 Hot sodium reacting with chlorine.

> **Tip**
>
> The first two sections of this chapter remind you of the model of ionic giant structures that you learned about in Year 1 of your chemistry course. In these sections, the 'Test yourself' questions help you to check your understanding of ionic compounds and enthalpy changes. From Section 13.1.3, the chapter goes on to show that this model can be tested quantitatively by studying the energy changes involved in the formation of crystals held together by ionic bonding.

In Figure 13.1.2, the electrons of one element are shown as dots and those of the other reactants are shown as crosses. Diagrams of this kind provide a useful balance sheet for keeping track of the electrons when ionic compounds form.

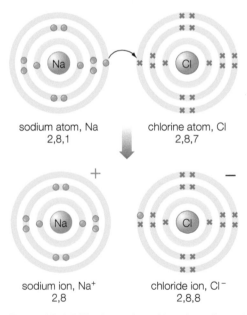

sodium atom, Na
2,8,1

chlorine atom, Cl
2,8,7

+

−

sodium ion, Na^+
2,8

chloride ion, Cl^-
2,8,8

Figure 13.1.2 The formation of ions in sodium chloride when sodium reacts with chlorine.

Very often it is sufficient to show simply the outer shell electrons in dot-and-cross diagrams and two of these simplified dot-and-cross diagrams are shown in Figure 13.1.3.

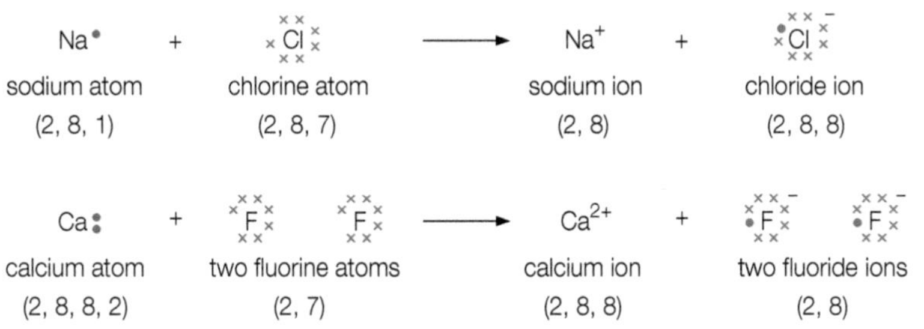

Na$^\bullet$ + $\overset{\times\,\times}{\underset{\times\,\times}{\times\,\text{Cl}\,\times}}$ $\longrightarrow$ Na$^+$ + $\overset{\times\,\times}{\underset{\times\,\times}{\times\,\text{Cl}\,\times}}^-$

sodium atom   chlorine atom   sodium ion   chloride ion

(2, 8, 1)   (2, 8, 7)   (2, 8)   (2, 8, 8)

Ca$\vdots$ + $\overset{\times\,\times}{\underset{\times\,\times}{\times\,\text{F}\,\times}}$ $\overset{\times\,\times}{\underset{\times\,\times}{\times\,\text{F}\,\times}}$ $\longrightarrow$ Ca^{2+} + $\overset{\times\,\times}{\underset{\times\,\times}{\bullet\,\text{F}\,\times}}^-$ $\overset{\times\,\times}{\underset{\times\,\times}{\bullet\,\text{F}\,\times}}^-$

calcium atom   two fluorine atoms   calcium ion   two fluoride ions

(2, 8, 8, 2)   (2, 7)   (2, 8, 8)   (2, 8)

Figure 13.1.3 Dot-and-cross diagrams for the formation of sodium chloride and calcium fluoride showing only the electrons in the outer shells of the reactants and products.

Ionic crystals

Ionic crystals consist of giant lattices containing billions of positive and negative ions packed together in a regular pattern (Figure 13.1.4). In the lattice, each Na$^+$ ion is surrounded by Cl$^-$ ions, and each Cl$^-$ ion is surrounded by Na$^+$ ions. The oppositely charged ions attract each other. At the same time, the chloride ions repel other chloride ions and sodium ions repel other sodium ions, but overall there are strong net electrostatic attractions between ions in all directions throughout the lattice. These electrostatic attractions between oppositely charged ions are described as ionic bonding.

Many other compounds have the same lattice structure as sodium chloride including the chlorides, bromides and iodides of lithium, sodium and potassium and the oxides and sulfides of magnesium, calcium and strontium.

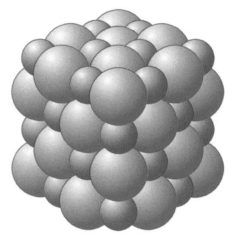

Figure 13.1.4 A three-dimensional model of the structure of sodium chloride. The smaller red spheres represent Na$^+$ ions. The larger green spheres represent Cl$^-$.

Test yourself

1 Draw dot-and-cross diagrams, similar to those in Figure 13.1.3 for:

a) potassium oxide

b) magnesium sulfide.

2 Why do metals form positive ions, whereas non-metals form negative ions?

3 Why do you think the melting point of magnesium oxide (2852 °C) is so much higher than that of sodium fluoride (993 °C)?

4 a) Why do ionic compounds conduct electricity when molten but not when solid?

b) Write equations for the reactions at the cathode and anode during electrolysis of molten magnesium chloride.

5 Why are chloride ions larger than sodium ions as shown in Figure 13.1.4?

6 Why are the electrostatic attractions between ions with opposite charges in an ionic lattice overall greater than the repulsive forces between ions with the same charge?

Tip

Electrostatic forces operate between the ions in a crystal. Oppositely charged ions attract each other while ions with like charges repel each other. The size of the electrostatic force, F, between two charges is given by the equation:

$$F \propto \frac{Q_1 \times Q_2}{d^2}$$

- The larger the charges, Q_1 and Q_2, the stronger the force.
- The greater the distance, d, between the charges, the smaller the force and this has a big effect because it is the square of the distance that matters.

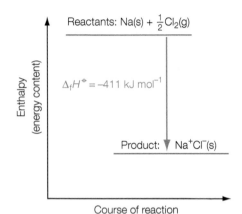

Reactants: $Na(s) + \frac{1}{2}Cl_2(g)$

$\Delta_fH^\ominus = -411 \text{ kJ mol}^{-1}$

Product: $Na^+Cl^-(s)$

Enthalpy (energy content)

Course of reaction

Figure 13.1.5 An energy level diagram for the formation of sodium chloride.

13.1.2 Energy changes and ionic bonding

Standard enthalpy changes

When sodium reacts with chlorine, a very exothermic reaction occurs and energy is given out to the surroundings. As the product, sodium chloride, cools down to room temperature, the system loses energy to its surroundings. This can be represented by an energy level diagram for the reaction (Figure 13.1.5).

In order to compare energy changes fairly and consistently it is important to make thermochemical measurements under the same conditions. The conditions chosen for comparing enthalpy changes and other thermochemical measurements are called standard conditions. These standard conditions are:

- a temperature of 25 °C (298 K)
- a pressure of 1×10^5 Pa = 100 kPa (this is very close to standard atmospheric pressure at sea level, which is 101.3 kPa)
- all reactants and products in their standard (stable) states at 25 °C and 1 atmosphere pressure
- any solutions at a concentration of 1 mol dm^{-3}.

The symbol for these standard enthalpy changes is $\Delta H^\ominus$, and $\Delta_fH^\ominus$ for standard enthalpy changes of formation.

The enthalpy change shown in Figure 13.1.5 relates to the formation of one mole of sodium chloride from its elements sodium and chlorine. If the measurements have been made at 25 °C (298 K) and 1 atmosphere pressure the result is described at the **standard enthalpy change of formation** of sodium chloride. This can be written either as:

$$Na(s) + \frac{1}{2}Cl_2(g) \rightarrow Na^+Cl^-(s) \quad \Delta_fH^\ominus = -411 \text{ kJ mol}^{-1}$$

or as:

$$\Delta_fH^\ominus[NaCl(s)] = -411 \text{ kJ mol}^{-1}$$

> **Tip**
>
> The superscript sign in $\Delta H^\ominus$ shows that the value quoted is for standard conditions. The symbol is pronounced 'delta *H* standard'.

> **Key term**
>
> The **standard enthalpy change of formation** of a compound, $\Delta_fH^\ominus$, is the enthalpy change when one mole of the compound forms from its elements under standard conditions with the elements and the compound in their standard (stable) states.

> **Test yourself**
>
> 7 Why does $\Delta_fH^\ominus = 0 \text{ kJ mol}^{-1}$ for an element?
>
> 8 Why are values for the standard enthalpy changes of formation of compounds containing carbon based on graphite and not diamond?
>
> 9 Write an equation for the reaction for which the enthalpy change is the standard enthalpy change of formation of calcium iodide.

13.1.3 Enthalpy changes when ions form

Figure 13.1.2 shows the formation of sodium chloride from its elements, but it simplifies the process in many ways. As far as sodium is concerned, Figure 13.1.2 ignores the following facts:

- Sodium starts as a giant lattice of metal atoms.
- Energy is required to separate the sodium atoms in the giant lattice to produce gaseous atoms. The energy change for this process is the standard enthalpy change of atomisation of sodium.
- Energy is also required to remove one electron from each gaseous sodium atom to form positive sodium ions, Na^+. This is the first ionisation energy of sodium.

As far as chlorine is concerned, Figure 13.1.2 ignores the following facts:

- Chlorine consists of Cl_2 molecules.
- Energy is required to break the bonds between Cl atoms in the Cl_2 molecules and form separate gaseous Cl atoms. The energy change, for each mole of chlorine atoms formed, is the standard enthalpy change of atomisation of chlorine.
- An energy change also occurs when one electron is added to each gaseous Cl atom forming chloride ions, Cl^-. The energy change for this process is called the first electron affinity of chlorine.

Finally, and equally importantly, Figure 13.1.2 ignores the fact that energy is given out when gaseous Na^+ and Cl^- ions come together forming a giant ionic lattice of sodium chloride, $Na^+Cl^-(s)$. The energy change for this process is called the lattice energy of sodium chloride.

Key terms

The **standard enthalpy change of atomisation** of an element is the energy change needed to produce one mole of gaseous atoms of the element.

For sodium this is:

$$Na(s) \rightarrow Na(g) \qquad \Delta_{at}H^{\ominus}[Na(s)] = +107\,kJ\,mol^{-1}$$

And for chlorine this is:

$$\tfrac{1}{2}Cl_2(g) \rightarrow Cl(g) \qquad \Delta_{at}H^{\ominus}[\tfrac{1}{2}Cl_2(g)] = +122\,kJ\,mol^{-1}$$

The first **ionisation energy** of an element is the energy needed to remove one electron from each atom in one mole of gaseous atoms of the element under standard conditions.

For sodium this is:

$$Na(g) \rightarrow Na^+(g) + e^- \qquad 1st\ IE[Na(g)] = +496\,kJ\,mol^{-1}$$

Tip

Note that the standard enthalpy of atomisation of chlorine is **per mole of atoms formed** – not per mole of molecules atomised. So the standard enthalpy change of atomisation of chlorine is half the size of the Cl–Cl bond energy.

Successive ionisation energies for the same element measure the energy needed to remove a second, third, fourth electron, and so on. For example, the third ionisation energy of sodium relates to the process:

$$Na^{2+}(g) \rightarrow Na^{3+}(g) + e^-$$

The first **electron affinity** of an element is the energy change when each atom in one mole of gaseous atoms gains one electron to form one mole of gaseous ions with a single negative charge.

The following two equations define the first and second electron affinities for oxygen:

$$O(g) + e^- \rightarrow O^-(g) \qquad \text{1st EA} = -141\,kJ\,mol^{-1}$$

$$O^-(g) + e^- \rightarrow O^{2-}(g) \qquad \text{2nd EA} = +798\,kJ\,mol^{-1}$$

The gain of the first electron is exothermic, but adding a second electron to a negatively charged ion is endothermic.

Lattice energies

The **lattice energy** of a compound is defined as the energy change when one mole of an ionic compound is formed from free gaseous ions. For sodium chloride, this is summarised by the equation:

$$Na^+(g) + Cl^-(g) \rightarrow Na^+Cl^-(s) \quad \Delta_{latt}H^\ominus[NaCl(s)] = -787\,kJ\,mol^{-1}$$

This is the lattice energy for the process shown diagrammatically in Figure 13.1.6.

Lattice energies are important because they can be used as a measure of the strength of the ionic bonding in different compounds.

The strength of ionic bonds, measured as lattice energies in $kJ\,mol^{-1}$, arises from the energy given out as billions upon billions of positive and negative ions come together to form a crystal lattice.

The overall force of attraction between the ions is stronger and this results in a more exothermic lattice energy if:

- the charges on the ions are large
- the ionic radii are small, allowing the ions to get closer to each other.

It is important to distinguish between the lattice energy of an ionic compound and its standard enthalpy change of formation. The lattice energy relates to the formation of one mole of a compound from its free gaseous ions, whereas the standard enthalpy change of formation relates to the formation of one mole of the compound from its elements in their stable states under standard conditions.

During the early part of the twentieth century, scientists found ways in which to measure enthalpy changes of formation and atomisation, ionisation energies and electron affinities of various elements. This led two German scientists, Max Born (1882–1970) and Fritz Haber (1868–1934), to analyse the energy changes in the formation of different ionic compounds. Their work resulted in Born–Haber cycles, which are thermochemical cycles for calculating lattice energies and for investigating the stability and bonding in ionic compounds.

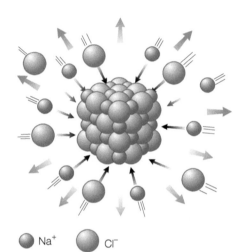

○ Na$^+$ ○ Cl$^-$

Figure 13.1.6 Lattice energy is the energy that would be given out to the surroundings (red arrows) if one mole of an ionic compound could be formed directly from free gaseous ions coming together (black arrows) and arranging themselves into a crystal lattice.

13.1.4 Born–Haber cycles

Born–Haber cycles are an application of Hess's law (Section 8.5 in Student Book 1). They make it possible to calculate lattice energies from other quantities that can be measured. They also enable chemists to test the ionic model of bonding in different substances.

A Born–Haber cycle identifies all the energy changes which contribute to the standard enthalpy change of formation of a compound.

These overall changes, shown in Figure 13.1.7, involve:

- the energy required to create free gaseous ions by atomising and then ionising the elements
- the energy given out (the lattice energy) when the ions come together to form a crystal lattice.

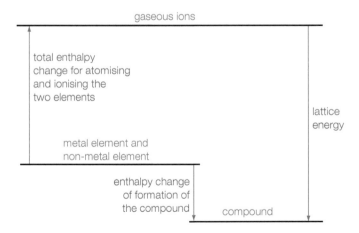

Figure 13.1.7 The overall structure of a Born–Haber cycle.

A Born–Haber cycle is often set out as an energy level diagram. All the processes in the cycle can be determined from experimental data except the lattice energy. So, by using Hess's law, it is possible to calculate the lattice energy. Figure 13.1.8 shows the Born–Haber cycle for sodium oxide, Na_2O.

Figure 13.1.8 The Born–Haber cycle for sodium oxide.

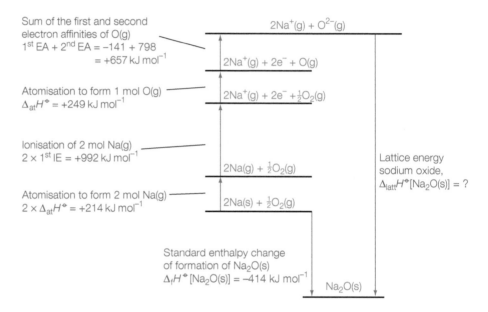

Starting with the elements sodium and oxygen, the measured value for the standard enthalpy change of formation of sodium oxide has been written beside a downwards arrow on the cycle, showing that it is exothermic. Above that, the terms and values for the atomisation and then ionisation of sodium are written beside arrows that point upwards as endothermic processes.

Notice also that the amount of sodium required is two moles because there are two moles of sodium in one mole of sodium oxide.

The terms and values for sodium are followed by those required for the conversion of half a mole of oxygen molecules, $\frac{1}{2}O_2(g)$, to one mole of oxide ions, $O^{2-}(g)$. This involves the atomisation of oxygen, followed by its first and second electron affinities. All these experimentally determined values make it possible to calculate the lattice energy.

Example

Calculate the lattice energy of sodium oxide, $\Delta_{latt}H^{\ominus}[Na_2O(s)]$, using the data in Figure 13.1.8.

Notes on the method

Apply Hess's law to the cycle in Figure 13.1.8 and remember that an endothermic change in one direction becomes an exothermic change with the opposite sign in the reverse direction.

Answer

$\Delta_{latt}H^{\ominus}[Na_2O(s)] = (-657 - 249 - 992 - 214 - 414)\,kJ\,mol^{-1}$

$= -2526\,kJ\,mol^{-1}$

Test yourself

17 Why are lattice energies:

a) always negative

b) impossible to measure directly?

18 Explain why a Born–Haber cycle is an application of Hess's law.

19 Look carefully at Figure 13.1.9, which is a Born–Haber cycle for magnesium chloride.

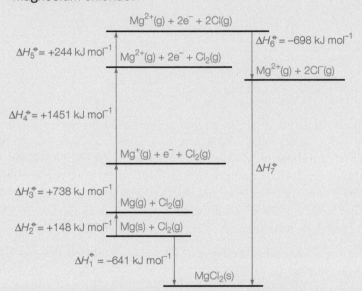

$Mg^{2+}(g) + 2e^- + 2Cl(g)$

$\Delta H_5^{\ominus} = +244 \text{ kJ mol}^{-1}$

$Mg^{2+}(g) + 2e^- + Cl_2(g)$

$\Delta H_6^{\ominus} = -698 \text{ kJ mol}^{-1}$

$Mg^{2+}(g) + 2Cl^-(g)$

$\Delta H_4^{\ominus} = +1451 \text{ kJ mol}^{-1}$

$Mg^+(g) + e^- + Cl_2(g)$

$\Delta H_7^{\ominus}$

$\Delta H_3^{\ominus} = +738 \text{ kJ mol}^{-1}$

$Mg(g) + Cl_2(g)$

$\Delta H_2^{\ominus} = +148 \text{ kJ mol}^{-1}$ $Mg(s) + Cl_2(g)$

$\Delta H_1^{\ominus} = -641 \text{ kJ mol}^{-1}$

$MgCl_2(s)$

Figure 13.1.9 A Born–Haber cycle for magnesium chloride.

a) Identify the energy changes $\Delta H_1^{\ominus}$, $\Delta H_2^{\ominus}$, $\Delta H_3^{\ominus}$, $\Delta H_4^{\ominus}$, $\Delta H_5^{\ominus}$, $\Delta H_6^{\ominus}$ and $\Delta H_7^{\ominus}$.

b) Calculate the lattice energy of magnesium chloride.

13.1.5 Testing the ionic model – ionic or covalent?

One way in which scientists can test their theories and models is by comparing the predictions from their theoretical models with the values obtained by experiment.

Born–Haber cycles are very helpful in this respect because they enable chemists to test the ionic model and check whether the bonding in a compound is truly ionic. The experimental lattice energy obtained from a Born–Haber cycle can be compared with a theoretical value calculated by applying the laws of electrostatics and assuming that the only bonding in the crystal is ionic.

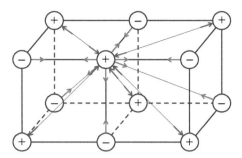

Figure 13.1.10 Some of the many attractions (red) and repulsions (blue) which must be taken into account in calculating a theoretical value for the lattice energy of an ionic crystal.

Using the laws of electrostatics, it is possible to calculate a theoretical value for the lattice energy of an ionic compound by summing up the effects of all the attractions and repulsions between the ions in the crystal lattice (Figure 13.1.10).

Table 13.1.1 shows both the experimentally determined lattice energies and the theoretical lattice energies for a number of compounds.

Table 13.1.1

Compound	Experimental lattice energy from a Born–Haber cycle/kJ mol^{-1}	Theoretical lattice energy calculated assuming that the only bonding is ionic/kJ mol^{-1}
NaCl	−780	−770
NaBr	−742	−735
NaI	−705	−687
KCl	−711	−702
KBr	−679	−674
KI	−651	−636
AgCl	−905	−833
MgI$_2$	−2327	−1944

Pure ionic bonding arises solely from the electrostatic forces between the ions in a crystal. Notice in Table 13.1.1 that there is close agreement between the experimental and theoretical values of the lattice energies for sodium and potassium halides. In all these compounds, the difference between the actual value found from experimental data and the theoretical value is less than 3%. This shows that ionic bonding can account almost entirely for the bonding in sodium and potassium halides.

But look at the experimental and theoretical lattice energies of silver chloride and magnesium iodide in Table 13.1.1. In these two compounds, the theoretical values based on the assumption that the bonding is purely ionic are much less exothermic than the experimental values. The actual bonding is clearly stronger than that predicted by a pure ionic model. This suggests that there is covalent bonding as well as ionic bonding in these substances.

Polarisation of ions

In ionic compounds, positive metal ions attracts the outermost electrons of negative ions. The attraction can pull these electrons into the space between the ions. This distortion of the electron clouds around anions by positively charged cations is an example of polarisation. As a result of polarisation, in some ionic compounds there is a significant degree of electron sharing, that is covalent bonding.

The contribution from covalent bonding makes the size of the lattice energy greater numerically than that expected from the purely ionic model. The values in Table 13.1.1 show clearly that both silver chloride and magnesium iodide, although mainly ionic, have significant extents of covalent bonding.

Figure 13.1.11 shows three examples of ionic bonding with increasing degrees of electron sharing as a positive cation polarises the neighbouring negative ion. In general, results show that:

- the polarising power of a cation is greater if it has a larger charge and a smaller radius
- the polarisability of an anion is greater if it has more electrons in shells and so a larger radius.

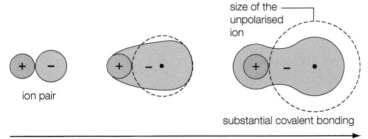

increasing polarisation of the negative ion by the positive ion

Figure 13.1.11 Ionic bonding with increasing degrees of electron sharing as a positive cation polarises the neighbouring negative anion. (Dotted circles show the size of unpolarised ions.)

In a larger negative anion with more electrons, the outermost electrons are further from the attraction of its positive nucleus. Consequently, its outermost electrons are more readily attracted to a neighbouring positive ion and are therefore more polarisable. Also, a negative ion with a 2^- charge is more polarisable than an ion with a 1^- charge.

This means that iodide ions are more polarisable than bromide ions, bromide ions are more polarisable than chloride ions, and fluoride ions are very difficult to polarise. In fact, fluorine, with its small singly charged fluoride ion, forms compounds that are more ionic than those of any other non-metal.

Key terms

The **polarising power** of a positive ion (cation) is its ability to distort the electron cloud of a neighbouring negative ion (anion).

Polarisability is an indication of the extent to which the electron cloud in a molecule, or an ion, can be distorted by a nearby electric charge.

Test yourself

20 Table 13.1.2 shows the ionic radii of some ions.

Ion	Li^+	Na^+	K^+	Mg^{2+}	Al^{3+}
Ionic radius/nm	0.074	0.102	0.138	0.072	0.053
Ion	N^{3-}	O^{2-}	F^-		
Ionic radius/nm	0.171	0.140	0.133		

Table 13.1.2

a) Why do the ionic radii decrease from N^{3-} through O^{2-} to F^-?

b) Use the data in Table 13.1.2 to explain why:

   i) the polarising power of Mg^{2+} is greater than that of Li^+

   ii) the polarising power of Li^+ is greater than that of K^+

   iii) the polarising power of Al^{3+} is much greater than that of Na^+

   iv) the polarisability of N^{3-} is greater than that of F^-.

21 The lattice energy of LiF is $-1031\,kJ\,mol^{-1}$ and that of LiI is $-759\,kJ\,mol^{-1}$.

a) Why is the lattice energy of LiI less exothermic than the lattice energy of LiF?

b) Which of these two compounds would you expect to have the closer agreement between the Born–Haber experimental value of its lattice energy and its theoretical value based on the ionic model?

c) Explain your answer to part (b).

22 Here are four values for lattice energy in $kJ\,mol^{-1}$: -3791, -3299, -3054 and -2725. The four ionic compounds to which these values relate are BaO, MgO, BaS and MgS. Match the formulae with the values and justify your choice.

The stability of ionic compounds

Almost all the compounds of metals with non-metals are regarded as ionic and these compounds have standard enthalpy changes of formation which are exothermic. This means that the compounds are at a lower energy level and therefore more stable than the elements from which they are formed.

The Born–Haber cycles in Figures 13.1.8 and 13.1.9, show that an ionic compound has an exothermic standard enthalpy of formation if its negative lattice energy outweighs the total energy needed to produce gaseous ions from the elements.

Using a Born–Haber cycle with a theoretically calculated value for the lattice energy, it is possible to estimate the standard enthalpy change of formation for compounds which do not normally exist. For example, consider the Born–Haber cycle for the hypothetical compound MgCl in Figure 13.1.12.

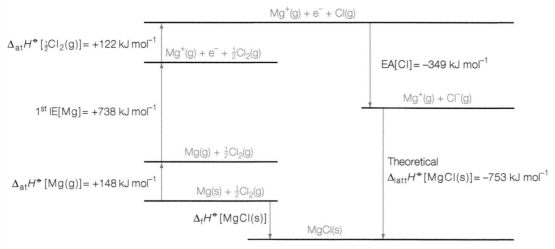

$\Delta_{at}H^{\ominus}[\frac{1}{2}Cl_2(g)] = +122\,kJ\,mol^{-1}$

$Mg^+(g) + e^- + Cl(g)$

$Mg^+(g) + e^- + \frac{1}{2}Cl_2(g)$

$EA[Cl] = -349\,kJ\,mol^{-1}$

$1^{st}\,IE[Mg] = +738\,kJ\,mol^{-1}$

$Mg^+(g) + Cl^-(g)$

$Mg(g) + \frac{1}{2}Cl_2(g)$

Theoretical
$\Delta_{latt}H^{\ominus}[MgCl(s)] = -753\,kJ\,mol^{-1}$

$\Delta_{at}H^{\ominus}[Mg(g)] = +148\,kJ\,mol^{-1}$

$Mg(s) + \frac{1}{2}Cl_2(g)$

$\Delta_f H^{\ominus}[MgCl(s)]$

$MgCl(s)$

Figure 13.1.12 A Born–Haber cycle for the hypothetical compound MgCl.

1 Use Figure 13.1.12 to calculate a value for the standard enthalpy change of formation of MgCl(s).

2 What does your answer to Question 1 suggest about the stability of MgCl(s)?

3 Using the Hess cycle in Figure 13.1.13, calculate the standard enthalpy change for the reaction:

$$2MgCl(s) \rightarrow MgCl_2(s) + Mg(s)$$

given that $\Delta_f H^{\ominus}[MgCl_2(s)] = -641\,kJ\,mol^{-1}$.

4 What does your result for Question 3 tell you about the stability of MgCl(s)?

5 A Born–Haber cycle for the hypothetical compound $MgCl_3$ suggests that $\Delta_f H^{\ominus}[MgCl_3(s)] = +3950\,kJ\,mol^{-1}$.

   a) What does the value of $\Delta_f H^{\ominus}[MgCl_3(s)]$ tell you about the stability of MgCl₃(s)?

   b) Suggest why the value of $\Delta_f H^{\ominus}[MgCl_3(s)]$ is so endothermic.

6 The estimated lattice energy of $MgCl_3(s)$ is $-5440\,kJ\,mol^{-1}$.

   a) Write an equation to summarise the lattice energy of $MgCl_3$.

   b) Why is the lattice energy of $MgCl_3$ more exothermic than that of $MgCl_2(s)$?

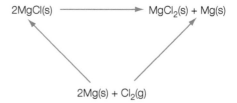

$2MgCl(s) \longrightarrow MgCl_2(s) + Mg(s)$

$2Mg(s) + Cl_2(g)$

Figure 13.1.13 A Hess cycle for the reaction $2MgCl(s) \rightarrow MgCl_2(s) + Mg(s)$.

13.1.6 Enthalpy changes during dissolving

Why do ionic crystals dissolve in water, even though ions in the lattice are strongly attracted to each other? What, in general, are the factors that determine the extent to which an ionic salt dissolves in water (Figure 13.1.14)?

Chemists look for answers to questions of this kind by analysing the energy changes that take place as crystals dissolve.

An ionic compound, such as sodium chloride, does not dissolve in a non-polar solvent like hexane, but it does dissolve in a polar solvent like water. When one mole of sodium chloride dissolves in a large volume of water to produce a very dilute solution, there is an enthalpy change of $+3.8\,kJ\,mol^{-1}$. This enthalpy change is described as the **enthalpy change of solution** of sodium chloride.

Figure 13.1.14 The concentration of sodium chloride in the Dead Sea is so high that salt crystallises out in some places.

The process can be summarised by the equation:

$$NaCl(s) + aq \rightarrow Na^+(aq) + Cl^-(aq) \quad \Delta_{sol}H^\ominus = +3.8\,kJ\,mol^{-1}$$

or simply as $\Delta_{sol}H^\ominus[NaCl(s)] = +3.8\,kJ\,mol^{-1}$.

In the equation above, '+ aq' is short for the addition of water.

Sodium chloride readily dissolves in water despite the fact that the process is slightly endothermic. As this example shows, the sign of ΔH is not a reliable guide to whether or not a process happens. This is particularly the case when the magnitude of ΔH is small (see Chapter 13.2).

It is not immediately obvious why the charged ions in a crystal such as sodium chloride separate and go into solution in water. Where does the energy come from to overcome the attractive forces between oppositely-charged ions?

When sodium chloride dissolves in water, the overall process can be pictured in two stages; these are shown in Figure 13.1.15.

- First of all, Na^+ and Cl^- ions must be separated from the solid NaCl crystals to form well-spaced ions in the gaseous state, $Na^+(g)$ and $Cl^-(g)$. This is the reverse of the lattice energy and labelled $-\Delta_{latt}H^\ominus = +787\,kJ\,mol^{-1}$ in Figure 13.1.15.
- In the second stage, gaseous $Na^+(g)$ and $Cl^-(g)$ ions are hydrated by polar water molecules forming a dilute solution of sodium chloride, $Na^+(aq) + Cl^-(aq)$. Under standard conditions, this process is the sum of the standard **enthalpy changes of hydration** of $Na^+(g)$ and $Cl^-(g)$. This is written as

$$\Delta_{hyd}H^\ominus[Na^+] + \Delta_{hyd}H^\ominus[Cl^-] = -784\,kJ\,mol^{-1} \text{ in Figure 13.1.15.}$$

It is now possible to see from Figure 13.1.15 why sodium chloride dissolves in water. The explanation is that the ions, Na^+ and Cl^-, are so strongly hydrated by the polar water molecules that the exothermic enthalpy changes of hydration nearly balance the energy required to separate the ions (the reverse lattice energy).

Key terms

The **enthalpy change of solution**, $\Delta_{sol}H^\ominus$, is the enthalpy change when one mole of a compound dissolves to form a solution of infinite dilution. An infinitely dilute solution is one where there is so much water that there is no further energy change if more water is added.

The **enthalpy change of hydration** is the enthalpy change when one mole of gaseous ions dissolve in water to give an infinitely dilute solution.

For example, for sodium ions:

$$Na^+(g) + aq \rightarrow Na^+(aq)$$
$$\Delta_{hyd}H^\ominus = -444\,kJ\,mol^{-1}$$

Enthalpy changes of hydration are sometimes just called hydration enthalpies.

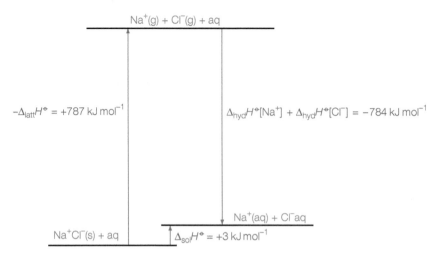

$Na^+(g) + Cl^-(g) + aq$

$-\Delta_{latt}H^\ominus = +787\,kJ\,mol^{-1}$

$\Delta_{hyd}H^\ominus[Na^+] + \Delta_{hyd}H^\ominus[Cl^-] = -784\,kJ\,mol^{-1}$

$Na^+(aq) + Cl^-aq$

$Na^+Cl^-(s) + aq$

$\Delta_{sol}H^\ominus = +3\,kJ\,mol^{-1}$

Figure 13.1.15 An energy level diagram for sodium chloride dissolving in water.

The enthalpy change of solution is the difference between the energy needed to separate the ions from the crystal lattice (the reverse of the lattice energy) and the energy given out as the ions are hydrated (the sum of the hydration enthalpies).

Figure 13.1.16 shows the structure of hydrated sodium and chloride ions. In water molecules, there is a δ^+ charge in the region between the hydrogen atoms and a δ^- charge on the oxygen atoms. This means that the polar water molecules are attracted to both positive cations and negative anions. The bond between the ions and the water molecules is an electrostatic attraction.

With cations, the electrostatic attraction involves the positive charge on the cations and the δ^- charges on the oxygen atoms of the water molecules. In contrast, with anions, the attraction involves the negative charge on the anions and the δ^+ charge between the hydrogen atoms in the water molecules.

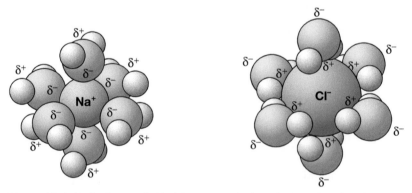

Figure 13.1.16 Sodium and chloride ions are hydrated when they dissolve in water. Polar water molecules are attracted to both cations and anions.

The effect of ionic charge on enthalpy change of hydration and lattice energy

Enthalpy changes of hydration and lattice energies both involve electrostatic attractions between opposite charges. Because of this, both processes are exothermic.

In addition, both processes become more exothermic as the charge on an ion increases because the charge density of the ion increases, and therefore its attraction for any opposite charge increases.

This is illustrated very well by the enthalpy changes of hydration for Na^+ and Mg^{2+} and the lattice energies of NaF and MgO in Table 13.1.3.

Enthalpy change of hydration/kJ mol^{-1}		Lattice energy/kJ mol^{-1}	
Na^+	−444	NaF	−918
Mg^{2+}	−2003	MgO	−3791
Li^+	−559	LiF	−1031
K^+	−361	KF	−817

Table 13.1.3 Comparing some enthalpy changes of hydration and lattice energies.

The effect of ionic radius on enthalpy change of hydration and lattice energy

Comparing ions with the same charge, the larger the radius of the ion the smaller its charge density. This results in a weaker attraction for oppositely charged ions and for the δ^+ or δ^- charges on polar molecules such as water. So, an increase in ionic radius leads to less exothermic values for enthalpy changes of hydration and lattice energies. This point is neatly illustrated by the enthalpy changes of hydration for Li^+ and K^+ and the lattice energies of LiF and KF in Table 13.1.3.

Trends in solubility

It is difficult to use enthalpy cycles, like that in Figure 13.1.15, to account for trends in the solubilities of ionic compounds for three reasons.

First, the enthalpy change of solution is generally a small difference between two large enthalpy changes, neither of which can be measured directly. So, even small errors in estimating trends in the values of lattice energies and hydration enthalpies can lead to large percentage errors in the predicted enthalpy changes of solution.

Second, both lattice energies and hydration enthalpies are affected in the same way by changes in the size of ions and their charges. This tends to reduce the likelihood of any clear trends in enthalpy changes of solution.

Finally, it is clear from the small endothermic value for sodium chloride that the sign and magnitude of the enthalpy change of solution is not a reliable guide as to whether or not a solid will dissolve. Other factors must be taken into account; these are considered in Chapter 13.2.

Test yourself

26 a) Use the data sheet headed 'Lattice energies and enthalpy changes of hydration', which you can access via the QR code for this chapter on page 320, to calculate the enthalpies of solution of lithium fluoride and lithium iodide.

 b) Account for the relative values of the lattice enthalpies and hydration enthalpies of the two compounds in terms of ionic radii.

 c) To what extent, if at all, can your answers to part (a) explain the differences in the solubilities of the two compounds?

 (Solubilities: LiF = 5×10^{-5} mol in 100 g water; LiI = 1.21 mol in 100 g water.)

27 Why do you think the lattice energy of magnesium oxide, $\Delta_{latt}H^{\ominus}[\text{MgO (s)}] = -3791\,\text{kJ mol}^{-1}$, is roughly four times more exothermic than that of sodium fluoride, $\Delta_{latt}H^{\ominus}[\text{NaF(s)}] = -918\,\text{kJ mol}^{-1}$?

Exam practice questions

1 a) i) Show that the formation of CaF_2 from its elements is a redox reaction. *(2)*

ii) Draw a dot-and-cross diagram to represent the bonding in calcium fluoride. *(2)*

b) i) State two characteristic physical properties of compounds with the types of structure and bonding found in calcium fluoride. *(2)*

ii) What is the model that chemists use to describe the structure and bonding of a compound such as calcium fluoride? How does this model account for the physical properties of such compounds? *(3)*

2 a) Define the term 'first electron affinity'. *(2)*

b) The equation below represents the change for which the energy change is the second electron affinity of nitrogen.

$$N^-(g) + e^- \rightarrow N^{2-}(g)$$

i) Explain why the second electron affinity for all elements is endothermic *(2)*

ii) Write equations for the first and third electron affinities of nitrogen. *(2)*

c) In magnesium iodide, MgI_2, the iodide ions are polarised, but in sodium iodide they are not polarised.

i) What is meant by the term 'polarisation' of an ion? *(2)*

ii) State two factors which help to explain the polarisation of iodide ions in magnesium iodide. *(2)*

3 The following data can be used in Born–Haber cycles for silver fluorides AgF and AgF_2

Enthalpy change of atomisation of fluorine,
$\Delta_{at}H^{\ominus}[\frac{1}{2}F_2(g)] = +79\,kJ\,mol^{-1}$

Enthalpy change of atomisation of silver,
$\Delta_{at}H^{\ominus}[Ag(s)] = +289\,kJ\,mol^{-1}$

First ionisation energy of silver,
1st $IE[Ag(g)] = +732\,kJ\,mol^{-1}$

Second ionisation energy of silver,
2nd $IE[Ag^+(g)] = +2070\,kJ\,mol^{-1}$

Electron affinity of fluorine,
$EA[F(g)] = -348\,kJ\,mol^{-1}$

Lattice energy of silver(II) fluoride, $AgF_2(s)$,
$\Delta_{latt}H^{\ominus}[AgF_2(s)] = -2650\,kJ\,mol^{-1}$

Lattice energy of silver(I) fluoride, $AgF(s)$,
$\Delta_{latt}H^{\ominus}[AgF(s)] = -955\,kJ\,mol^{-1}$

a) Use the outline of the Born–Haber cycle for silver(II) fluoride shown below to:

i) write the formulae and state symbols of the species that should appear in boxes A, B and C *(3)*

ii) name the enthalpy changes D, E and F. *(3)*

b) Define the term 'lattice energy'. *(2)*

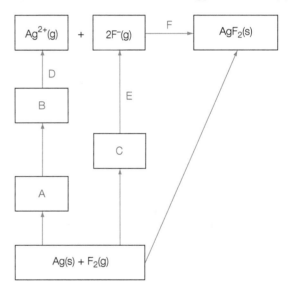

c) Use the diagram and the data supplied to calculate the enthalpy change of formation of silver(II) fluoride. Give a sign and units in your answer. *(3)*

d) Draw a similar Born–Haber cycle for silver(I) fluoride and use the data supplied to calculate the enthalpy change of formation of silver(I) fluoride. Give a sign and units in your answer. *(6)*

e) i) Use the values of enthalpy change of formation from parts (c) and (d) to calculate an enthalpy change for the reaction:

$$AgF_2(s) \rightarrow AgF(s) + \tfrac{1}{2}F_2(g)$$ *(2)*

ii) Comment on the relative stability of $AgF_2(s)$ compared to its elements and also compared to $AgF(s)$ *(2)*

4 The lattice energy of rubidium iodide, RbI, can be determined indirectly using a Born–Haber cycle.

a) Use the data in the table below to construct a Born–Haber cycle for rubidium iodide. *(6)*

Enthalpy change	Energy/kJ mol^{-1}
Formation of rubidium iodide	−334
Atomisation of rubidium	+81
Atomisation of iodine	+107
First ionisation energy of rubidium	+403
First electron affinity of iodine	−295

b) Determine a value for the lattice energy of rubidium iodide. *(2)*

c) Explain why the lattice energy of lithium iodide, LiI, is more exothermic than that of rubidium iodide. *(3)*

d) A theoretical value for the lattice energies of LiI and RbI can be calculated assuming the compounds are purely ionic. When this is done, the experimentally determined value of the lattice energy of lithium iodide is greater than the calculated value by 21 kJ mol^{-1}, whereas the experimental value for rubidium iodide is greater than the calculated value by only 11 kJ mol^{-1}. Explain the difference in these values. *(6)*

5 When calcium chloride dissolves in water, the process can be represented by the equation:

$$CaCl_2(s) + aq \rightarrow Ca^{2+}(aq) + 2Cl^-(aq)$$

The enthalpy change for this process is called the enthalpy change of solution. Its value can be calculated from a Born–Haber cycle using the following data.

$$\Delta_{latt}H^\ominus[CaCl_2(s)] = -2258 \text{ kJ mol}^{-1}$$

$$\Delta_{hyd}H^\ominus[Ca^{2+}(g)] = -1657 \text{ kJ mol}^{-1}$$

$$\Delta_{hyd}H^\ominus[Cl^-(g)] = -340 \text{ kJ mol}^{-1}$$

a) Draw and label the Born–Haber cycle linking the enthalpy change of solution of calcium chloride with the enthalpy changes in the data above. *(4)*

b) Use your Born–Haber cycle to calculate the enthalpy change of solution of calcium chloride. *(3)*

c) What factors affect the size of the enthalpy change of hydration of $Ca^{2+}(g)$ compared with that of $Li^+(g)$? *(2)*

d) Why are the hydration enthalpies of both anions and cations negative? *(2)*

6 The table shows the values of the first five ionisation energies of boron.

Ionisation energies of boron	Value/kJ mol^{-1}
First	+800
Second	+2 400
Third	+3 700
Fourth	+25 000
Fifth	+32 800

a) i) Give the electron configuration of a boron atom. *(1)*

   ii) Use your answer to (i) to explain the pattern of the five ionisation energies of boron. *(3)*

b) Boron forms an oxide with the formula B_2O_3. Use the table of ionisation energies and this data to calculate a value for the lattice energy of boron oxide.
Enthalpy change of atomisation of oxygen, $\Delta_{at}H^\ominus[\frac{1}{2}O_2(g)] = +250 \text{ kJ mol}^{-1}$
Enthalpy change of atomisation of boron, $\Delta_{at}H^\ominus[B(s)] = +590 \text{ kJ mol}^{-1}$
First electron affinity of oxygen, 1st EA$[O(g)] = -140 \text{ kJ mol}^{-1}$
Second electron affinity of oxygen, 2nd EA$[O^-(g)] = +790 \text{ kJ mol}^{-1}$
Standard enthalpy change of formation of boron oxide, $\Delta_f H^\ominus$[boron oxide] $= -1270 \text{ kJ mol}^{-1}$ *(6)*

c) Predict whether or not the value for the lattice energy for boron oxide calculated in (b) is in good agreement with a value calculated in theory by applying the laws of electrostatics to boron oxide assuming that has an ionic lattice. Give reasons to explain your prediction. *(2)*

d) One mole of boron oxide slowly dissolves in water to given a product that dissolves in water forming a solution that is slightly acidic. One mole of the oxide reacts with three moles of water.

   i) Write a balanced equation for the reaction of boric oxide with water. *(1)*

   ii) Comment on this property of boron oxide in relation to your answer to (b). *(2)*

e) Boron forms a chloride which melts at 166 K and boils at 286 K. Suggest the type of bonding in this chloride and predict how it reacts when added to water. *(3)*

13.2.1 Enthalpy changes and the direction of change

Chemists have devised a range of ways for predicting the direction and extent of change. They use equilibrium constants (Chapters 11 and 12) and electrode potentials (Chapter 14) to explain why some reactions go while others do not. These quantities are related and there is a more fundamental concept which links them together; however this concept is not the enthalpy change for reactions, but the entropy change. Many exothermic reactions with a negative enthalpy change of reaction do tend to go naturally (Figure 13.2.1), but change can also happen in directions that are endothermic if there are other changes in the surroundings (Figure 13.2.2).

Figure 13.2.1 A cheetah hunting its prey in Kenya. The cheetah gets its energy from respiration, taking advantage of the natural direction of change. Carbohydrates react with oxygen in muscle cells to form carbon dioxide and water, releasing energy.

Figure 13.2.2 Energy can drive change in the direction opposite to the natural direction of change. Photosynthesis effectively reverses the changes of respiration. Leaves harness energy from the Sun to convert carbon dioxide and water into carbohydrates.

Spontaneous changes

A spontaneous reaction is a reaction which tends to go without being driven by any external agency. Spontaneous reactions are the chemical equivalent of water flowing downhill (Figure 13.2.3). Any reaction which naturally tends to happen is spontaneous in this sense even if it is very slow, just as water has a tendency to flow down a valley even when held up behind a dam. The chemical equivalent of a dam is a high activation energy for a reaction.

In practice, chemists also use the word 'spontaneous' in its everyday sense to describe reactions which not only tend to go, but go fast on mixing the reactants at room temperature. Here is a typical example:

'The hydrides of silicon catch fire spontaneously in air, unlike methane which has an ignition temperature of about 500 °C.'

Figure 13.2.3 Metals such as magnesium, iron and aluminium react spontaneously with oxygen. They are ingredients of fireworks. They burn, when heated, by the spontaneous reactions between sulfur, carbon and potassium nitrate in gunpowder.

Figure 13.2.4 Theory shows that the change of diamond to graphite is spontaneous. Graphite is more stable than diamond. Fortunately for owners of valuable jewellery the change is very, very slow.

Key term

A **feasible** reaction is one that naturally tends to happen, even if it is very slow because the activation energy is high.

Figure 13.2.5 Antacid tablets that contain citric acid and sodium hydrogencarbonate fizz when added to water. The reaction of the acid and the hydrogencarbonate is a spontaneous endothermic reaction.

Figure 13.2.6 The endothermic reaction between the solids barium hydroxide and ammonium chloride cools the flask so much that the water between the flask and the wooden block freezes.

The reaction of methane with oxygen is also spontaneous in the thermodynamic sense, even at room temperature. However, the activation energy for the reaction is so high that nothing happens until the gas is heated with a flame. There are many examples of reactions that theory shows are spontaneous, but which in practice happen very slowly (Figure 13.2.4).

The possible ambiguity in the use of the term 'spontaneous' means that chemists generally prefer an alternative term. They describe a reaction as feasible if it tends to go naturally.

Test yourself

1 Classify these changes as feasible/fast, feasible/slow or not feasible:
   a) ice melting at 5 °C
   b) ammonia gas condensing to a liquid at room temperature
   c) diamond reacting with oxygen at room temperature
   d) water splitting up into hydrogen and oxygen at room temperature
   e) sodium reacting with water at room temperature.

Enthalpy changes and feasible reactions

Most reactions that are feasible are also exothermic. They have a negative standard enthalpy change of reaction. This is such a common pattern that chemists often use the sign of ΔH to decide whether or not a reaction is likely to go.

However, some endothermic processes are feasible too. This shows the limitation of using the enthalpy change to decide the likely direction of change. One example is the reaction of ethanoic acid with ammonium carbonate. The mixture fizzes vigorously while getting colder and colder. ΔH for the reaction is positive. There is a similar reaction between citric acid and sodium hydrogencarbonate (Figure 13.2.5).

Another example is the reaction between solid barium hydroxide and solid ammonium chloride. When these chemicals are mixed in a flask, the endothermic reaction means that the temperature inside the beaker drops dramatically. If the reaction flask is sitting on a damp wooden block, the temperature drop is enough to freeze the water, sticking the flask and block together (Figure 13.2.6).

13.2.2 Feasible by chance alone

Diffusion

Open a bottle of perfume in the corner of a room, and it is not long before everyone else in the room can smell it. The perfume molecules spread out naturally and mix with the air molecules. This is diffusion (Figures 13.2.7 and 13.2.8).

The opposite never happens. If there is a bad smell in a room it is impossible to get rid of it by persuading all the smelly molecules to collect together in a small bottle before putting in a stopper.

Figure 13.2.7 The right-hand bottle contains bromine vapour. The left-hand bottle contains air. The two are separated by a barrier.

Figure 13.2.8 After removing the barrier, the bromine diffuses from the right-hand bottle until the bromine molecules are evenly spread between the two bottles. Once mixed, the air and bromine never unmix. When the concentrations are the same in both bottles, molecules continue to diffuse between the bottles, but overall there seems to be no change. This is an example of dynamic equilibrium.

Diffusion happens by chance alone. This is shown by the very simple example in Figure 13.2.9. This shows an imaginary situation with just six molecules of bromine in the right-hand jar (jar R). The left-hand jar is empty and there is a barrier between the jars. This situation can be represented as RRRRRR.

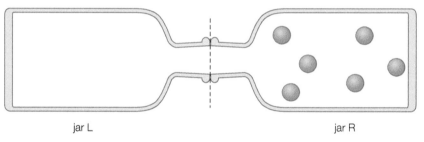

jar L   jar R

Figure 13.2.9 Two gas jars separated by a barrier with six molecules of bromine in the right-hand jar. The molecules are in rapid, random motion (RRRRRR).

The molecules in jar R are moving around randomly, bumping into each other and the sides of the jar. Figure 13.2.10 shows what happens immediately after removing the barrier. One molecule has moved into the left–hand jar. This can be represented as RRRRRL.

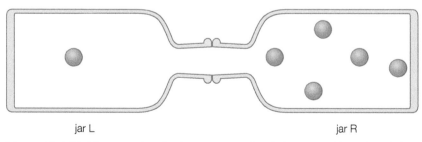

jar L   jar R

Figure 13.2.10 After removing the barrier one molecule has moved into the left-hand jar (RRRRRL).

As the molecules move around randomly in the two jars, there are many possible arrangements. Each molecule can either be R or L. For six molecules moving between the jars, there are $2 \times 2 \times 2 \times 2 \times 2 \times 2 = 2^6 = 64$ possible arrangements.

Only one of these arrangements is RRRRRR, so there is a 1 in 64 chance that all the molecules will end up back in jar R. There is a much bigger chance that the molecules will be arranged in some other way.

In practice the number of molecules is always many more than six. In Figure 13.2.7 there are about 10^{22} molecules of bromine in the right-hand jar. The number of ways of distributing the molecules between the two jars is $2^{10^{22}}$. This is a huge number – too big to be imagined. Only one of all this number of arrangements has all the molecules back in the right-hand jar. The chance of this happening is impossibly small.

It is overwhelmingly likely, by chance alone, that bromine molecules will spread out and mix with air molecules unless there is a barrier to stop them. This is a spontaneous change. In general, gas molecules naturally tend to mix up and disperse themselves randomly.

Molecules and energy

It is not only the arrangement of molecules that matters. Even more important is the way that energy is spread out between molecules. The molecules in bromine gas are moving around, spinning and vibrating. The energies of the tiny molecules are quantised, in a similar way to the energies of electrons in atoms. All the time the molecules are bumping into each other; as they do so, they lose and gain energy quanta to and from other molecules.

Taking a very simple situation, Table 13.2.1 shows the number of ways that two molecules can share four energy quanta. In all there are five different ways. The more molecules, and the more energy quanta, the more ways there are of sharing the energy between the molecules, as shown in Table 13.2.2.

Table 13.2.1 The number of ways that two molecules can share four energy quanta.

Number of quanta	
Molecule 1	Molecule 2
2	2
3	1
1	3
0	4
4	0

Table 13.2.2 The number of ways of sharing energy between molecules.

Number of molecules	Number of quanta	Number of ways of sharing the molecules between the quanta
10	100	$\approx 10^{12}$
100	10	$\approx 10^{13}$
100	100	$\approx 10^{60}$
200	110	$\approx 10^{86}$

Increasing the temperature of a substance increases the number of energy quanta in the system and so increases the number of possible ways that the energy quanta can be shared out between the molecules.

13.2.3 Entropy changes

Entropy

Random changes, which happen by chance, always tend to go in the direction that increases the number of ways of distributing the molecules and energy quanta. This is a fundamental principle.

However, the number of ways, W, of distributing the energy quanta between the molecules in a mole of gas at room temperature is huge. The numbers involved are very hard to deal with. Fortunately, the Austrian physicist Ludwig Boltzmann (1844–1906) showed how the established laws of thermodynamics could be explained in terms of molecules and their energies.

Physicists had already developed the concept of entropy to account for the way that steam engines work. What Boltzmann was able to show was that there is a relationship between this quantity entropy and the number of ways that any chemical system could distribute its molecules and energy. He demonstrated the truth of this formula:

entropy, $S = k \ln W$

where S is the entropy of the system, k is a constant named after Boltzmann and $\ln W$ is the natural logarithm of the number of ways of arranging the particles and energy in the system.

Figure 13.2.11 The reaction mixture in the flask is 'the system'. The air and everything else around the flask makes up the 'surroundings'. It is the total entropy in the system **and** the surroundings which determines whether or not the reaction is feasible.

> ### Tip
>
> Find out more about natural logarithms, ln, in Section 3 of 'Mathematics in A Level chemistry', which you can access via the QR code for Chapter 13.2 on page 320.

Now chemists can use this quantity called entropy, S, to decide whether or not a reaction is feasible. The formula shows that as W increases, S increases. So change happens in the direction which leads to a total increase in entropy.

Chemists sometimes describe entropy as a measure of disorder or randomness. These descriptions have to be interpreted with care because the disorder refers not only to the arrangement of the particles in space, but much more significantly to the numbers of ways of distributing the energy of the system across all the available energy levels.

When considering chemical reactions it is essential to calculate the total entropy change in two parts: the entropy change of the system and the entropy change of the surroundings (Figure 13.2.11).

$$\Delta S_{total} = \Delta S_{system} + \Delta S_{surroundings}$$

Standard molar entropies

In a perfectly ordered crystal at $0\,K$ the entropy is zero. The entropy of a chemical rises as the temperature rises. This is because increasing the temperature raises the number of energy quanta to share between the atoms and molecules. There are also jumps in the entropy of the chemical wherever there is a change of state. This is because energy is added to change a solid to a liquid or a liquid to a gas (Figure 13.2.12). Also there is an increase in the number of ways of arranging the atoms or molecules as a solid changes to a liquid and then to a gas.

Standard molar entropy values are quoted for pure chemicals under standard conditions (298 K and 1 atmosphere pressure). Gases generally have higher entropies than comparable liquids, which have higher entropies than similar solids (Figures 13.2.13–13.2.15).

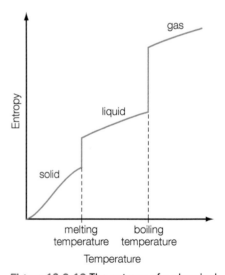

Figure 13.2.12 The entropy of a chemical increases as its temperature rises. There is a jump in the entropy values when the chemical melts or boils.

> ### Key term
>
> **Standard molar entropy, S,** is the entropy per mole for a substance under standard conditions. Chemists use values for standard molar entropies to calculate entropy changes and so to predict the direction and extent of chemical change.

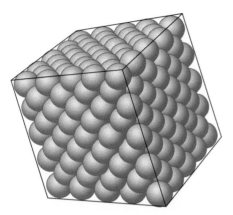

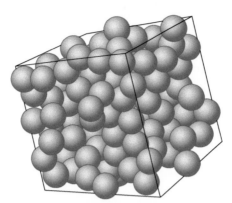

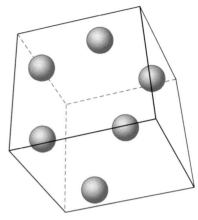

Figure 13.2.13 Structure of a solid. Solids have relatively low values for standard molar entropies. In diamond the carbon atoms are held firmly in place by strong, highly directional covalent bonds. The standard molar entropy of diamond is low. Lead has a higher value for its standard molar entropy because metallic bonds are not directional. The heavier, larger atoms can vibrate more freely and share out their energy in more ways than can carbon atoms in diamond.

Figure 13.2.14 Structure of a liquid. In general, liquids have higher standard molar entropies than comparable solids because the atoms or molecules are free to move. There are many more ways of distributing the particles and energy – there is more disorder. The standard molar entropy of mercury is higher than that of lead. Molecules with more atoms have higher standard molar entropies because they can vibrate, rotate and arrange themselves in yet more ways.

Figure 13.2.15 Atoms of a noble gas. Gases have even higher standard molar entropies than comparable liquids because the atoms or molecules are not only free to move but also widely spaced. There are even more ways of distributing the particles and energy – the disorder is even greater. The standard molar entropy of argon is higher than that of mercury. As with liquids, molecules with more atoms have even higher standard molar entropies because they can vibrate, rotate and arrange themselves in more ways.

> **Tip**
>
> The units for standard molar entropy are joules per kelvin per mole ($J\,mol^{-1}\,K^{-1}$). Note that the units are joules and not kilojoules.

Table 13.2.3 Standard molar entropies for selected solids, liquids and gases.

Solids	$S^{\ominus}/$ $J\,mol^{-1}\,K^{-1}$	Liquids	$S^{\ominus}/$ $J\,mol^{-1}\,K^{-1}$	Gases	$S^{\ominus}/$ $J\,mol^{-1}\,K^{-1}$
Carbon (diamond)	2.4	Mercury	76.0	Argon	155
Magnesium oxide	26.9	Water	69.9	Ammonia	192
Copper	33.2	Ethanol	160	Carbon dioxide	214
Lead	64.8	Benzene	173	Propane	270

> **Test yourself**
>
> 2 Refer to Table 13.2.3. Why is the value of the standard molar entropy of:
>   a) mercury higher than the value for copper
>   b) ammonia higher than the value for water
>   c) propane higher than the value for argon?

3 Which substance in each of the following pairs is expected to have the higher standard molar entropy at 298 K:

a) $Br_2(l)$ or $Br_2(g)$

b) $H_2O(s)$ or $H_2O(l)$

c) $HF(g)$ or $NH_3(g)$

d) $CH_4(g)$ or $C_2H_6(g)$

e) $NaCl(s)$ or $NaCl(aq)$

f) ethane or poly(ethene)

g) pentene gas or cyclopentane gas?

The entropy change of the system

Tables of standard molar entropies make it possible to calculate the entropy change of the system of chemicals during a reaction.

$\Delta S^{\ominus}_{system}$ = the sum of the standard molar entropies of the products
− the sum of the standard molar entropies of the reactants

This can be shortened to:

$\Delta S^{\ominus}_{system} = \Sigma S^{\ominus}[products] - \Sigma S^{\ominus}[reactants]$

> **Tip**
>
> The symbol Σ is the Greek capital letter for sigma. It is used in science and maths to mean 'sum of'.

Example

Calculate the entropy change for the system, $\Delta S^{\ominus}_{system}$, for the synthesis of ammonia from nitrogen and hydrogen. Comment on the value.

Notes on the method

Write the balanced equation for the reaction.

Look up the standard molar entropies on the data sheet headed 'Thermodynamic properties of selected elements and compounds', which you can access via the QR code for Chapter 13.2 on page 320. Take careful note of the units.

Answer

$N_2(g) + 3H_2(g) \rightarrow 2NH_3(g)$

Sum of the standard molar entropies of the products

$= 2S^{\ominus}[NH_3(g)]$

$= 2 \times 192.4 \, J \, mol^{-1} \, K^{-1} = 384.8 \, J \, mol^{-1} \, K^{-1}$

Sum of the standard molar entropies of the reactants

$= S^{\ominus}[N_2(g)] + 3S^{\ominus}[H_2(g)]$

$= 191.6 \, J \, mol^{-1} \, K^{-1} + (3 \times 130.6 \, J \, mol^{-1} \, K^{-1}) = 583.4 \, J \, mol^{-1} \, K^{-1}$

$\Delta S^{\ominus}_{system} = 384.8 \, J \, mol^{-1} \, K^{-1} - 583.4 \, J \, mol^{-1} \, K^{-1}$

$= -198.6 \, J \, mol^{-1} \, K^{-1}$

This shows that the entropy of the system decreases when nitrogen and hydrogen combine to form hydrogen. This is not surprising since the change halves the number of molecules so the amount of gas decreases.

4 Without doing any calculations, predict whether the entropy of the system increases or decreases as a result of these changes:

a) $KCl(s) + aq \rightarrow KCl(aq)$ b) $H_2O(l) \rightarrow H_2O(g)$

c) $Mg(s) + Cl_2(g) \rightarrow MgCl_2(s)$ d) $N_2O_4(g) \rightarrow 2NO_2(g)$

e) $NaHCO_3(s) + HCl(aq) \rightarrow NaCl(aq) + H_2O(l) + CO_2(g)$

f) $2CH_3COOH(l) + (NH_4)_2CO_3(s) \rightarrow 2CH_3COONH_4(s) + H_2O(l) + CO_2(g)$

5 Refer to the data sheet headed 'Thermodynamic properties of selected elements and compounds', which you can access via the QR code for Chapter 13.2 on page 320. Use values from the data sheet to calculate the entropy change for the system, $\Delta S^\ominus$, for the catalytic reaction of ammonia with oxygen to form NO and steam. Comment on the value.

The entropy change of the surroundings

It is not enough to consider only the entropy of the system. What matters is the total entropy change, which is the sum of the entropy changes of the system and the entropy change in the surroundings.

It turns out that the entropy change of the surroundings during a chemical reaction is determined by the size of the enthalpy change, ΔH, and the temperature, T. The relationship is:

$$\Delta S_{surroundings} = -\frac{\Delta H}{T}$$

The minus sign is included because the entropy change becomes more positive, the more energy that is transferred to the surroundings. For an exothermic reaction, which transfers energy to the surroundings, ΔH is negative so $-\Delta H$ is positive.

What this relationship shows is that the more energy transferred to the surroundings by an exothermic process, the larger the increase in the entropy of the surroundings (Figure 13.2.16). It also shows that, for a given quantity of energy, the increase in entropy is greater when the surroundings are cool than when they are hot. Adding energy to molecules in a cool system has a proportionately greater effect on the number of ways of distributing matter and energy than adding the same quantity of energy to a system that is already very hot.

Figure 13.2.16 The thermite reaction between iron(III) oxide and aluminium metal is highly exothermic. The molten iron formed can be used to weld railway lines. The reaction gives out a great deal of energy to its surroundings. The entropy change in the surroundings is large and positive.

Total entropy changes

A reaction is only feasible if the total entropy change, ΔS_{total}, is positive. This means that:

● Most exothermic reactions tend to go because around room temperature the value of $-\Delta H/T$ is much larger and more positive than ΔS_{system}, which means that ΔS_{total} is positive.
● An endothermic reaction can be feasible so long as the increase in the entropy of the system is greater than the decrease in the entropy of the surroundings.
● A reaction that does not tend to go at room temperature may become feasible as the temperature rises, because $\Delta S_{surroundings}$ decreases in magnitude as T increases.

6 Consider the reaction of magnesium with oxygen:

$$2Mg(s) + O_2(g) \rightarrow 2MgO(s) \qquad \Delta H^\ominus = -602\,kJ\,mol^{-1}$$
$$\Delta S^\ominus_{system} = -217\,J\,mol^{-1}K^{-1}$$

a) Why does the entropy of the system decrease?

b) Show why the reaction of magnesium with oxygen is feasible at 298 K despite the decrease in the entropy.

7 Consider the reaction of ammonium chloride with barium hydroxide:

$$2NH_4Cl(s) + Ba(OH)_2(s) \rightarrow BaCl_2.2H_2O(s) + 2NH_3(g)$$

Refer to the data sheet headed 'Thermodynamic properties of selected elements and compounds', which you can access via the QR code for Chapter 13.2 on page 320.

a) Use the data from the data sheet to calculate the entropy change for the system and comment on the value you get.

b) Calculate the enthalpy change for the reaction.

c) Calculate the entropy change of the surroundings at 298 K.

d) Work out the total entropy change for the reaction and decide whether or not it is feasible under standard conditions. Comment on your answer.

Entropy changes during dissolving

Sodium chloride dissolves readily in water, despite the fact that the process is slightly endothermic. This is another example that shows that the sign of ΔH is not a reliable guide to whether or not a process will happen. This is particularly the case when the magnitude of ΔH is small.

When sodium chloride dissolves, the disorder increases as the ions leave the regular lattice and mix with the molecules of liquid water. This means that the entropy of the system increases as a salt dissolves. The entropy change in the surroundings is negative for this endothermic change, but the increase in the entropy of the system is more than enough to compensate for this and so the total entropy change is positive.

Dissolving depends on the balance between the change in entropy of the solution and the change in entropy of the surroundings. This balance starts to alter as soon as a salt starts to dissolve and the concentration of the solution rises. In a more concentrated solution the ions are closer together and there are fewer free water molecules to hydrate the ions. These changes modify the values of the hydration energies and the entropies of the ions in solution. Overall, this means that once a certain amount of a salt has dissolved, the processes ceases to be feasible and no more solid dissolves. At this point the solution is saturated. The salt crystals and the saturated solution are in equilibrium.

Table 13.2.4

	$S^{\ominus}/\text{J mol}^{-1}\text{K}^{-1}$
NaCl(s)	+72.1
Na$^+$(aq)	+321
Cl$^-$(aq)	+56.5

Example

Use the values in Table 13.2.4 to calculate the total entropy change when sodium chloride dissolves in water under standard conditions.

$$\text{NaCl(s)} \rightarrow \text{Na}^+\text{(aq)} + \text{Cl}^-\text{(aq)} \quad \Delta H^{\ominus}_{\text{system}} = +3.8\,\text{kJ mol}^{-1}$$

Notes on the method

Calculate the entropy change of the surroundings using the formula:

$$\Delta S^{\ominus}_{\text{surroundings}} = -\frac{\Delta H^{\ominus}}{T}$$

Remember to convert the value of the enthalpy change to J mol^{-1}.

Answer

$$\Delta S^{\ominus}_{\text{surroundings}} = -\frac{3800\,\text{J mol}^{-1}}{298\,\text{K}} = -12.8\,\text{J mol}^{-1}\,\text{K}^{-1}$$

$$\Delta S^{\ominus}_{\text{system}} = \Sigma S^{\ominus}[\text{products}] - \Sigma S^{\ominus}[\text{reactants}]$$

$$= 321\,\text{J mol}^{-1}\text{K}^{-1} + 56.5\,\text{J mol}^{-1}\text{K}^{-1} - 72.1\,\text{J mol}^{-1}\text{K}^{-1}$$

$$= +305\,\text{J mol}^{-1}\text{K}^{-1}$$

$$\Delta S^{\ominus}_{\text{total}} = \Delta S^{\ominus}_{\text{system}} + \Delta S^{\ominus}_{\text{surroundings}}$$

$$= +305\,\text{J mol}^{-1}\text{K}^{-1} - 12.8\,\text{J mol}^{-1}\text{K}^{-1} = +292\,\text{J mol}^{-1}\text{K}^{-1}$$

The total entropy change is positive. Sodium chloride dissolving in water is a feasible process.

Table 13.2.5

	$S^{\ominus}/\text{J mol}^{-1}\text{K}^{-1}$
NH$_4$NO$_3$(s)	151
NH$_4^+$(aq)	113
NO$_3^-$(aq)	146

Test yourself

8 a) Use the values in Table 13.2.5 to calculate the total entropy change when ammonium nitrate dissolves in water under standard conditions.

$$\text{NH}_4\text{NO}_3\text{(s)} \rightarrow \text{NH}_4^+\text{(aq)} + \text{NO}_3^-\text{(aq)} \quad \Delta H^{\ominus} = +28.1\,\text{kJ mol}^{-1}$$

b) Comment on the fact that ammonium nitrate dissolves in water even though the process is endothermic.

13.2.4 Free energy

A chemical change is feasible if the total entropy change is positive. There is no doubt about this. The problem is that using entropy to decide on the direction change involves three steps: working out the entropy change of the system, working out the entropy change of the surroundings, and then putting the two together to calculate the total entropy change. This can be laborious and so chemists are grateful to the American physicist Willard Gibbs, who discovered an easier way of unifying all that chemists know about predicting the extent and direction of change.

Free energy and entropy

Willard Gibbs (1839–1903) was the first to define the thermochemical quantity free energy. The symbol for a free energy change is ΔG. If ΔG is negative the reaction is feasible and tends to go.

The advantage of $\Delta G^{\ominus}$ values for chemists is that tables of standard free energies of formation can be used to calculate the standard free energy change for any reaction. The calculations follow exactly the same steps as the calculations to calculate standard enthalpy changes for reactions from standard enthalpies of formation.

The quantity 'free energy' is closely related to the idea of entropy and can be thought of as the 'total entropy change' in disguise. Willard Gibbs defined free energy as:

$$\Delta G^{\ominus} = -T\Delta S^{\ominus}_{total}$$

Given that:

$$\Delta S^{\ominus}_{total} = \Delta S^{\ominus}_{system} + \Delta S^{\ominus}_{surroundings}$$

And that for a change at constant temperature and pressure:

$$\Delta S^{\ominus}_{surroundings} = \frac{-\Delta H^{\ominus}}{T}$$

It follows that:

$$\Delta S^{\ominus}_{total} = \Delta S^{\ominus}_{system} + \left(\frac{-\Delta H^{\ominus}}{T}\right)$$

Hence: $-T\Delta S^{\ominus}_{total} = -T\Delta S^{\ominus}_{system} + \Delta H^{\ominus}$

From Gibbs' definition this becomes: $\Delta G^{\ominus} = \Delta H^{\ominus} - T\Delta S^{\ominus}_{system}$

The great advantage of this equation is that all the terms refer to the system and so it is no longer necessary to calculate changes in the surroundings. Given that this is the case, the equation is usually written as here, with the understanding that the entropy change is $\Delta S^{\ominus}_{system}$.

$$\Delta G^{\ominus} = \Delta H^{\ominus} - T\Delta S^{\ominus}$$

Table 13.2.6 summarises the implications of this important relationship.

Table 13.2.6

Enthalpy change	Entropy change of the system	Is the reaction feasible?
Exothermic (ΔH negative)	Increase (ΔS positive)	Yes, ΔG is negative
Exothermic (ΔH negative)	Decrease (ΔS negative)	Yes, if the number value of ΔH is greater than the magnitude of $T\Delta S$
Endothermic (ΔH positive)	Increase (ΔS positive)	Yes, if the magnitude of $T\Delta S$ is greater than the number value of ΔH
Endothermic (ΔH positive)	Decrease (ΔS negative)	No, ΔG is positive

> **Key term**
>
> The **free energy change**, ΔG, is the thermochemical quantity used by chemists to decide whether a reaction tends to go and how far it will go. ΔG is the test for the feasibility of a reaction. If ΔG is negative the reaction is feasible.

The possibilities listed in Table 13.2.6 show why chemists sometimes say that the feasibility of a reaction depends on the balance between the enthalpy changes and the entropy changes for the process.

Table 13.2.6 also shows that a change that is not feasible at a lower temperature may become feasible if the temperature rises. Generally the values of $\Delta H^{\ominus}$ and $\Delta S^{\ominus}$ do not change markedly with temperature, so it is possible to estimate the higher temperature at which a reaction that is not feasible at room temperature becomes feasible.

Full data are not available for all reactions, so it is not always possible to calculate the free energy change. At relatively low temperatures the $T\Delta S^{\ominus}$ term is often relatively small compared to the enthalpy change, so that $\Delta G^{\ominus} \approx \Delta H^{\ominus}$.

This means that chemists can often use the sign of $\Delta H^{\ominus}$ as a guide to feasibility. This can be misleading if the magnitude of the entropy change for the reaction system is large. Also, the approximation becomes less justified at higher temperatures when T is bigger and so the magnitude of $T\Delta S^{\ominus}$ is bigger.

Example

Calculate $\Delta S^{\ominus}$ and $\Delta H^{\ominus}$ for the synthesis of methanol from carbon monoxide and hydrogen. Work out the temperature at which the synthesis ceases to be feasible.

Notes on the method

Start by writing the equation for the reaction.

Look up the standard enthalpies changes of formation on the data sheet headed 'Thermodynamic properties of selected elements and compounds', which you can access via the QR code for this chapter on page 320.

Remember that all temperatures are measured on the Kelvin scale. Note too that the values for the enthalpy change and entropy change must be converted to be in the same units.

From the equation $\Delta G^{\ominus} = \Delta H^{\ominus} - T\Delta S^{\ominus}$, it follows that $\Delta G^{\ominus}$ becomes positive and an exothermic reaction ceases to be feasible when the temperature is high enough for $-T\Delta S^{\ominus}$ to be positive and large enough to balance the negative value for the enthalpy change.

Answer

$$CO(g) + 2H_2(g) \rightarrow CH_3OH(l)$$

$$\Delta_r H^{\ominus} = \Sigma \Delta_f H^{\ominus}[\text{products}] - \Sigma \Delta_f H^{\ominus}[\text{reactants}]$$

$$\Delta H^{\ominus} = -239\,\text{kJ}\,\text{mol}^{-1} - (-110\,\text{kJ}\,\text{mol}^{-1}) = -129\,\text{kJ}\,\text{mol}^{-1}$$

$$\Delta S^{\ominus}_{\text{system}} = \Sigma S^{\ominus}[\text{products}] - \Sigma S^{\ominus}[\text{reactants}]$$

$$\Delta S^{\ominus} = 240\,\text{J}\,\text{mol}^{-1}\text{K}^{-1} - (260\,\text{J}\,\text{mol}^{-1}\text{K}^{-1} + 198\,\text{J}\,\text{mol}^{-1}\text{K}^{-1})$$

$$= -218\,\text{J}\,\text{mol}^{-1}\text{K}^{-1} = -0.218\,\text{kJ}\,\text{mol}^{-1}\text{K}^{-1}$$

The reaction ceases to be feasible when $-T\Delta S$ becomes more positive than $-129\,\text{kJ}\,\text{mol}^{-1}$.

$$-T\Delta S^{\ominus} = +129\,\text{kJ}\,\text{mol}^{-1}$$

$$T = 129\,\text{kJ}\,\text{mol}^{-1} \div 0.218\,\text{kJ}\,\text{mol}^{-1}\text{K}^{-1} = 592\,\text{K}$$

The synthesis ceases to be feasible when the temperature is above 592 K.

9 Can an exothermic reaction which is not feasible at room temperature become feasible at a higher temperature if the entropy change for the reaction is negative?

10 Consider the reduction of iron(III) oxide by carbon:

$$2Fe_2O_3(s) + 3C(s) \rightarrow 4Fe(s) + 3CO_2(g) \qquad \Delta H^\ominus = +468\,kJ\,mol^{-1}$$
$$\Delta S^\ominus = +558\,J\,mol^{-1}\,K^{-1}$$

a) Why does the entropy of the system increase for this reaction?

b) Show that the reaction is not feasible at room temperature (298 K).

c) Assuming that $\Delta H^\ominus$ and $\Delta S^\ominus$ do not vary with temperature, estimate the temperature at which the reaction becomes feasible.

Free energy and equilibrium constants

Chapter 11 shows that the direction and extent of change can be described in terms of equilibrium constants. This chapter has introduced the idea that the feasibility of a reaction can be determined by calculating its standard free energy change. It turns out that these two methods for determining the direction and extent of chemical change are directly related. The theory of thermochemistry shows that, for reactions involving gases, they are related by this formula, where R is a constant (the same constant as in the ideal gas equation):

$$\Delta G^\ominus = -RT \ln K$$

This equation applies to reactions involving gases and gives the value of K_p (see Section 11.4).

This relationship shows that the values for $\Delta G^\ominus$ and value for K offer alternative ways of answering the same questions for a reaction:

- Will the reaction go?
- How far will it go?

The question that **cannot** be answered by these thermodynamic quantities is: 'How fast will it go?'.

It is very important to keep in mind that even if a reaction is feasible, it may be very slow. This is fortunate, otherwise it would not be possible to fill a car's petrol tank in the presence of air.

Tip

For guidance about the meaning and use of natural logarithms, ln, refer to Section 1 in 'Mathematics for A Level chemistry', which you can access via the QR code for Chapter 13.2 on page 320.

Tip

Chapter 14 introduces a third method of predicting the direction and extent of change for redox reactions based on standard electrode potentials. In practice, chemists use the quantity that it is more convenient to measure. Knowing the value of one of the two quantities, it is possible to calculate the others.

Nitrogen can react with hydrogen to make ammonia, $\Delta G^{\ominus} = -32.9\,kJ\,mol^{-1}$.

$$N_2(g) + 3H_2(g) \rightleftharpoons 2NH_3(g)$$

Calculate the equilibrium constant, K, for this reaction at 298 K. Comment on the value of K.

Notes on the method

The value of the gas constant $R = 8.31\,J\,K^{-1}\,mol^{-1}$.

Work in consistent units by converting kJ to J.

Rearrange $\Delta G^{\ominus} = -RT\ln K$ to find $\ln K$.

Refer to Section 3 in 'Mathematics in A Level chemistry', which you can access via the QR code for this chapter on page 320, to find out how to calculate the value of K from $\ln K$.

Logarithms do not have units, so the calculation based on this thermodynamic equation gives a value for K that does not have units.

Answer

$$\ln K = \frac{-\Delta G^{\ominus}}{RT}$$

$$\ln K = \frac{-(-32\,900\ \text{J mol}^{-1})}{8.31\,\text{J K}^{-1}\text{mol}^{-1} \times 298\,\text{K}} = 13.3$$

$$K = e^{13.3} \approx 6 \times 10^5$$

The value of K is very large. This shows that, at room temperature, the equilibrium is well over to the product side. This is consistent with a negative value of $\Delta G^{\ominus}$, which shows that the reaction is feasible under these conditions. However, at room temperature, in the absence of a catalyst, the reaction is very, very slow.

Test yourself

11 $\Delta G^{\ominus} = +163\,kJ\,mol^{-1}$ for the conversion of oxygen to ozone at 298 K.

$$\tfrac{3}{2}O_2(g) \rightleftharpoons O_3(g)$$

a) Calculate a value for the equilibrium constant, K, of this reaction at 298 K.

b) Comment on the values of $\Delta G^{\ominus}$ and K.

Stable or inert?

The study of energetics (thermochemistry) and rates of reaction (kinetics) helps to explain why some chemicals are stable (Figure 13.2.17) while others react rapidly.

Compounds are stable if they have no tendency to decompose into their elements or into other compounds. Magnesium oxide, for example, has no tendency to split up into magnesium and oxygen. However, a compound that is stable at room temperature and pressure may become more or less stable as conditions change.

Key term

A chemical or mixture of chemicals is thermodynamically **stable** if there is no tendency for a reaction. A positive free energy change, ΔG, indicates that the reaction does not tend to occur.

Chemists often use standard enthalpy changes as an indicator of stability. Strictly they should use standard free energy, $\Delta_f G^{\ominus}$, values but in many cases $\Delta_f G^{\ominus} \approx \Delta_f H^{\ominus}$.

A chemical is inert if it has no tendency to react even when the reaction is feasible (Figure 13.2.18). Nitrogen, for example, is a relatively unreactive gas that can be used to create an 'inert atmosphere' free of oxygen (which is much more reactive). However, nitrogen is not inert in all circumstances. It reacts with hydrogen in the Haber process to form ammonia and with oxygen at high temperatures in motor engines and power stations to form nitrogen oxides.

Figure 13.2.17 The Giant Buddha in a temple at Bangkok Thailand. Gold is stable in air and water. It has no tendency to react and tarnish.

A compound such as the gas N_2O, for example, is thermodynamically unstable but it continues to exist at room temperature because it is kinetically inert ($\Delta_f H^{\ominus} = +82 \, kJ \, mol^{-1}$ and $\Delta_f G^{\ominus} = +104 \, kJ \, mol^{-1}$). The free energy change for the decomposition reaction is negative so the compound tends to decompose into its elements, but the rate is very slow under normal conditions.

Examples of kinetic inertness are:

- a fuel tank containing petrol and air
- mixture of hydrogen and oxygen at room temperature
- a solution of hydrogen peroxide in the absence of a catalyst
- aluminium metal in dilute hydrochloric acid.

'Kinetic stability' is a term sometimes used for 'kinetic inertness'. It helps, however, to make a sharp distinction between two quite different types of explanation. For clarity, chemists refer to:

- systems with no tendency to react as 'stable' and
- systems that should react but do not do so for a rate (kinetic) reason as 'kinetically inert'.

Figure 13.2.18 Distorted reflections of surrounding park and buildings in an aluminium street sculpture in Los Angeles. Aluminium is a reactive metal, but it is inert in air and water because it is protected from corrosion by a thin layer of the metal oxide on its surface.

Table 13.2.7 Sometimes there is no tendency for a reaction to go because the reactants are stable. This is so if the free energy change for the reaction is positive. Sometimes there is no reaction even though thermochemistry suggests that it should go. The free energy change is negative, so the change is feasible, but a high activation energy means that the rate of reaction is very, very slow.

$\Delta G^{\ominus}$ $(\approx \Delta H^{\ominus})$	Activation energy	Change observed	Stability
Positive	High	No reaction	Reactants stable relative to products
Negative	High	No reaction	Reactants unstable relative to products but kinetically inert
Positive	Low	No reaction	Reactants stable relative to products
Negative	Low	Fast reaction	Reactants unstable relative to products

Test yourself

12 Suggest examples of reactions to illustrate each of the four possibilities in Table 13.2.7.

13 Draw a reaction profile to show the energy changes from reactants to products for reactants that are thermodynamically unstable relative to the products, but kinetically inert.

Activity

The thermal stability of Group 2 carbonates

The decomposition of Group 2 metal carbonates is used on a large scale to make oxides such as magnesium and calcium oxides.

$$MgCO_3(s) \rightarrow MgO(s) + CO_2(g)$$
$$\Delta H^{\ominus} = +117 \, kJ \, mol^{-1}, \ \Delta S^{\ominus} = +175 \, J \, mol^{-1} \, K^{-1}$$

$$BaCO_3(s) \rightarrow BaO(s) + CO_2(g)$$
$$\Delta H^{\ominus} = +268 \, kJ \, mol^{-1}, \ \Delta S^{\ominus} = +172 \, J \, mol^{-1} \, K^{-1}$$

The carbonates of Group 2 metals do not decompose at room temperature. They do decompose on heating.

1 Why does the entropy increase when a Group 2 carbonate decomposes?

2 a) Calculate the free energy change for the decomposition of:
   i) magnesium carbonate
   ii) barium carbonate.
   b) Are these two compounds stable or unstable relative to decomposition into their oxide and carbon dioxide at 298 K?

3 Assuming that $\Delta H^{\ominus}$ and $\Delta S^{\ominus}$ for the reactions do not vary with temperature, estimate the temperatures at which the two decomposition reactions become feasible.

4 Down Group 2, do the metal carbonates become more or less stable relative to decomposition into the oxide and carbon dioxide?

Figure 13.2.19 Crystals of chalcopyrite on dolomite with a large calcite crystal. Dolomite is a calcium magnesium carbonate rock. Calcite is pure calcium carbonate.

One of the ways that chemists explain the trend in thermal stability of the Group 2 carbonates is to analyse the related energy changes.

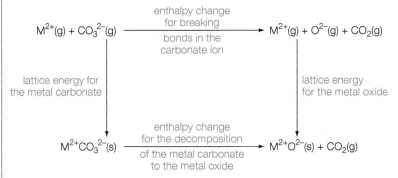

Figure 13.2.20 An energy cycle for the decomposition of the carbonate of a Group 2 metal, M.

Figure 13.2.21 Decomposition of the Group 2 carbonate into its oxide.

5 Which of the enthalpy changes in Figure 13.2.20 are exothermic and which are endothermic?

6 Why is the lattice energy of magnesium oxide more negative than the lattice energy of barium oxide?

7 Why is the lattice energy of magnesium oxide more negative than the lattice energy of magnesium carbonate?

8 Why is the difference between the lattice energies of the metal carbonates and oxides significant in explaining the trend in thermal stability of the Group 2 carbonates?

9 How does the trend in thermal stability of the metal carbonates down Group 2 relate to the polarising power of the metal ions?

Exam practice questions

1 The graph below shows how the entropy of water changes from 0 K to 450 K.

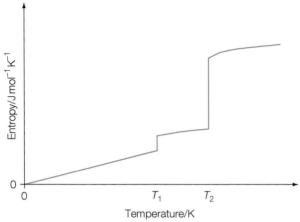

a) Explain why the molar entropy of water is zero at 0 K. *(2)*

b) Account for the entropy changes at temperatures T_1 and T_2. Explain why one is bigger than the other. *(3)*

c) i) Why does $\Delta G = 0\,\text{kJ}\,\text{mol}^{-1}$ for the change of water to steam at 373 K and 1 atmosphere pressure? *(2)*

    ii) The enthalpy change of vaporisation of water is $+41.1\,\text{kJ}\,\text{mol}$ at 373 K. Calculate the value of ΔS for the conversion of water into steam at 373 K and 1 atmosphere pressure. *(4)*

2 Write equations, with state symbols, for these changes. Without doing any calculations, state whether the entropy change for the system is positive or negative during each change. Explain your answers.

a) Ammonium nitrate decomposing to form dinitrogen oxide and steam. *(2)*

b) Dissolving potassium chloride in water. *(2)*

c) Reacting nitrogen and oxygen to form nitrogen dioxide. *(2)*

d) Adding a catalyst to an aqueous solution of hydrogen peroxide. *(2)*

e) Mixing ammonia and hydrogen iodide to form ammonium iodide. *(2)*

3 Electrolysis is usually used to extract aluminium metal. In principle, it should be possible to extract the metal by heating its oxide with carbon. Use the data in the table to answer the questions.

Substance	$Al_2O_3(s)$	$C(s)$	$Al(s)$	$CO(g)$
$\Delta_f H^{\ominus}/\text{kJ}\,\text{mol}^{-1}$	−1669	0	0	−111
$S^{\ominus}\,\text{J}\,\text{mol}^{-1}\,\text{K}^{-1}$	50.9	5.7	28.3	198

a) Write the equation for the reduction of aluminium oxide by carbon, assuming that the carbon is converted to carbon monoxide. *(1)*

b) Calculate the standard enthalpy change for the reduction reaction. *(3)*

c) Calculate the standard entropy change for the reduction reaction. *(3)*

d) i) Calculate the standard free energy change for the reduction at 298 K. *(2)*

    ii) What does you answer to (d)(i) tell you about the feasibility of the reaction at 298 K? *(1)*

e) Calculate the minimum temperature at which the reduction of aluminium oxide by carbon becomes feasible. *(2)*

f) Suggest a reason why the reaction between aluminium oxide and carbon does not happen to a significant extent until the temperature is about 1000 degrees higher than your answer to (e). *(1)*

4 The equation shows the reaction of carbon dioxide and water to form glucose:

$$6CO_2(g) + 6H_2O(l) \rightarrow C_6H_{12}O_6(s) + 6O_2(g)$$

$$\Delta H^{\ominus} = +2879\,\text{kJ}\,\text{mol}^{-1}$$

$$\Delta S^{\ominus} = -256\,\text{J}\,\text{K}^{-1}\,\text{mol}^{-1}$$

a) Use the equation and the data in the table to calculate the standard entropy of glucose. *(2)*

Molecule	$CO_2(g)$	$H_2O(l)$	$O_2(g)$
$S^{\ominus}/\text{J}\,\text{K}^{-1}\,\text{mol}^{-1}$	214	70	205

b) Calculate $\Delta G^{\ominus}$ for the reaction. *(2)*

c) i) Discuss whether or not this reaction is feasible at any temperature. *(3)*

    ii) Account for the fact that plants are able to make glucose from carbon dioxide and water by photosynthesis. *(1)*

5 Carbon monoxide is one of the products when methane burns in a limited supply of air.

$$CH_4(g) + \tfrac{3}{2}O_2(g) \rightarrow CO(g) + 2H_2O(g)$$
$$\Delta H^{\ominus} = -519\,kJ\,mol^{-1}$$
$$\Delta S^{\ominus} = +82\,J\,mol^{-1}\,K^{-1}$$

a) Calculate $\Delta G^{\ominus}$ for the reaction. *(2)*

b) i) Plot a graph to show how ΔG for this reaction varies with temperature using your answer to (a) and the values below. *(2)*

Temperature/K	1500	3000	4000
ΔG/kJ mol⁻¹	−640	−765	−845

ii) From the shape of the line on the graph, what can you conclude about the variation with temperature of ΔS for this reaction? *(2)*

c) i) Is the reaction of methane with oxygen to form carbon monoxide and steam feasible in the temperature range 200 K to 3000 K? *(2)*

ii) Why does methane not burn in air at 298 K? *(2)*

d) Sooty carbon is another of the products when methane burns in a limited supply of air.

$$CH_4(g) + O_2(g) \rightarrow C(s) + 2H_2O(g)$$
$$\Delta S^{\ominus} = -8\,J\,mol^{-1}\,K^{-1}$$

How do you account in the difference in the values for $\Delta S^{\ominus}$ for the reaction producing carbon monoxide and the reaction that forms soot? *(4)*

6 a) Explain why the standard entropy of bromine is greater than that of iodine. *(2)*

b) The enthalpy change of vaporisation of ethoxyethane is $+26.0\,kJ\,mol^{-1}$ at its boiling temperature, which is 35 °C. What is the value of $\Delta S^{\ominus}_{system}$ for the change shown in this equation? *(4)*

$$C_2H_5OC_2H_5(l) \rightarrow C_2H_5OC_2H_5(g)$$

c) Predict (without any calculations) the sign (positive or negative) of the values of $\Delta H^{\ominus}$, $\Delta S^{\ominus}$ and $\Delta G^{\ominus}$ for each of these changes.

i) The decomposition of water into hydrogen and oxygen. *(3)*

ii) The combustion of a hydrocarbon such as octane in petrol. *(3)*

d) The reaction of carbon with steam produces synthesis gas.

$$C(s) + H_2O(g) \rightleftharpoons CO(g) + H_2(g)$$
$$\Delta H^{\ominus} = +132\,kJ\,mol^{-1}$$
$$\Delta S^{\ominus} = +233\,J\,mol^{-1}\,K^{-1}$$

i) Calculate the value of $\Delta G^{\ominus}$ for the reaction. *(2)*

ii) Calculate the value of the equilibrium constant K for the reaction. *(3)* The gas constant $R = 8.31\,J\,K^{-1}\,mol^{-1}$.

iii) Comment on the values of $\Delta G^{\ominus}$ and K. *(2)*

iv) At what temperature does the reaction become feasible? *(2)*

Redox II

14.1 Redox reactions

Redox reactions are very important in the natural environment, in living things and in modern technology. It is no surprise that the Earth, with its oxygen atmosphere, has an extensive range of redox chemistry. Every year, oxidation of ions such as Fe^{2+} in weathered rocks, and oxidation of molecules such as hydrogen sulfide, carbon monoxide and methane in volcanic gases (Figure 14.1), removes about one thousand billion (10^{12}) moles of oxygen from the atmosphere.

Redox reactions are also involved in the metabolic pathways of respiration. These pathways produce the molecule adenosine triphosphate (ATP). ATP transfers the energy released when food is oxidised to make possible the movement, growth and all the other activities in living things that need a source of energy.

In addition, the voltages of chemical cells are obtained from the energy of redox reactions, and redox is also involved in manufacturing processes that use electrolysis to make products such as chlorine and aluminium.

Figure 14.1 Volcanoes release millions of tonnes of reducing gases into the atmosphere where they react with oxygen. The April 2010 eruption of the Eyjafjallajokull volcano in Iceland created an ash cloud that grounded air traffic around the world.

> **Tip**
>
> This first section of this chapter revisits ideas that you met when studying Chapter 3 in Student Book 1. The 'Test yourself' questions in this section are to help you revise your understanding of the theory of redox reactions.

Definitions of redox

Descriptions and theories of oxidation and reduction have developed over the years. Although there are now several definitions of redox, oxidation and reduction always go together.

Oxidation originally meant addition of oxygen or removal of hydrogen, but the term now covers all reactions in which atoms or ions lose electrons. Chemists have further extended the definition of oxidation to include molecules by defining oxidation as a change that makes the oxidation number of an element more positive, or less negative.

Similarly, reduction originally meant removal of oxygen or addition of hydrogen, but the term now covers all reactions in which atoms, molecules or ions gain electrons. Defining reduction as a change in which the oxidation number of an element decreases further extends the concept of reduction.

Key terms

Redox stands for Reduction + Oxidation.

Oxidation involves the loss of electrons or an increase in oxidation number.

An **oxidation number** is a number assigned to an atom or ion to describe its relative state of oxidation or reduction.

Reduction involves the gain of electrons or a decrease in oxidation number.

Oxidation states and oxidation numbers

Elements in the s block of the periodic table have only one oxidation state in their compounds: this is +1 for Group 1 elements and +2 for Group 2 elements. However, most elements in the p block and d block form compounds in which their atoms have different oxidation states. Displaying the compounds of an element on an oxidation state diagram provides a 'map' of its chemistry and shows the different oxidation numbers that it can have (Figure 14.2).

There are strict rules for assigning oxidation numbers; these are shown in Figure 14.3.

> **Tip**
>
> Remember the mnemonic **OIL RIG** – **O**xidation **I**s **L**oss; **R**eduction **I**s **G**ain of electrons.

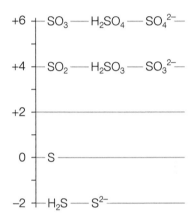

Figure 14.2 The oxidation numbers of sulfur in its different oxidation states.

Oxidation number rules

1. The oxidation number of the atoms in uncombined elements is zero.
2. In simple ions, the oxidation number of the element is the charge on the ion.
3. In neutral molecules, the sum of the oxidation numbers of the constituent elements is zero.
4. In ions containing two or more elements, the charge on the ion is the sum of the oxidation numbers.
5. In any compound, the more electronegative element has a negative oxidation number and the less electronegative element has a positive oxidation number.
6. The oxidation number of hydrogen in all its compounds is +1, except in metal hydrides in which it is −1.
7. The oxidation number of oxygen in all its compounds is −2, except in peroxides in which it is −1 and OF_2 in which it is +2.
8. The oxidation number of chlorine in all its compounds is −1, except in compounds with oxygen and fluorine in which it is positive.

Figure 14.3 Rules for assigning oxidation numbers.

Half-equations

Half-equations are ionic equations used to describe either the gain or the loss of electrons during a redox process. Half-equations help to show what is happening during a redox reaction. Two half-equations can be combined to give the full equation for a redox reaction.

For example, zinc metal can reduce Cu^{2+} ions in copper(II) sulfate solution, forming copper metal and Zn^{2+} ions in zinc sulfate solution. This can be shown as two half-equations:

- electron loss (oxidation) $Zn(s) \rightarrow Zn^{2+}(aq) + 2e^-$
- electron gain (reduction) $Cu^{2+}(aq) + 2e^- \rightarrow Cu(s)$

This leads to the full equation by balancing the number of electrons lost by Zn with the number gained by $Cu^{2+}(aq)$:

$$Zn(s) + Cu^{2+}(aq) \rightarrow Zn^{2+}(aq) + Cu(s)$$

1 Describe, in terms of gain or loss of electrons, the redox reactions in these examples:

a) the reaction of calcium with iodine to form calcium iodide

b) the changes at the electrodes during the manufacture of aluminium from molten (liquid) aluminium oxide

c) the reaction of iron with chlorine to form iron(III) chloride.

2 Explain why the oxidation number of oxygen is:

a) +2 in OF_2

b) −1 in peroxides such as Na_2O_2.

3 State the changes in oxidation number when concentrated sulfuric acid reacts with potassium bromide.

$$2KBr(s) + 3H_2SO_4(l)$$
$$\rightarrow 2KHSO_4(s) + Br_2(l) + SO_2(g) + 2H_2O(l)$$

4 What is the oxidation number of each element in:

a) KIO_3

b) N_2O_5

c) H_2O_2

d) SF_6

e) NaH?

5 Which sulfur molecules or ions in Figure 14.2 can:

a) act as an oxidising agent or as a reducing agent depending on the conditions

b) only act as an oxidising agent

c) only act as a reducing agent?

6 Draw charts similar to Figure 14.2 to show that:

a) nitrogen can exist in the −3, 0, +1, +2, +3, +4 and +5 oxidation states

b) chlorine can exist in the −1, 0, +1, +3, +5 and +7 oxidation states.

7 Are the named elements below oxidised or reduced in the following conversions?

a) magnesium to magnesium sulfate

b) iodine to aluminium iodide

c) hydrogen to lithium hydride

d) iodine to iodine monochloride, ICl

14.2 Balancing equations for redox reactions

Half-equations and stoichiometry

One way to arrive at the balanced equation for a redox reaction is to combine two half-equations. In the reaction, the number of electrons given up in one half-equation must equal the number taken in the other. The amounts of different substances involved in a balanced chemical equation is sometimes described as the reaction **stoichiometry**.

For example, when hydrogen peroxide in acid solution reacts with iron(II) ions, the half-equations are:

$$H_2O_2(aq) + 2H^+(aq) + 2e^- \rightarrow 2H_2O(l)$$

$$Fe^{2+}(aq) \rightarrow Fe^{3+}(aq) + e^-$$

Multiplying the second equation by 2 balances the number of electrons given up with those taken, and the overall equation can be obtained by adding the two half-equations together.

$$H_2O_2(aq) + 2H^+(aq) + 2e^- \rightarrow 2H_2O(l)$$

$$2Fe^{2+}(aq) \rightarrow 2Fe^{3+}(aq) + 2e^-$$

Overall: $H_2O_2(aq) + 2H^+(aq) + 2Fe^{2+}(aq) \rightarrow 2H_2O(l) + 2Fe^{3+}(aq)$

Key term

The **stoichiometry** of a reaction is the amounts in moles of the reactants and products as shown in the balanced equation for a reaction. A stoichiometric reaction is one that uses up reactants and produces products in amounts exactly as predicted by the balanced equation.

Tip

When balancing ionic equations for redox reactions, always check that the overall charges balance.

Oxidation numbers and stoichiometry

It is also possible to use oxidation numbers to balance equations for redox reactions. This is done by matching the decrease in oxidation number of the element that is reduced to the increase in oxidation number of the element that is oxidised.

The oxidation of iron(II) ions by manganate(VII) ions in acid solution shows how this can be done to balance an equation when oxoanions are involved.

Example

What is the balanced equation for the reaction in which manganate(VII) ions, MnO_4^-, in acid solution are reduced to manganese(II) ions as they oxidise iron(II) ions to iron(III) ions?

Notes on the method

When oxoanions such as MnO_4^- (and $Cr_2O_7^{2-}$) react in acid solution, the oxygen atoms are converted to water molecules by H^+ ions. The atoms of the metal in the oxoanions become stable simple ions.

The initial and final states of the metal element must be known before it is possible to write the balanced equation.

Answer

Step 1 Write the formulae of the atoms, ions and molecules involved in the reaction.

$MnO_4^- + H^+ + Fe^{2+} \rightarrow Mn^{2+} + H_2O + Fe^{3+}$

Step 2 Identify the elements which change in oxidation number and the extent of change.

$$\overbrace{MnO_4^- \ + \ H^+ \ + \ \underbrace{Fe^{2+} \ \rightarrow \ Mn^{2+}}_{\text{change of +1 in Fe}} \ + \ H_2O \ + \ Fe^{3+}}^{\text{change of -5 in Mn}}$$

Step 3 Balance the equation so that the decrease in oxidation number of one element equals the increase in oxidation number of the other element.
In this example, the decrease of −5 in the oxidation number of manganese is balanced by five Fe^{2+} ions, each increasing their oxidation number by +1.

$MnO_4^- + H^+ + 5Fe^{2+} \rightarrow Mn^{2+} + H_2O + 5Fe^{3+}$

Step 4 Balance for oxygen and hydrogen.
In this example, the four oxygen atoms of the MnO_4^- ion join with eight hydrogen ions to form four water molecules.

$MnO_4^- + 8H^+ + 5Fe^{2+} \rightarrow Mn^{2+} + 4H_2O + 5Fe^{3+}$

Step 5 Finally, check that the overall charges on each side of the equation balance and then add state symbols.
The net charge on the left is 17+, which is the same as that on the right. So, the equation for the reaction is:

$MnO_4^-(aq) + 8H^+(aq) + 5Fe^{2+}(aq) \rightarrow Mn^{2+}(aq) + 4H_2O(l) + 5Fe^{3+}(aq)$

Oxidising agents and reducing agents

> **Key term**
>
> **Oxidising agents** (oxidants) take electrons and are reduced when they react.

Oxidising agents (oxidants) are chemical reagents that can oxidise other substances. They do this either by taking electrons away from these substances, or by increasing their oxidation number. Common oxidising agents include oxygen, chlorine, bromine, hydrogen peroxide, the manganate(VII) ion in potassium manganate(VII) and the dichromate(VI) ion in potassium or sodium dichromate(VI) (Figure 14.4).

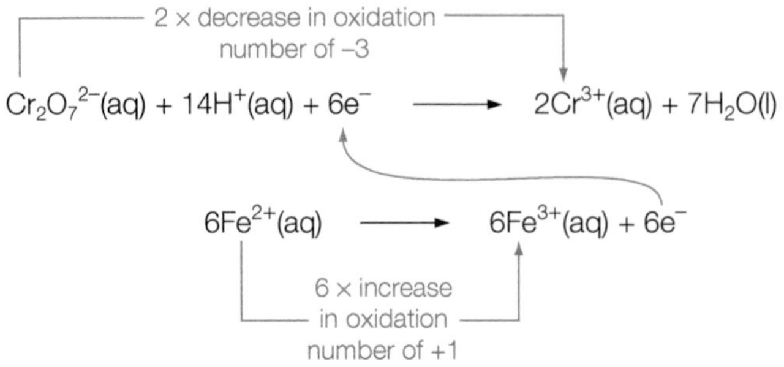

2 × decrease in oxidation number of −3
$Cr_2O_7^{2-}(aq) + 14H^+(aq) + 6e^- \rightarrow 2Cr^{3+}(aq) + 7H_2O(l)$
$6Fe^{2+}(aq) \rightarrow 6Fe^{3+}(aq) + 6e^-$
6 × increase in oxidation number of +1

Figure 14.4 Dichromate(VI) ions act as oxidising agents by taking electrons from iron(II) ions in acid solution. An oxidising agent is itself reduced when it reacts.

Some reagents change colour when they are oxidised which makes them useful for detecting oxidising agents. In particular, a colourless solution of iodide ions is oxidised to iodine, which turns the solution to a yellow-brown colour, so long as excess iodide ions are present. Iodine is only very slightly soluble in water, but it dissolves in a solution containing iodide ions to form the tri-iodide ion, $I_3^-(aq)$.

$$I_2(s) + I^-(aq) \rightarrow I_3^-(aq)$$

A reagent labelled 'iodine solution' is normally $I_2(s)$ in KI(aq) which forms $KI_3(aq)$. The $I_3^-(aq)$ ion is yellow-brown, which explains the colour change when iodine is produced from iodide ions.

$$2I^-(aq) \rightarrow I_2(aq) + 2e^-$$

electrons taken by oxidising agent

This can be a very sensitive test for oxidising agents if starch is also present because starch forms an intense blue-black colour with iodine. Moistened starch-iodide paper is a version of this test which can detect oxidising gases such as chlorine and bromine vapour.

> **Key term**
>
> **Reducing agents** (reductants) give up electrons and are oxidised when they react.

Reducing agents (reductants) are chemical reagents that can reduce other substances. They do this either by giving electrons to these substances or by decreasing their oxidation number. Common reducing agents include metals such as zinc and iron, often with acid, sulfite ions (SO_3^{2-}), iron(II) ions and iodide ions (Figure 14.5).

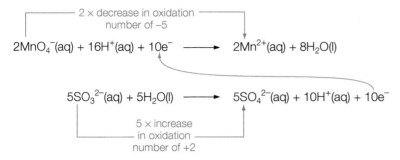

$$2MnO_4^-(aq) + 16H^+(aq) + 10e^- \longrightarrow 2Mn^{2+}(aq) + 8H_2O(l)$$

2 × decrease in oxidation number of −5

$$5SO_3^{2-}(aq) + 5H_2O(l) \longrightarrow 5SO_4^{2-}(aq) + 10H^+(aq) + 10e^-$$

5 × increase in oxidation number of +2

Figure 14.5 Sulfite ions act as reducing agents by giving electrons to manganate(VII) ions. A reducing agent is itself oxidised when it reacts.

Some reagents change colour when they are reduced, which makes them useful for detecting reducing agents (Figure 14.6 and Figure 14.7).

Figure 14.6 A test for reducing agents. *The test*: add a solution of purple potassium manganate(VII) acidified with dilute sulfuric acid to the reducing agent. *The result*: the purple solution turns colourless as purple MnO_4^- ions are reduced to very pale pink Mn^{2+} ions.

Figure 14.7 Another test for reducing agents. *The test*: add orange dichromate(VI) solution acidified with dilute sulfuric acid to the reducing agent. *The result*: the orange solution turns green as orange $Cr_2O_7^{2-}$ ions are usually reduced to green Cr^{3+} ions but sometimes to blue Cr^{2+} ions.

Test yourself

8 Write half-equations to show what happens when the following act as oxidising agents:

a) $Fe^{3+}(aq)$ b) $Br_2(aq)$

c) $H_2O_2(aq)$ in acid solution.

9 Write half-equations to show what happens when the following act as reducing agents:

a) $Zn(s)$ b) $I^-(aq)$

c) $Fe^{2+}(aq)$.

10 Explain why moist starch-iodide paper can be used as a very sensitive test for chlorine.

11 a) Write the half-equations involved in the reaction between Fe^{3+} ions in a solution of iron(III) chloride and iodide ions in potassium iodide solution.

   b) Write a full redox equation for the reaction involved.

   c) Describe the change you would see in the solution when the reaction occurs.

   d) How many moles of Fe^{3+} ions react with one mole of I^- ions?

   e) Why does iron(III) iodide not exist?

12 Write a balanced equation for each of these redox reactions:

   a) manganese(IV) oxide with hydrochloric acid to form manganese(II) ions and chlorine

   b) copper metal with nitrate ions in nitric acid to form copper(II) ions and nitrogen dioxide gas.

14.3 Redox titrations

In a redox titration, an oxidising agent reacts with a reducing agent. During the titration, the aim is to measure the volume of a standard solution of an oxidising agent or a reducing agent that reacts exactly with a measured volume of the other reagent.

Measuring reducing agents – potassium manganate(VII) titrations

Potassium manganate(VII) is often chosen to measure reducing agents because it can be obtained as a pure, stable solid which reacts in acid solution exactly as in the following equation:

$$MnO_4^-(aq) + 8H^+(aq) + 5e^- \rightarrow Mn^{2+}(aq) + 4H_2O(l)$$

Because of these properties, potassium manganate(VII) is a **primary standard** used to make up **standard solutions**.

No indicator is required in potassium manganate(VII) titrations. On adding the potassium manganate(VII) solution from a burette, the purple MnO_4^- ions change rapidly to the very pale, pink Mn^{2+} ions which look colourless in dilute solution. At the end–point, one drop of excess MnO_4^- is sufficient to produce a permanent pale pink colour.

> ### Key terms
>
> A **primary standard** is a chemical which can be weighed out accurately to make up a standard solution. A primary standard must:
> - be very pure
> - not gain or lose mass when exposed to the air
> - have a relatively high molar mass so weighing errors are minimised
> - react exactly and rapidly as described by the chemical equation.
>
> A **standard solution** is a solution with an accurately known concentration. The method of preparing a standard solution is to dissolve a weighed sample of a primary standard in water and then make the solution up to a definite volume in a graduated flask.

Example

Two iron tablets (mass 1.30 g) containing iron(II) sulfate were dissolved in dilute sulfuric acid and made up to 100 cm³ (Figure 14.8). 10.0 cm³ of this solution required 12.00 cm³ of a standard solution of 0.00500 mol dm⁻³ KMnO₄ to produce a faint red colour. What is the percentage of iron in the iron tablets? (Fe = 55.8)

Notes on the method

1 Write the half-equations and work out the amounts in moles of Fe^{2+} and MnO_4^- that react.

2 Calculate the amount of MnO_4^- that reacts in the titration, and hence the amount of Fe^{2+} which reacts.

3 Work out the amount of Fe^{2+} in the whole solution, and hence in the tablets dissolved.

4 Calculate the percentage of iron in the tablets.

Answer

1 The half-equations for the reaction are:

$$MnO_4^- + 8H^+ + 5e^- \rightarrow Mn^{2+} + 4H_2O \text{ and}$$

$$(Fe^{2+} \rightarrow Fe^{3+} + e^-) \times 5$$

Therefore 5 mol Fe^{2+} react with 1 mol MnO_4^-.

2 Amount of MnO_4^- reacting in the titration

$$= \frac{12.00}{1000} \text{ dm}^3 \times 0.00500 \text{ mol dm}^{-3}$$

Therefore the amount of Fe^{2+} reacting in the titration

$$= \frac{12.00}{1000} \times 0.00500 \times 5 \text{ mol}$$

3 Amount of Fe^{2+} in 100 cm³ of solution (2 tablets)

$$= \frac{12.00}{1000} \times 0.00500 \times 5 \times 10.0 \text{ mol}$$

$$= \frac{3.00}{1000} \text{ mol}$$

4 Mass of Fe^{2+} in 2 tablets

$$= \frac{3.00}{1000} \text{ mol} \times 55.8 \text{ g mol}^{-1}$$

$$= 0.1674 \text{ g}$$

Therefore the percentage of iron in the tablets

$$= \frac{0.1674}{1.30} \times 100 = 12.9\%$$

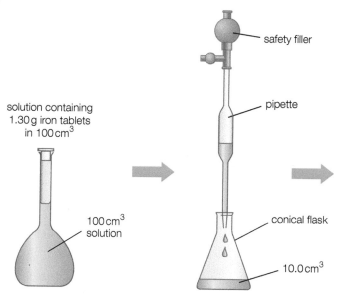

solution containing 1.30 g iron tablets in 100 cm³

100 cm³ solution

safety filler

pipette

conical flask

10.0 cm³

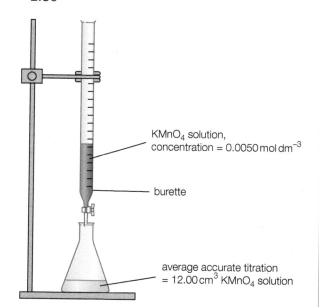

KMnO₄ solution, concentration = 0.0050 mol dm⁻³

burette

average accurate titration = 12.00 cm³ KMnO₄ solution

Figure 14.8 Finding the percentage of iron in iron tablets.

Measuring oxidising agents – iodine/thiosulfate titrations

Many oxidising agents rapidly convert iodide ions to iodine by taking away electrons.

$$2I^-(aq) \rightarrow I_2(aq) + 2e^-$$

The iodine which forms can be titrated with thiosulfate ions and be reduced back to iodide ions.

$$I_2(aq) + 2e^- \rightarrow 2I^-(aq)$$

$$\underset{\text{thiosulfate ion}}{2S_2O_3^{2-}(aq)} \rightarrow \underset{\text{tetrathionate ion}}{S_4O_6^{2-}(aq)} + 2e^-$$

These two half-equations show that 2 mol of thiosulfate ions reduce 1 mol of iodine molecules.

This system can be used to investigate quantitatively any oxidising agent that can oxidise iodide ions to iodine. Oxidising agent that can be estimated by this method include iron(III) ions, copper(II) ions, chlorine, and manganate(VII) ions in acid.

The procedure is:

- Add excess potassium iodide to a measured quantity of the oxidising agent, which then converts iodide ions to iodine.
- Titrate the iodine formed with a standard solution of sodium thiosulfate.

The greater the amount of oxidising agent added, the more iodine is formed and the more thiosulfate is needed from the burette to react with it.

Iodine is not soluble in water. It dissolves in potassium iodide solution to form a solution that is dark brown when concentrated but pale yellow when dilute.

At the end-point, the pale yellow iodine colour disappears to give a colourless solution. Adding a few drops of starch solution just before the end-point makes the colour change much sharper. Starch gives a deep blue colour with iodine that disappears at the end-point.

Tip

When using thiosulfate to titrate iodine formed by an oxidising agent, iodide ions are first oxidised to iodine and then reduced back to iodide. This means you do not need to include the iodine in the calculation; it is present as a 'go between'. You can relate the thiosulfate directly to the oxidising agent with the help of the half-equations.

The following practical illustrates the use of iodine/thiosulfate titrations to determine the concentration of supermarket bleaches.

> **Tip**
>
> Refer to Practical skills sheet 7, 'Measuring chemical amounts by titration', which you can access via the QR code for Chapter 14 on page 321.

Core practical 11

A redox titration

The active reagent in household bleaches is sodium chlorate(I), NaClO (Figure 14.9). To increase the cleaning power of these bleaches manufacturers usually add detergents, and to improve their smell they add perfumes. Sodium chlorate(I) is a strong oxidising agent which bleaches by oxidising coloured materials to colourless or white substances.

The half-equation when sodium chlorate(I) acts as an oxidising agent is:

$$ClO^-(aq) + 2H^+(aq) + 2e^- \rightarrow Cl^-(aq) + H_2O(l)$$

A student is asked to determine the concentration of sodium chlorate(I) in a supermarket bleach. The bleach is too concentrated to be titrated directly, so it first has to be diluted.

Using a measuring cylinder, $100\,cm^3$ of the bleach is added to a graduated flask and made up to a volume of $1000\,cm^3$. $10.0\,cm^3$ of the diluted solution is then pipetted into a conical flask, followed by the addition of excess potassium iodide.

The iodine produced is finally titrated with $0.100\,mol\,dm^{-3}$ sodium thiosulfate solution, giving an average accurate titre of $26.60\,cm^3$.

1 Write a half-equation for the oxidation of iodide ions to iodine.
2 Write a balanced equation for the reaction of chlorate(I) ions with iodide ions in acid solution to form iodine, chloride ions and water.
3 Write a balanced equation for the reaction of iodine with thiosulfate ions during the titration.
4 Using the equations from Questions 2 and 3, work out the number of moles of thiosulfate that react with the iodine produced by 1 mol of chlorate(I) ions.
5 Calculate the number of moles of thiosulfate in the average accurate titration, and hence the amount in moles of sodium chlorate(I) in $10.0\,cm^3$ of the diluted bleach.
6 Calculate the mass of sodium chlorate(I) in $100\,cm^3$ of undiluted bleach. (Na = 23.0, Cl = 35.5, O = 16.0)

Figure 14.9 Pouring concentrated bleach into a bucket before use for cleaning.

7 Give examples of random and systematic errors that can affect the results from this procedure and explain how can they be minimised.
8 With the help of Practical skills sheet 5, 'Identifying errors and estimating uncertainties', which you can access via the QR code for this chapter on page 321, calculate:
  a) the uncertainty and percentage uncertainty in:
    i) the volume of undiluted bleach taken
    ii) the volume of diluted bleach pipetted
    iii) the volume of thiosulfate titrated
    iv) the concentration of the thiosulfate solution
  b) the total percentage uncertainty in the mass of sodium chlorate(I) in $100\,cm^3$ of undiluted bleach.
9 Finally, write your result for the mass of sodium chlorate(I) in undiluted bleach in the form $x \pm y\,g$ per $100\,cm^3$.

14.4 Electrode potentials

Electrochemical cells

Redox reactions, like all reactions, tend towards a state of dynamic equilibrium. Redox reactions involve electron transfer and chemists have developed **electrochemical cells** based on redox changes. Some of these cells have great practical and technological importance while others, particularly fuel cells, are becoming a serious alternative to oil-based fuels for vehicles. Measurements of the voltages of cells help to assess the feasibility and likelihood of redox reactions.

Redox reactions involve the transfer of electrons from a reducing agent to an oxidising agent. The electron transfer can be shown by writing half-equations. So, for example, when zinc is added to copper(II) sulfate solution, Zn atoms give up electrons to form Zn^{2+} ions. At the same time, the electrons are transferred to Cu^{2+} ions, which form Cu atoms.

The two half-equations for the reaction are:

$$Zn(s) \rightarrow Zn^{2+}(aq) + 2e^-$$

and $\quad Cu^{2+}(aq) + 2e^- \rightarrow Cu(s)$

The overall balanced equation is:

$$Zn(s) + Cu^{2+}(aq) \rightarrow Zn^{2+}(aq) + Cu(s)$$

Instead of mixing two reagents, it is possible to carry out a redox reaction in an electrochemical cell so that electron transfer takes place along a wire connecting the two electrodes. This harnesses the energy from the redox reaction to produce an electrical potential difference (voltage).

> ### Key term
>
> **Electrochemical cells** produce an electric potential difference (voltage) from a redox reaction. In an electrochemical cell, the two half reactions happen in separate half cells. The electrons flow from one cell to the other through a wire connecting the electrodes. The electric circuit is completed by a salt bridge connecting the two solutions.

Test yourself

15 Write two ionic half-equations and the overall balanced equation for each of the following redox reactions. In each example, state which atom, ion or molecule is oxidised and which is reduced:

a) magnesium metal with copper(II) sulfate solution

b) aqueous chlorine with a solution of potassium bromide

c) a solution of silver nitrate with copper metal.

One of the first useable cells was based on the reaction of zinc metal with aqueous copper(II) ions (Figure 14.10). In the cell, zinc is oxidised to zinc(II) ions as copper(II) ions are reduced to copper metal.

In electrochemical cells, the two half-reactions happen in separate half-cells. The electrons flow from one cell to the other through a wire connecting the electrodes. A salt bridge connecting the two solutions completes the electrical circuit.

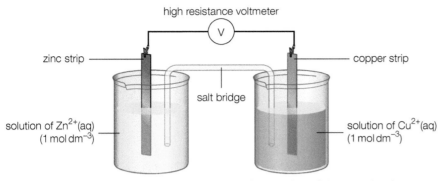

Figure 14.10 An electrochemical cell based on the reaction of zinc metal with aqueous copper(II) ions. In this cell, electrons tend to flow from the negative zinc electrode to the positive copper electrode through the external circuit.

The salt bridge makes an electrical connection between the two halves of the cell by allowing ions to flow while preventing the two solutions from mixing. At its simplest, a salt bridge consists of a strip of filter paper soaked in saturated potassium nitrate solution and folded over each of the two beakers.

All potassium salts and all nitrates are soluble so the salt bridge does not react to produce precipitates with any of the ions in the half-cells. In more permanent cells, a salt bridge may consist of a porous solid such as sintered glass.

Chemists measure the tendency for the current to flow in the external circuit by using a high-resistance voltmeter to measure the maximum cell e.m.f. when no current is flowing.

In Figure 14.10, electrons tend to flow out of the zinc electrode (negative) through the external circuit to the copper electrode (positive). The maximum voltage of the cell, usually called its **electromotive force (e.m.f.)**, or cell potential, is 1.10 V under standard conditions.

> **Key term**
>
> The **electromotive force (e.m.f.)** of a cell measures the maximum 'voltage' produced by an electrochemical cell. The symbol for e.m.f. is E and its SI unit is the volt (V). The e.m.f. is the energy transferred in joules per coulomb of charge flowing through the circuit connected to a cell. Cell e.m.f.s are at a maximum when no current flows because under these conditions no energy is lost due to the internal resistance of the cell as the current flows.

> **Test yourself**
>
> 16 Identify the oxidation and reduction reactions that take place in the cell in Figure 14.10 when a current flows and state where these processes take place.
>
> 17 Why is the copper strip in Figure 14.10 the positive electrode and the zinc strip the negative electrode?

Standard conditions

In order to compare the voltages (e.m.f.s) developed by different electrochemical cells, scientists carry out the measurements under standard conditions. These standard conditions for electrochemical measurements are the same as those for thermochemical measurements. They are:

- temperature 298 K (25 °C)
- gases at a pressure of 100 kPa
- solutions at a concentration of $1.0 \, mol \, dm^{-3}$.

Chemists have also developed a convenient shorthand called a cell diagram for describing cells. The cell diagram for the cell in Figure 14.10 is shown in Figure 14.11 with an explanation below each entry. Under standard conditions, the symbol for the e.m.f. of the cell is $E_{cell}^{\ominus}$ and this is called the standard e.m.f. of the cell.

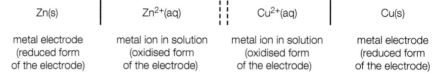

Zn(s)		Zn^{2+}(aq)			Cu^{2+}(aq)		Cu(s)
metal electrode (reduced form of the electrode)		metal ion in solution (oxidised form of the electrode)		metal ion in solution (oxidised form of the electrode)		metal electrode (reduced form of the electrode)

Figure 14.11 A cell diagram for the cell composed of the Zn^{2+}(aq) | Zn(s) and Cu^{2+}(s) | Cu(s) half-cells. The solid vertical lines, |, separate the different physical states of each half-electrode. The double dotted line, ⁞⁞, represents the salt bridge. Note that the reduced form of each electrode appears towards the outside of the cell diagram. This is the general rule.

If the cell e.m.f. is positive, the reaction in the cell tends to go according to the cell diagram reading from left to right. As a current flows in the external circuit connecting the two electrodes in Figure 14.11, zinc atoms turn into zinc ions and go into solution, while copper ions turn into copper atoms and deposit on the copper electrode (Figure 14.12).

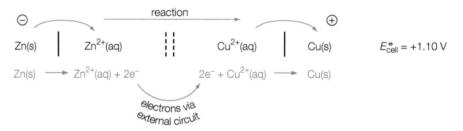

Figure 14.12 The direction of change in an electrochemical cell.

Test yourself

18 Consider a cell based on the redox reaction below that tends to go in the direction shown.

$$Mg(s) + Zn^{2+}(aq) \rightarrow Mg^{2+}(aq) + Zn(s)$$

The potential difference between the electrodes is +1.61 V.

a) Write the half-equations for the electrode processes when the cell supplies a current.

b) Write the conventional cell diagram for the cell, including the value for $E_{cell}^{\ominus}$.

Standard electrode potentials

The study of many cells has shown that a half-electrode, such as the Cu^{2+}(aq) | Cu(s) electrode, makes the same contribution to the cell e.m.f. in **any** cell, so long as the measurements are made under the same conditions. But there is no way of measuring the e.m.f. of an isolated, single electrode because it has only one terminal.

Chemists have solved this problem by selecting a standard electrode system as a reference electrode against which they can compare all other electrode systems. The chosen reference electrode is the standard hydrogen electrode.

By convention, the electrode potential of the standard hydrogen electrode is zero. This is represented as:

$$Pt[H_2(g)] \mid 2H^+(aq) \| \qquad\qquad E^{\ominus} = 0.00\,V$$

The standard electrode potential for any half-cell is measured relative to a standard hydrogen electrode under standard conditions (Figure 14.13). A standard hydrogen electrode sets up an equilibrium between hydrogen ions in solution ($1.00\,mol\,dm^{-3}$) and hydrogen gas ($100\,kPa$ pressure) at $298\,K$ on the surface of a platinum electrode coated with platinum black.

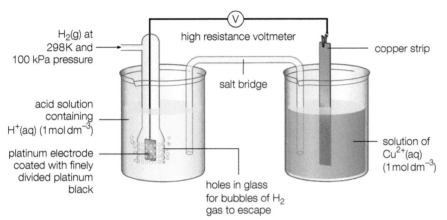

Figure 14.13 The e.m.f. of this cell under standard conditions is, by definition, the standard electrode potential of the $Cu^{2+}(aq) \mid Cu(s)$ electrode for which $E^{\ominus} = +0.34\,V$.

By convention, when a standard hydrogen electrode is the left-hand electrode in an electrochemical cell, the cell e.m.f. is the electrode potential of the right-hand electrode.

So, the conventional cell diagram for the cell which defines the standard electrode potential of the $Cu^{2+}(aq) \mid Cu(s)$ electrode is:

$$Pt[H_2(g)] \mid 2H^+(aq) \| Cu^{2+}(aq) \mid Cu(s) \qquad E^{\ominus} = +0.34\,V$$

The electrode and its standard electrode potential are often represented more simply as:

$$\underset{\text{oxidised form}}{Cu^{2+}(aq)} + 2e^- \rightleftharpoons \underset{\text{reduced form}}{Cu(s)} \qquad E^{\ominus} = +0.34\,V$$

This also serves to emphasise that standard electrode potentials represent reduction processes.

A hydrogen electrode is difficult to set up and maintain, so it is much easier to use a secondary standard such as a silver/silver chloride electrode or a calomel electrode as a reference electrode. These electrodes are available commercially and are reliable to use. They have been calibrated against a standard hydrogen electrode. Calomel is an old-fashioned name for mercury(I) chloride. The cell reaction and electrode potential relative to a hydrogen electrode for a calomel electrode are:

$$Hg_2Cl_2(s) + 2e^- \rightleftharpoons 2Hg(l) + 2Cl^-(aq) \qquad E^{\ominus} = +0.27\,V$$

Key terms

A **standard hydrogen electrode** is a half-cell in which a $1.00\,mol\,dm^{-3}$ solution of hydrogen ions is in equilibrium with hydrogen gas at $100\,kPa$ pressure on the surface of a platinum electrode coated with platinum black at $298\,K$.

The **standard electrode (reduction) potential**, $E^{\ominus}$, of a standard half-cell is the e.m.f. of that half-cell relative to a standard hydrogen electrode under standard conditions. Standard electrode potentials are sometimes called standard redox potentials.

A **reference electrode** is used to measure electrode potentials in place of the standard hydrogen electrode.

When used as a reference electrode, on the left of the cell, the reverse reaction, $E^{\ominus}$ has the opposite sign. This gives:

$$2Hg(l) + 2Cl^-(aq) \rightleftharpoons Hg_2Cl_2(s) + 2e^- \qquad E^{\ominus} = -0.27\,V$$

Figure 14.13 shows how to measure the standard electrode potential of metals in contact with their ions in aqueous solution. However, it is also possible to measure the standard electrode potentials of electrode systems in which both the oxidised and reduced forms are ions in solution, such as ions of the same element in different oxidation states. In these cases, the electrode in the system is platinum (Figures 14.14 and 14.15).

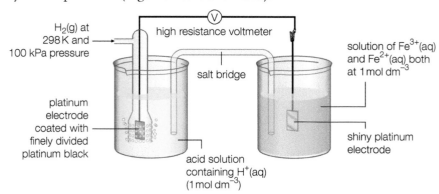

Figure 14.14 The apparatus in a cell for measuring the standard electrode potential of the redox reaction $Fe^{3+}(aq) + e^- \rightleftharpoons Fe^{2+}(aq)$.

$Pt[H_2(g)]$		$2H^+(aq)$		$Fe^{2+}(aq)$	,	$Fe^{2+}(aq)$		$Pt(s)$
hydrogen gas on Pt electrode coated with Pt black (reduced form of the electrode)		hydrogen ion in solution (oxidised form of the electrode)		metal ion in solution (oxidised form of the electrode)		metal ion in solution (reduced form of the electrode)		shiny platinum (inert electrode)

Figure 14.15 The cell diagram for the cell in Figure 14.14. Here both the reduced and oxidised forms of the chemicals in right-hand half-cell are in solution. As in Figure 14.11, the reduced form of each electrode system appears towards the outside of the cell diagram.

Test yourself

19 Suggest why a hydrogen electrode is difficult to set up and maintain.

20 Why is it not possible to measure the electrode potential for the $Na^+(aq) \mid Na(s)$ system using the method illustrated in Figure 14.14?

21 Why do you think that platinum metal is used as the electrode for systems in which both the oxidised and reduced forms are ions in solution, such as $Fe^{3+}(aq)$ and $Fe^{2+}(aq)$?

22 What are the half-equations and standard electrode potentials of the right-hand electrode in each of the following cells?

a) $Pt[H_2(g)] \mid 2H^+(aq) \parallel Sn^{2+}(aq) \mid Sn(s)$
   $E^{\ominus} = -0.14\,V$

b) $Pt[H_2(g)] \mid 2H^+(aq) \parallel Br_2(aq), 2Br^-(aq) \mid Pt(s)$
   $E^{\ominus} = +1.07\,V$

c) $Pt \mid [2Hg(l) + 2Cl^-(aq)], Hg_2Cl_2(s) \parallel Cr^{3+}(aq) \mid Cr(s)$
   $E^{\ominus} = -1.01\,V$

23 The standard electrode potential for the $Cu^{2+}(aq) \mid Cu(s)$ electrode is +0.34 V. For the cell, $Cu(s) \mid Cu^{2+}(aq) \parallel Pb^{2+}(aq) \mid Pb(s)$, the standard cell e.m.f., $E_{cell}^{\ominus}$, equals −0.47 V.

What is the standard electrode potential for the $Pb^{2+}(aq) \mid Pb(s)$ electrode?

24 a) What is the e.m.f. when a standard calomel electrode is connected to a standard $Cu^{2+}(aq) \mid Cu(s)$ electrode?

b) Write half-equations for the reactions at the electrodes.

14.5 Cell e.m.f.s and the direction of change

Chemists use standard electrode potentials:

- to calculate the e.m.f.s (standard potentials) of electrochemical cells
- to predict the direction (feasibility) of redox reactions.

The data sheet entitled 'Standard electrode potentials', which you can access via the QR code for Chapter 14 on page 321, lists redox half-reactions in order of their standard electrode (reduction) potentials from the most negative to the most positive.

The size and sign of a standard electrode potential shows how likely it is that a half-reaction will occur. The more positive the standard electrode potential, the more likely it is that the half-reaction will occur.

So, the half-reaction $H_2O_2(aq) + 2H^+(aq) + 2e^- \rightarrow 2H_2O(l)$ with a standard electrode potential of +1.77 V near the bottom of the table has a stronger tendency to change in the direction shown by the equation than the half-reaction $Li^+(aq) + e^- \rightarrow Li(s)$ with a standard electrode potential of −3.03 V at the top of the list. Indeed, this second half-reaction is much more likely to occur in the opposite direction, as it does when it contributes a voltage of +3.03 V to any electrochemical cell (Figure 14.16).

Using a table of standard electrode potentials, it is possible to calculate cell e.m.f.s ($E_{cell}^{\ominus}$ values) by combining the standard electrode potentials of the two half-cells that make up the full cell. This provides a prediction of the expected direction of chemical change for the redox reaction in the cell.

Look again at Figure 14.12 and the value of the cell e.m.f., $E_{cell}^{\ominus}$. The value of +1.10 V arises from the sum of the E values for two half-reactions:

$$Cu^{2+}(aq) + 2e^- \rightarrow Cu(s) \qquad E^{\ominus} = +0.34 \text{ V}$$

and the reverse of:

$$Zn^{2+}(aq) + 2e^- \rightarrow Zn(s) \qquad E^{\ominus} = -0.76 \text{ V}$$

Combining these two half-equations gives:

$$Cu^{2+}(aq) + 2e^- \rightarrow Cu(s) \qquad\qquad E^{\ominus} = +0.34 \text{ V}$$
$$\underline{\qquad\qquad Zn(s) \rightarrow Zn^{2+}(aq) + 2e^- \qquad\qquad E^{\ominus} = +0.76 \text{ V}}$$
Overall: $\quad Zn(s) + Cu^{2+}(aq) \rightarrow Zn^{2+}(aq) + Cu(s) \qquad E_{cell}^{\ominus} = +1.10 \text{ V}$

Key term

Feasibility: a feasible reaction is one that naturally tends to happen, even if it is very slow because it has a high activation energy.

Figure 14.16 Lithium batteries come in a range of sizes.

This shows that:

$$E^\ominus_{cell} = -E^\ominus_{(\text{left-hand electrode})} + E^\ominus_{(\text{right-hand electrode})}$$

$$= E^\ominus_{(\text{right-hand electrode})} - E^\ominus_{(\text{left-hand electrode})}$$

Positive values of $E^\ominus_{cell}$ indicate that the sign of the right–hand electrode is positive and that a reaction is tends to go in the direction of the overall equation. The more positive the value of $E^\ominus_{cell}$, the greater the tendency for the reaction to happen. On the other hand, negative values for $E^\ominus_{cell}$ indicate that reactions tends to go in the opposite direction to that shown in the cell diagram.

Example

Write the cell diagram for a cell based on the two half-equations below. Work out the e.m.f. of the cell and write the overall equation for the reaction which is likely to occur (the feasible reaction).

$Fe^{3+}(aq) + e^- \rightleftharpoons Fe^{2+}(aq)$ $E^\ominus = +0.77\,V$

$Cu^{2+}(aq) + 2e^- \rightleftharpoons Cu(s)$ $E^\ominus = +0.34\,V$

Notes on the method

Write the cell diagram with the more positive electrode on the right.

Then use the equation:

$$E^\ominus_{cell} = E^\ominus_{(\text{right-hand electrode})} - E^\ominus_{(\text{left-hand electrode})}$$

to calculate the cell e.m.f.

Answer

The $Fe^{3+}(aq)$, $Fe^{2+}(aq)$ electrode is the more positive so it should be on the right-hand side of the cell diagram. Both the oxidised and reduced forms are in solution so a shiny platinum electrode is needed.

$Fe^{3+}(aq)$, $Fe^{2+}(aq) \mid Pt(s)$

The left-hand electrode is $Cu^{2+}(aq) \mid Cu(s)$ so the reduced from is copper metal which can also be the conducting electrode. The cell diagram is therefore:

$Cu(s) \mid Cu^{2+}(aq) \parallel Fe^{3+}(aq)$, $Fe^{2+}(aq) \mid Pt(s)$

and $E^\ominus_{cell} = (+0.77\,V) - (+0.34\,V) = +0.43\,V$

$E^\ominus_{cell}$ is positive and so the reaction tends to go from left to right in the direction of the cell diagram. Balancing the two half-equations in this direction gives the overall equation:

$Cu(s) + 2Fe^{3+}(aq) \rightarrow Cu^{2+}(aq) + 2Fe^{2+}(aq)$

Test yourself

25 Write the cell diagram for a cell based on each of the following pairs of half-equations. For each example, find the standard electrode potentials from the data sheet for this chapter accessed via the QR code on page 321. Work out the e.m.f. of the cell and write the overall equation for the reaction that tends to happen (the feasible reaction):

a) $V^{3+}(aq) + e^- \rightleftharpoons V^{2+}(aq)$
$Zn^{2+}(aq) + 2e^- \rightleftharpoons Zn(s)$

b) $Br_2(aq) + 2e^- \rightleftharpoons 2Br^-(aq)$
$I_2(aq) + 2e^- \rightleftharpoons 2I^-(aq)$

c) $Cl_2(aq) + 2e^- \rightleftharpoons 2Cl^-(aq)$
$PbO_2(s) + 4H^+(aq) + 2e^- \rightleftharpoons Pb^{2+}(aq) + 2H_2O(l)$

Investigating some electrochemical cells

The unlabelled diagram in Figure 14.17 can be used to investigate cells made using strips of metals dipping into solutions of their own ions.

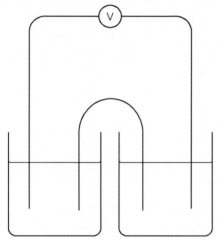

Figure 14.17 Outline of an apparatus for setting up chemical cells.

1 Copy Figure 14.17 and label it to show how to investigate a cell combining an $Ag^+(aq) \mid Ag(s)$ electrode and a $Zn^{2+}(aq) \mid Zn(s)$ electrode under standard conditions.
2 What is used to make the part of the apparatus that links the solutions in the two beakers? Explain its purpose.

Table 14.1 shows the results of measuring the cell e.m.f.s of three cells. The concentration of the silver nitrate solution used was $0.10 \, mol \, dm^{-3}$ because of the high cost of the silver salt.

Table 14.1

Cell	Negative electrode	Positive electrode	Cell e.m.f./V
A	$Cu^{2+}(aq) \mid Cu(s)$	$Ag^+(aq) \mid Ag(s)$	0.40
B	$Zn^{2+}(aq) \mid Zn(s)$	$Cu^{2+}(aq) \mid Cu(s)$	1.06
C	$Zn^{2+}(aq) \mid Zn(s)$	$Ag^+(aq) \mid Ag(s)$	1.48

3 For cell A:
   a) write the half-equation for the reaction taking place at the copper electrode and explain why this electrode is negative
   b) write the half-equation for the reaction taking place at the silver electrode and explain why this electrode is positive.
4 Write half-equations for the reactions at each of the electrodes in cells B and C.
5 Write the overall cell reaction for each of the three cells.
6 Give the conventional cell diagram for each cell in Table 14.1. Use Table 14.2 on page 104 to work out the expected e.m.f. of each cell under standard conditions and comment on any differences with the experimental values.
7 Explain the observations in Figures 14.18 and 14.19 and show that they are consistent with the results shown in Table 14.1.

Figure 14.18 The strip of zinc in this test tube was dipped into a solution of copper(II) sulfate solution that was an even darker blue at the start.

Figure 14.19 The copper wire in this test tube was originally added to a colourless solution of silver nitrate solution.

Table 14.2 The standard electrode potentials of some common metals.

Metal ion/metal electrode	Standard electrode potential, $E^{\ominus}/V$
$Li^+(aq) \mid Li(s)$	−3.03
$K^+(aq) \mid K(s)$	−2.92
$Na^+(aq) \mid Na(s)$	−2.71
$Al^{3+}(aq) \mid Al(s)$	−1.66
$Zn^{2+}(aq) \mid Zn(s)$	−0.76
$Fe^{2+}(aq) \mid Fe(s)$	−0.44
$Pb^{2+}(aq) \mid Pb(s)$	−0.13
$Cu^{2+}(aq) \mid Cu(s)$	+0.34
$Ag^+(aq) \mid Ag(s)$	+0.80

> **Tip**
>
> In whatever order electrode (reduction) potentials are tabulated, it is always true that:
>
> - the half-cell with the most positive electrode potential has the greatest tendency to gain electrons, so the species on the left-hand side of the half-equation is the most powerfully oxidising
> - the half-cell with the most negative electrode potential has the greatest tendency to give up electrons, so the species on the right-hand side of the half-equation is the most powerfully reducing.

The electrochemical series

A list of electrode systems set out in order of their electrode potentials (as on the data sheet 'Standard electrode potentials', which you can access via the QR code for this chapter on page 321) is a useful guide to the behaviour of oxidising and reducing agents. It is an electrochemical series.

The metal ion | metal electrodes with highly negative electrode potentials involve half-reactions for Group 1 metal ions and metals (Table 14.2). Lithium is the most reactive of these metals when it reacts as a reducing agent forming metal ions. Consequently, the reverse reaction of Li^+ ions forming Li metal is least likely and this results in the most negative standard electrode potential.

> **Tip**
>
> The data in tables such as Table 14.2 show reduction potentials for changes that can be represented as:
>
> oxidised form (oxidising agent) + electron(s) → reduced form (reducing agent).

The metal ion | metal electrodes with positive electrode potentials involve half-reactions for d-block metal ions and metals low in the reactivity series, such as copper and silver. These metals are relatively unreactive as reducing agents and they do not react with dilute acids to form hydrogen gas. However, their ions are readily reduced to the metal, which results in positive standard electrode potentials.

The order of metal ion/metal systems in Table 14.2 closely corresponds to the reactivity series for metals and the reactions shown by metal/metal ion displacement reactions (Figures 14.18 and 14.19).

The electrode potentials of the half-equations involving halogen molecules and halide ions are positive. The $F_2(aq) \mid 2F^-(aq)$ system is the most positive, showing that fluorine is the most reactive of the halogens as an oxidising agent. The next most reactive halogen is chlorine, then bromine and finally iodine is the least reactive. This corresponds to the order of reactivity of the halogens and the results of their displacement reactions.

An electrochemical series based on electrode potentials can be used to predict the direction of change in redox reactions. This is an alternative approach to the use of cell diagrams to make predictions.

> **Test yourself**
>
> For Questions 26 and 27, refer to the standard electrode potential values from the data sheet for this chapter, which you can access via the QR code on page 321.
>
> 26 Using the standard electrode potential values, arrange the following sets of metals in order of decreasing strength as reducing agents:
>   a) Ca, K, Li, Mg, Na
>   b) Cu, Fe, Pb, Sn, Zn.

27 Using the standard electrode potential values, arrange the following sets of molecules or ions in order of decreasing strength as oxidising agents in acid solution:

  a) $Cr_2O_7^{2-}$, Fe^{3+}, H_2O_2, MnO_4^-

  b) Br_2, Cl_2, ClO^-, H_2O_2, O_2.

Example

Use electrode potentials to predict what happens when chlorine is added to a solution of iodide ions.

Notes on the method

The first step is to identify the two half-equations. Write them down one above the other.

The more positive half-reaction tends to go from left to right, taking in electrons, while the more negative half-reaction goes from right to left.

Answer

Figure 14.20 shows how to predict the direction of change for the two half-equations involved when chlorine reacts with iodide ions.

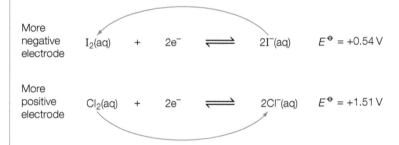

More negative electrode $I_2(aq)$ + $2e^-$ ⇌ $2I^-(aq)$ $E^{\ominus}$ = +0.54 V

More positive electrode $Cl_2(aq)$ + $2e^-$ ⇌ $2Cl^-(aq)$ $E^{\ominus}$ = +1.51 V

Figure 14.20 Chlorine is a stronger oxidising agent than iodine. Iodide ions are stronger reducing agents than chloride ions.

As expected, the electrode potentials predict that chlorine displaces iodine from a solution of aqueous iodide ions.

Disproportionation reactions

Electrode potentials to predict whether or not **disproportionation reactions** are likely to occur. During a disproportionation reaction, the same element both increases and decreases its oxidation number. Figure 14.21 on the next page shows that copper(I) ions do tend to disproportionate in aqueous solution while iron(II) ions do not.

Key term

A **disproportionation reaction** is a change in which the same element both increases and decreases its oxidation number. Some of the element is oxidised while the rest of it is reduced.

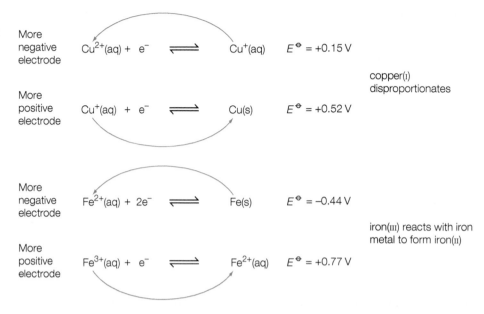

The limitations of predictions from $E^{\ominus}$ data

Tip

Note that strictly the term 'disproportionation' refers to a particular element in a compound not to the compound as a whole. However, as in Question 28, it is common to refer to the disproportionation of compounds.

Test yourself

28 Using the standard electrode potential values from the data sheet for this chapter, which you can access via the QR code on page 321, show that:

a) hydrogen peroxide tends to disproportionate to give oxygen and water under acid conditions

b) nitrous acid, HNO_2, tends to disproportionate to give nitrate(v) ions and nitrogen monoxide under acid conditions.

The limitations of predictions from $E^{\ominus}$ data

Although the electrode potentials for a redox reaction suggest that a reaction should take place, in practice the reaction may be too slow for any change to be observed. In other words, the $E^{\ominus}$ values show whether or not a reaction is thermodynamically feasible, but they do not give any indication about the rate of the reaction. There may be something to inhibit the reaction kinetically, so that the reaction mixture is inert.

For example, $E^{\ominus}$ values predict that $Cu^{2+}(aq)$ should oxidise $H_2(g)$ to H^+ ions:

$$Cu^{2+}(aq) + H_2(g) \rightarrow Cu(s) + 2H^+(aq) \qquad E^{\ominus}_{cell} = +0.34\,V$$

However, nothing happens when hydrogen is bubbled into copper(ɪɪ) sulfate solution because the activation energy is so high that the reaction rate is effectively zero.

A second important point about $E^{\ominus}$ values is that they relate only to standard conditions. Changes in concentration, temperature and pressure affect electrode potentials. In particular, all electrode (reduction) potentials become more positive if the concentration of reactant ions is increased and less positive if their concentration is reduced. This means that some reactions which are not possible under standard conditions occur under non-standard conditions, and vice versa.

For example, under standard conditions, MnO_2 does not oxidise $1.0\,mol\,dm^{-3}$ HCl(aq) to Cl_2.

$$MnO_2(s) + 4H^+(aq) + 2e^- \rightleftharpoons Mn^{2+}(aq) + 2H_2O \qquad E^{\ominus} = +1.23\,V$$

$$Cl_2(aq) + 2e^- \rightleftharpoons 2Cl^-(aq) \qquad E^{\ominus} = +1.36\,V$$

Chlorine is the stronger oxidising agent under standard conditions.

But if MnO_2 is heated with **concentrated** HCl, the electrode potentials of both half-equations change in such a way that chlorine is produced, as predicted by equilibrium theory. This is because, in hydrochloric acid, the hydrogen ion concentration is about $12\,mol\,dm^{-3}$. Under these non-standard conditions manganese(IV) oxide is a stronger oxidising agent with a more positive electrode potential. The higher chloride ion concentration in the concentrated acid means that chlorine is less powerfully oxidising, and so its electrode potential is less positive. These changes are sufficient to reverse the predicted direction of change.

So, predictions from cell e.m.f.s about the feasibility of redox reactions may not occur in practice due to kinetic effects (slow reaction rates) or non-standard conditions of concentration and temperature.

Test yourself

For these questions, refer to the standard electrode potential values from the data sheet for this chapter, which you can access via the QR code on page 321.

29 a) Using the standard electrode potential values, show that aluminium is expected to react with dilute hydrochloric acid.

b) Suggest a reason why there is very little change at first when a piece of aluminium foil is added to $1.0\,mol\,dm^{-3}$ hydrochloric acid at room temperature.

30 Using the standard electrode potential values, explain why when copper reacts with dilute nitric acid the product is nitrogen dioxide and not hydrogen.

14.6 How far and in which direction?

Chapter 11 shows that the direction and extent of a chemical reaction can be described by its equilibrium constant, K_c. The larger the value of the equilibrium constant, the greater is the proportion of products to reactants at equilibrium.

Chapter 13.2 introduces the connection between the total entropy change of a reaction, $\Delta S^{\ominus}_{total}$, and its free energy change, $\Delta G^{\ominus}$. The two are related by this formula: $\Delta G^{\ominus} = -T\Delta S^{\ominus}_{total}$. The connection between the free energy change and the equilibrium constant for a reaction involving gases is summarised in this equation:

$$\Delta G^{\ominus} = -RT\ln K$$

where R is the gas constant $= 8.31\,J\,K^{-1}\,mol^{-1}$.

This chapter has now shown that for redox reactions, electrode potentials offer yet another way of deciding the direction and extent of a reaction. A positive value for the e.m.f. of a cell means that the reaction that it is based on is feasible. The more positive $E^{\ominus}_{cell}$, the greater the tendency for the reaction to go. So it is not surprising that scientists have found that the total entropy change of a redox reaction, its equilibrium constant and $E^{\ominus}_{cell}$ are also closely related.

Thus:

$$E^{\ominus}_{cell} \propto \Delta S^{\ominus}_{total}$$

$$E^{\ominus}_{cell} \propto \ln K$$

What this all shows is that ΔS_{total}, $\Delta G^{\ominus}$, K_c and $E^{\ominus}_{cell}$ are different ways of presenting what is essentially the same information. Given the value for one of these quantities it is possible, in principle, to calculate any of the others. They are quantities that can all answer two key questions for any reaction:

- Will the reaction go? and
- How far will it go?

In practice, chemists use the quantity that is most easily determined by direct experiment or by calculation from experimental data. For example, they use equilibrium constants to explain the behaviour of weak acids; they use standard electrode potentials to explain what happens during redox reactions; they use free energy changes to determine the conditions needed to extract metals from oxide ores.

Table 14.3 shows how the values of these predictors are related to the extent of a reaction.

Table 14.3 Predicting the direction and extent of chemical reactions from the values of $\Delta G^{\ominus}$, $\Delta S^{\ominus}_{total}$, $E^{\ominus}_{cell}$ and K_c.

$\Delta G^{\ominus}$ kJ mol^{-1}	$\Delta S^{\ominus}_{total}/$ J mol^{-1}K^{-1}	$E^{\ominus}_{cell}/V$	K_c (units depend on the reaction)	Extent of reaction
More negative than −60	More positive than +200	More positive than +0.6	Greater than 10^{10}	Goes to completion
≈ −10	≈ +40	≈ +0.1	≈ 10^2	Equilibrium with more products than reactants
≈ 0	≈ 0	≈ 0	≈ 1	Roughly equal amounts of reactants and products
≈ +10	≈ −40	≈ −0.1	≈ 10^{-2}	Equilibrium with more reactants than products
More positive than +60	More negative than −200	More negative than −0.6	Less than 10^{-10}	No reaction

The fact that ΔS_{total}, $\Delta G^{\ominus}$, K_c and E_{cell} are all related to each other has had a profound impact on scientific thinking and on chemists in particular. It has brought together concepts of entropy, equilibrium and electrochemistry, showing that ideas developed in different areas and different contexts of chemistry are all related to the over-riding concept of the thermodynamic feasibility of chemical reactions.

14.7 Modern storage cells

Mobile phones, tablets and laptop computers depend on the existence of chemical cells that can be recharged. These are storage cells because they store the electricity - usually from the mains power supply. Often a single cell does not have a voltage that is high enough, and so cells are linked together in series to make a battery of cells.

When a storage cell is recharged, an electric current passes through it in the opposite direction to the current that the cell produces. Recharging is an example of electrolysis as chemical reactions occur to reform the chemicals that make up the electrodes.

Key terms

A **storage cell** is an electrochemical cell that is based on reversible chemical changes so that it can be recharged by an external electricity supply.

A **battery** is two or more electrochemical cells connected in series.

Lead–acid cells

Vehicles that run on petrol or diesel need storage batteries for the starter motor and for the lights when the engine is not running. Most of batteries in these vehicles are composed of six lead–acid cells in series, giving a total battery potential of 12 volts.

Lead–acid cells are also used to provide power for the motor in battery-operated vehicles such as electric wheelchairs, bicycles and scooters (Figure 14.22). Another important use of lead–acid batteries is to provide back-up power to computer systems, emergency lighting and hospital equipment in case power cuts interrupt the mains supply.

The negative terminal in a lead–acid cell is lead. The electrolyte is fairly concentrated sulfuric acid (about $6 \, mol \, dm^{-3}$). The lead gives up electrons, forming lead(II) ions when the cell is working normally (discharging). In the presence of sulfate ions, the lead(II) ions precipitate to form lead(II) sulfate.

$$Pb(s) + SO_4^{2-}(aq) \rightarrow PbSO_4(s) + 2e^-$$

The positive terminal is lead coated with lead(IV) oxide. During discharge, the lead(IV) oxide reacts with H^+ ions in the sulfuric acid electrolyte and takes electrons. Here, too, the lead(II) ions precipitate as lead(II) sulfate.

$$PbO_2(s) + SO_4^{2-}(aq) + 4H^+(aq) + 2e^- \rightarrow PbSO_4(s) + 2H_2O(l)$$

The formation of insoluble lead(II) sulfate creates a problem for lead–acid cells. If the cells are discharged for long periods, the precipitate of lead(II) sulfate becomes coarser and thicker and the process cannot be reversed when the cells are recharged.

When a lead–acid cell is recharged, the current is reversed and the reactions at each terminal are reversed. This turns Pb^{2+} ions back to lead metal at one terminal and back to PbO_2 at the other, with sulfate ions going back into the electrolyte.

Figure 14.22 Most battery-operated wheelchairs are powered by lead-acid cells.

Lithium cells

Modern mobile phones and laptop computers use lithium batteries. One advantage of electrodes based on lithium is that the metal has a low density, so that cells based on lithium electrodes can be relatively light. Also, lithium is very reactive, which means that the electrode potential of a lithium half-cell is relatively high and each cell has a large e.m.f.

The difficulty to overcome is that lithium is so reactive that it readily combines with oxygen in the air, forming a layer of non-conducting oxide on the surface of the metal. The metal also reacts rapidly with water. Research workers have solved these technical problems by developing electrodes with lithium atoms and ions inserted into the crystal lattices of other materials. In addition, the electrolyte is a polymeric material rather than an aqueous solution (Figure 14.23).

Figure 14.23 A schematic diagram of a lithium battery discharging. The electrode processes are reversible, so the battery can be recharged.

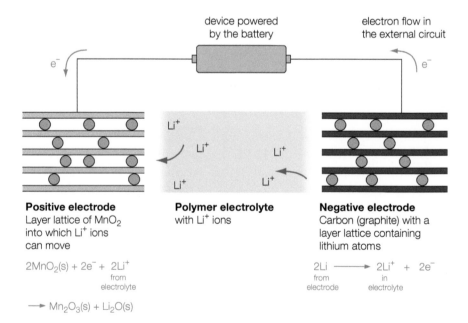

device powered by the battery

electron flow in the external circuit

Positive electrode
Layer lattice of MnO_2 into which Li^+ ions can move

$2MnO_2(s) + 2e^- + 2Li^+$ (from electrolyte)

$\longrightarrow Mn_2O_3(s) + Li_2O(s)$

Polymer electrolyte with Li^+ ions

Negative electrode
Carbon (graphite) with a layer lattice containing lithium atoms

$2Li \longrightarrow 2Li^+$ (from electrode) $+ 2e^-$ (in electrolyte)

14.8 Fuel cells

Fuel cells are electrochemical cells in which the chemical energy of a fuel is converted directly into electrical energy. Fuel cells differ from typical electrochemical cells such as lithium cells and lead–acid cells in having a continuous supply of reactants from which to produce a steady electric current. Fuel cells use a variety of fuels including hydrogen, hydrocarbons (such as methane) and alcohols. Inside a fuel cell, energy from the redox reaction between a fuel and oxygen is used to create a potential difference (voltage). Hydrogen and alcohol fuel cells have been used in the development of electric cars and in space exploration.

One of the most important fuel cells is the hydrogen–oxygen fuel cell (Figure 14.24).

> **Key term**
>
> A **fuel cell** is an electrochemical cell which is continuously supplied with fuel and oxidising agent. A fuel cell produces electric power from a fuel, directly, without having to burn it.

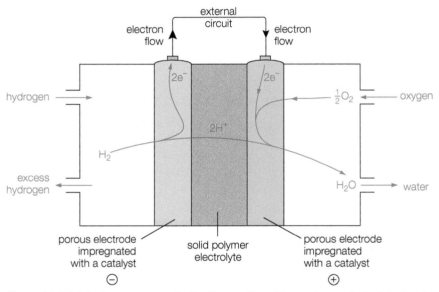

Figure 14.24 A hydrogen-oxygen fuel cell operating with a polymer electrolyte that is permeable to protons.

Figure 14.24 shows a hydrogen–oxygen fuel cell working in acid conditions. Fuel cells of this kind are used in buses and to provide back-up power for telecommunications and in data centres.

Hydrogen gas flows into the negative terminal where the H_2 molecules split into single H atoms in the presence of the catalyst. The H atoms then lose electrons and form H^+ ions.

> Negative terminal: $H_2(g) \rightarrow 2H^+ + 2e^-$ 　　　　Equation 1

The electrons flow into the external circuit as an electric current while the hydrogen ions migrate through the electrolyte.

Oxygen flows onto the positive terminal, where the nickel/nickel(II) oxide catalyses the splitting into single oxygen atoms. These oxygen atoms then combine with hydrogen ions from the electrolyte and electrons to form water.

Positive terminal: $\frac{1}{2}O_2(g) + 2H^+ + 2e^- \rightarrow H_2O(l)$ Equation 2

The overall reaction in the hydrogen–oxygen fuel cell (obtained by adding Equations 1 and 2) is:

$$H_2(g) + \frac{1}{2}O_2(g) \rightarrow H_2O(l)$$

NASA has used hydrogen–oxygen fuel cells in space missions since the 1960s to provide both electricity and drinking water. Typically, these fuel cells used aqueous potassium hydroxide as the electrolyte so they operated under alkaline conditions.

The two reduction half-equations with electrode potentials for an alkaline fuel cell using hydrogen are:

$$2H_2O(l) + 2e^- \rightarrow H_2(g) + 2OH^-(aq) \qquad E^\ominus = -0.83\,V$$

$$\frac{1}{2}O_2(g) + H_2O(l) + 2e^- \rightarrow 2OH^-(aq) \qquad E^\ominus = +0.40\,V$$

When an alcohol such as methanol is used as the energy source in place of hydrogen in a cell such as that shown in Figure 14.24, the following half-reactions occur at the terminals:

Negative terminal: $CH_3OH(l) + H_2O(l) \rightarrow CO_2(g) + 6H^+(aq) + 6e^-$

Positive terminal: $\frac{3}{2}O_2(g) + 6H^+(aq) + 6e^- \rightarrow 3H_2O(l)$

Fuel cells are no different in principle from more familiar electrochemical cells. The innovation is that new reactants (such as H_2, or CH_3OH, and O_2) are constantly fed into the cell and the products (H_2O and sometimes CO_2 or other products) are drawn off. This continuous flow of materials allows the cell potential to remain constant and the power output is uninterrupted.

The great advantage of all fuel cells is that they convert energy from chemical changes directly into electricity and in doing so achieve a remarkable efficiency of about 70%. In comparison, modern power plants and petrol engines using fossil fuels have a conversion efficiency for the energy from chemical reactions to electrical energy or kinetic energy of only about 40%.

Test yourself

32 a) What are the changes at the negative and positive terminals of a hydrogen fuel cell of the type used by NASA in spacecraft?

b) What is the overall equation for the reactions taking place in this type of fuel cell?

c) What is the cell e.m.f.?

33 a) What is the overall reaction for a fuel cell based on methanol?

b) Suggest one advantage and one disadvantages of using methanol rather than hydrogen.

Exam practice questions

1 a) For each of these reactions, identify the changes of oxidation number and state the element that is oxidised and the element that is reduced.

   i) $2NH_3(g) + 3Cl_2(g)$
$$\rightarrow N_2(g) + 6HCl(g) \quad (2)$$

   ii) $Cu_2O(s) + H_2SO_4(aq)$
$$\rightarrow CuSO_4(aq) + Cu(s) + H_2O(l) \quad (2)$$

   iii) $2KNO_3(s) \rightarrow 2KNO_2(s) + O_2(g)$ *(2)*

b) i) Write the half-equations involved when dichromate(VI) ions in acid solution react with sulfite ions, SO_3^{2-}, to form chromium(III) ions and sulfate ions.   *(2)*

   ii) Write a full, balanced redox equation for the reaction.   *(1)*

   iii) Describe the change you would see in the solution when the reaction occurs.   *(2)*

   iv) How many moles of sulfite ions react with one mole of dichromate(VI) ions? *(1)*

2 A $20.0\,cm^3$ sample of water was taken from a swimming pool which had recently been disinfected with chlorine. The sample was added to excess potassium iodide solution. The iodine formed was titrated with $0.00500\,mol\,dm^{-3}$ sodium thiosulfate solution. The volume of sodium thiosulfate solution needed to reach the end-point was $19.4\,cm^3$.

a) Write ionic equations for:

   i)   the reaction of chlorine with iodide ions   *(1)*

   ii) the reaction of iodine molecules with thiosulfate ions.   *(2)*

b) Name the indicator used to detect the end-point of the titration.   *(1)*

c) Describe the colour changes observed at each stage of the analysis.   *(3)*

d) Calculate the concentration of chlorine in the swimming pool water in $mol\,dm^{-3}$.   *(4)*

3 Use the standard electrode potentials below to answer the questions that follow.

$$Fe^{3+}(aq) \mid Fe^{2+}(aq) \qquad E^{\ominus} = +0.77V$$

$$Cu^{2+}(aq) \mid Cu(s) \qquad E^{\ominus} = +0.34V$$

An electrochemical cell was set up with $Fe^{3+}(aq) \mid Fe^{2+}(aq)$ as the right–hand electrode

a) Write half-equations for the reactions that occur in each half-cell when a current flows. State which half-equation involves oxidation and which involves reduction.   *(3)*

b) Calculate the change in oxidation number of the oxidised and reduced elements in each half-cell.   *(2)*

c) Write the conventional cell diagram and give the e.m.f. of the cell.   *(3)*

4 A student set up the electrochemical cell shown in the diagram.

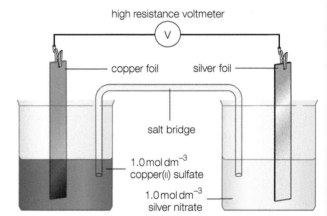

The standard electrode potentials are:

$$Cu^{2+} \mid Cu \qquad\qquad E^{\ominus} = +0.34V$$

$$Ag^+ \mid Ag \qquad\qquad E^{\ominus} = +0.80V$$

a) How could the student have made a salt bridge?   *(1)*

b) Write half-equations to show the reactions that occurred in:

   i)   the $Cu^{2+} \mid Cu$ half-cell   *(1)*

   ii) the $Ag^+ \mid Ag$ half-cell.   *(1)*

c) Write an equation for the overall cell reaction.   *(1)*

d) Calculate the e.m.f. (potential difference) for this cell.   *(2)*

e) At which electrode does reduction occur? Explain your answer.   *(2)*

f) The student found that the cell e.m.f. was less than the calculated value. Suggest two reasons for this.   *(2)*

5 a) Fuel cells can be made using these electrode systems.

$$2H^+(aq) + 2e^- \rightarrow H_2(g) \quad E^\ominus = 0.00\,V$$
$$O_2(g) + 4H^+(aq) + 4e^- \rightarrow 2H_2O(l)$$
$$E^\ominus = +0.40\,V$$

  i) Construct an overall equation for the cell reaction. Show your working. *(2)*

  ii) From which half-cell do electrons flow into the external circuit? *(1)*

b) What are the principal differences between a hydrogen–oxygen fuel cell and a storage cell, such as a lead–acid cell? *(3)*

c) Outline and explain two advantages that are gained by generating electricity using fuel cells rather than in thermal power stations. *(4)*

d) Discuss the advantages and disadvantages of using hydrogen fuel cells as the power supply for motor vehicles rather than using engines that burn fuels. *(6)*

6 Work out the results for each of the following redox titrations. In each case give the equation for any reactions, show the steps of your working in full and give an answer to an appropriate number of significant figures.

a) Sodium ethanedioate, $Na_2C_2O_4$, is a primary standard that can be used to standardise solutions of potassium manganate(VII). Under acidic conditions, $KMnO_4$ oxidises ethanedioate ions on heating to carbon dioxide. In a titration it was found that $28.85\,cm^3$ of a solution of $KMnO_4$ oxidised $25.00\,cm^3$ of a solution containing $7.445\,g\,dm^{-3}$ of sodium ethanedioate. What was the concentration of the potassium manganate(VII) solution?
($Na = 23.0$, $C = 12.0$, $O = 16.0$) *(5)*

b) Iron in $1.34\,g$ of iron ore was dissolved in acid and reduced to iron(II) ions. The solution was then titrated with $0.0200\,mol\,dm^{-3}$ potassium manganate(VII) solution. The titre was $26.75\,cm^3$. Calculate the percentage by mass of iron in the ore. ($Fe = 55.8$) *(5)*

c) A $25.00\,cm^3$ sample of a solution containing iron(II) and iron(III) ions was titrated with potassium manganate(VII) in the presence of acid. The titre was $18.00\,cm^3$ with a $0.0200\,mol\,dm^{-3}$ solution of $KMnO_4(aq)$. A second $25.00\,cm^3$ sample of the original solution was reduced with zinc and acid. After filtering off the zinc, the solution was titrated with the same solution of $KMnO_4(aq)$. This time the mean titre was $22.50\,cm^3$. Calculate the concentrations of the iron(II) and iron(III) ions in the original solution. *(5)*

7 Process the results for each of the following redox titrations. In each case give the equation for any reactions, show the steps of your working in full and give an answer to an appropriate number of significant figures.

a) Copper(II) ions oxidise iodide ions to iodine. A pale, off-white precipitate of a copper compound forms at the same time. $3.405\,g$ of $CuSO_4.5H_2O$ was dissolved in water and made up to $250\,cm^3$. Excess potassium iodide was added to $25.0\,cm^3$ of the copper(II) sulfate solution. In a titration, $18.00\,cm^3$ of a solution of $0.0760\,mol\,dm^{-3}$ sodium thiosulfate was required to react with the iodine formed. What is the oxidation number of the copper in the precipitated copper compound formed during the reaction of copper(II) ions with iodide ions? *(5)*

b) $0.275\,g$ of an alloy containing copper was dissolved in nitric acid and then diluted with water, producing a solution of copper(II) nitrate. An excess of potassium iodide was then added. The copper(II) ions reacted with the iodide ions to form a precipitate of a copper iodide and iodine. In a titration, the iodine reacted with $22.50\,cm^3$ of $0.140\,mol\,dm^{-3}$ sodium thiosulfate solution. ($Cu = 63.5$) Calculate the percentage by mass of copper in the alloy. *(5)*

8 The list below shows three standard electrode potentials.

I $\quad I_2(aq) + 2e^- \rightleftharpoons 2I^-(aq) \qquad E^\ominus = +0.54\,V$

II $\quad Cd^{2+}(aq) + 2e^- \rightleftharpoons Cd(s) \qquad E^\ominus = -0.40\,V$

III $\quad Fe^{2+}(aq) + 2e^- \rightleftharpoons Fe(s) \qquad E^\ominus = -0.44\,V$

a) Using the standard electrode potentials I, II and III above, suggest:

  i) one equation for a reaction which would go to completion, and explain your choice *(2)*

ii) one equation for a reaction which might occur, but only to an equilibrium position, and explain your choice. *(2)*

b) i) Draw a labelled diagram of the cell formed by connecting half-cells I and III. *(5)*

ii) Indicate on your diagram the direction of electron flow in the external circuit. *(1)*

iii) Calculate the e.m.f. (standard cell potential) of this cell. *(2)*

iv) Suggest what would be the effect on the cell e.m.f. of decreasing the concentration of $Fe^{2+}(aq)$. Explain your answer. *(3)*

c) The electrode potentials, E, of a number of copper/copper(II) sulfate half-cells with different $Cu^{2+}(aq)$ concentrations were measured against a standard hydrogen electrode at 298 K. The results are shown in the graph below.

i) From the graph, determine the value of E when $\log[Cu^{2+}(aq)]$ is zero. *(1)*

ii) What is the significance of the value of E when $\log[Cu^{2+}(aq)]$ is zero? *(2)*

iii) Suggest a formula to describe the shape of the graph and explain your answer. *(3)*

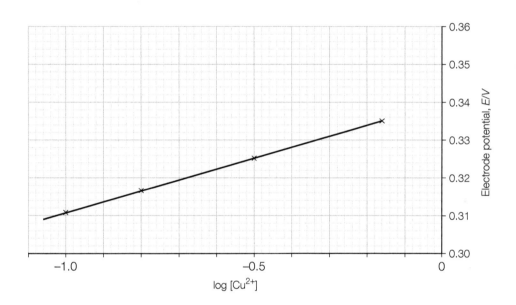

Transition metals

15.1 The atoms and ions of transition elements

The transition metals are vital to life and bring colour to our lives. They are also metals of great engineering and industrial importance. Chemically, these elements, which occupy the d block of the periodic table, are more alike than might be expected. Across the ten d-block metals from scandium to zinc in Period 4, the similarities are as striking as the differences. Chemists explain the characteristics of transition metals in terms of the electronic configurations of their atoms. Transition metal chemistry is colourful because of the range of oxidation states and complex ions. Transition metals matter because their properties are fundamental, not only to life, but also to modern technology (Figure 15.1).

Figure 15.1 Specimens of some d-block elements. The chemistry of an element is determined to a large extent by its outer shell electrons because they are the first to get involved in reactions. All the d-block elements have their outer electrons in the 4s sub-shell.

> **Tip**
>
> Chapter 1 of Student Book 1 introduces the description of atomic energy levels in terms of s, p and d energy levels. Then Chapter 4 shows how an understanding of atomic structure can explain the arrangement of elements in the periodic table with particular reference to elements in the s and p blocks of the table. This chapter builds on these ideas to explain the chemistry of the metals in the d block.

Electronic configurations

As the shells of electrons around the nuclei of atoms get further from the nucleus, they become closer in energy. Therefore, the difference in energy between the second and third shells is less than that between the first and second. When the fourth shell is reached, there is an overlap between the orbitals of highest energy in the third shell (the 3d orbitals) and that of lowest energy in the fourth shell (the 4s orbital) (Figure 15.2).

The 3d sub-shell is on average nearer the nucleus than the 4s sub-shell, but at a higher energy level. So, once the 3s and 3p sub-shells are filled, the next

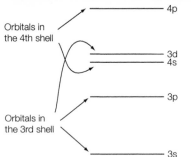

Figure 15.2 The relative energy levels of orbitals in the third and fourth shells.

electrons go into the 4s sub-shell because it occupies a lower energy level than the 3d sub-shell.

This means that potassium and calcium have the electron structure $[Ar]4s^1$ and $[Ar]4s^2$ respectively (Table 15.1).

Table 15.1 Electron configurations from potassium to zinc in Period 4 of the periodic table. ([Ar] represents the electronic configuration of argon.) Note the way that the electron configurations for chromium and copper atoms do not fit the general pattern.

Element	Symbol	Electronic structure	
		s,p,d,f notation	Electrons-in-boxes notation
Potassium	K	$[Ar]4s^1$	[Ar]
Calcium	Ca	$[Ar]4s^2$	[Ar]
Scandium	Sc	$[Ar]3d^14s^2$	[Ar]
Titanium	Ti	$[Ar]3d^24s^2$	[Ar]
Vanadium	V	$[Ar]3d^34s^2$	[Ar]
Chromium	Cr	$[Ar]3d^54s^1$	[Ar]
Manganese	Mn	$[Ar]3d^54s^2$	[Ar]
Iron	Fe	$[Ar]3d^64s^2$	[Ar]
Cobalt	Co	$[Ar]3d^74s^2$	[Ar]
Nickel	Ni	$[Ar]3d^84s^2$	[Ar]
Copper	Cu	$[Ar]3d^{10}4s^1$	[Ar]
Zinc	Zn	$[Ar]3d^{10}4s^2$	[Ar]

Look carefully at Table 15.1. In Period 4, the d-block elements run from scandium ($1s^22s^22p^63s^23p^63d^14s^2$) to zinc ($1s^22s^22p^63s^23p^63d^{10}4s^2$). But, notice that the electronic configurations of chromium and copper do not fit the general pattern. The explanation of these irregularities lies in the stability associated with half-filled and filled sub-shells. So, the electronic structure of chromium, $[Ar]3d^54s^1$, with half-filled sub-shells and an equal distribution of charge around the nucleus, is more stable than the electronic structure $[Ar]3d^44s^2$.

Similarly, the electronic structure of copper, $[Ar]3d^{10}4s^1$, with a filled 3d sub-shell and a half-filled 4s sub-shell is more stable than $[Ar]3d^94s^2$.

Along the series of d-block elements from scandium to zinc, the number of protons in the nucleus increases by one from one element to the next. However, the added electrons go into an inner d sub-shell, but the outer electrons are always in the 4s sub-shell. This means that there are clear similarities amongst the transition elements. Changes in their chemical properties across the series are much less marked than the big changes across a series of p-block elements such as aluminium to argon.

In this way, the energy-level model for electronic structure can help to account for the similarities in properties of transition metals. Later in this chapter, Section 15.5 shows that there are limitations to this energy-level model so that more sophisticated explanations are needed.

Ions of the transition metals

When transition metals form their ions, electrons are lost initially from the 4s sub-shell and not the 3d sub-shell. This may seem somewhat illogical because, prior to holding any electrons, the 4s level is more stable than the 3d level. But once the 3d sub-shell is occupied by electrons, these 3d electrons, being closer to the nucleus, repel the 4s electrons to a higher energy level. The 4s electrons are, in fact, repelled to an energy level higher than those occupying the 3d sub-shell. So, when transition metals form ions, they lose electrons from the 4s before the 3d level. This further emphasises the fact that transition metals have similar chemical properties dictated by the behaviour of the 4s electrons in their outer shells.

Test yourself

1 Write the full s,p,d electronic configuration of:

  a) a scandium atom

  b) a scandium(III) ion

  c) a manganese atom

  d) a manganese(II) ion.

2 Look at the electronic structures of iron and copper in Table 15.1.

  a) Write the electronic structure of an iron(II) ion.

  b) Write the electronic structure of an iron(III) ion.

  c) Which ion, Fe^{2+} or Fe^{3+}, would you expect to be the more stable? Explain your choice.

  d) Write the formula for the ion of copper that you would expect to be the more stable. Explain your choice.

15.2 Defining the transition metals

The simplest and neatest way to define the transition metals would be to say that they are the elements in the d block of the periodic table. But this simple definition leads to the inclusion of scandium and zinc as transition metals and ignores the fact that these two metals have some clear differences from the metals between them in the periodic table from titanium to copper. For instance:

- Scandium and zinc have only one oxidation state in their compounds (scandium +3, zinc +2), whereas the elements from titanium to copper have two or more.
- The compounds of scandium and zinc are usually white, unlike those of transition metals which are generally coloured.
- Scandium, zinc and their compounds show little catalytic activity.

Studying the ionisation energies of transition metals

Experimental evidence for the electronic configurations of transition metals atoms and ions can be obtained from the ionisation energies of the elements concerned.

Look carefully at Figure 15.3, which shows graphs of the first, second and third ionisation energies of the elements from scandium to zinc.

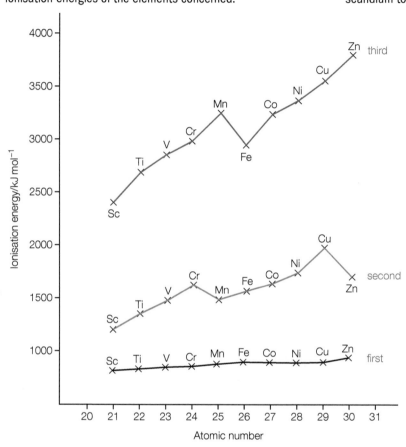

Figure 15.3 Graphs of the first, second and third ionisation energies of the elements from scandium to zinc in the periodic table.

1 Write the electronic structures of the following atoms and ions using [Ar] for the electronic structure of argon.
   a) Zn   b) Cu⁺   c) Zn⁺   d) Cr⁺   e) Mn⁺

2 Write an equation for:
   a) the second ionisation energy of chromium
   b) the third ionisation energy of iron.

3 Explain the general trend in ionisation energies as atomic number increases.

4 a) How does the first ionisation energy of zinc compare with those of the other d-block elements in Period 4?
   b) What does this tell you about zinc relative to the other elements?
   c) How does this relate to the electronic configuration of zinc atoms?

5 a) What does the high second ionisation energy of copper, relative to its neighbours in the periodic table, tell you about copper?

   b) How does this relate to the electronic configuration of copper atoms and ions?

6 a) What does the high second ionisation energy of chromium, relative to its neighbours in the periodic table, tell you about chromium?
   b) How does this relate to the electronic configuration of chromium atoms and ions?

7 a) Which elements have relatively high third ionisation energies compared with their neighbours in the d-block elements of Period 4?
   b) How do these relatively high third ionisation energies provide further evidence for the proposed electronic configurations of the elements concerned?

Key term

A **transition metal** is an element that has one or more stable ions with incompletely filled d orbitals.

As scandium and zinc do not show the typical properties of a **transition metal**, chemists looked for a more satisfactory definition. This definition should exclude scandium and zinc, but include all the elements from titanium to copper. In order to achieve this, chemists describe transition metals as those elements that form one or more stable ions with incompletely filled d orbitals.

Characteristics of the transition metals

In general, transition elements share a number of common properties (see the Data sheet 'Properties of selected elements – d-block metals', which you can access via the QR code for Chapter 15 on page 321).

- They are hard metals with useful mechanical properties, high melting and high boiling temperatures.
- They show variable oxidation numbers in their compounds.
- They form coloured ions in solution.
- They can act as catalysts both as the elements and as their compounds.
- They form complex ions involving monodentate, bidentate and polydentate ligands (Section 15.8).

15.3 The transition elements as metals

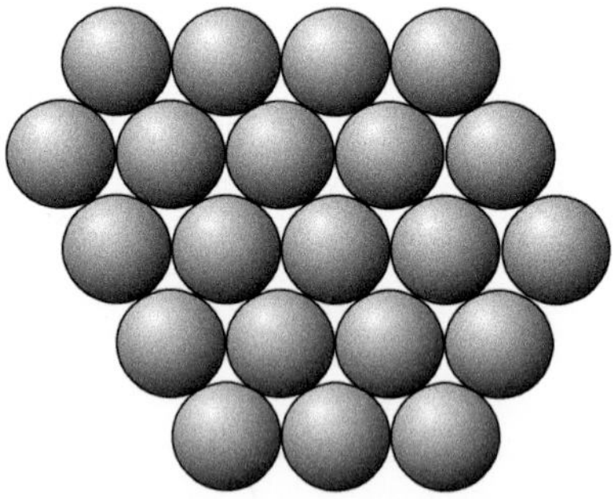

Most of the transition elements have a close-packed structure in which each atom has 12 nearest neighbours (Figure 15.4). In addition, transition elements have relatively low atomic radii because an increasingly large nuclear charge is attracting electrons that are being added to an inner sub-shell. The dual effect of close packing and small atomic radii results in strong metallic bonding. So, transition metals have higher melting temperatures, higher boiling temperatures, higher densities and higher tensile strengths than s-block metals such as calcium and p-block metals such as aluminium and lead. A plot of physical properties against atomic number often has two peaks or two troughs associated with a half-filled and then a filled d sub-shell (Figure 15.5).

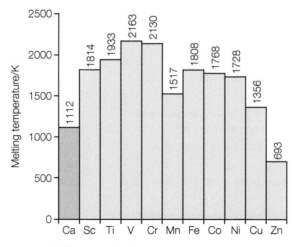

Figure 15.4 Close packing of atoms in one layer of a metal crystal. Each atom is in contact with six atoms in the same layer, three atoms in the layer above and three atoms in the layer below, so 12 in all.

Figure 15.5 A plot of melting temperature against atomic number for the elements calcium to zinc in the periodic table.

The transition metals are much less reactive than the s-block metals. However, the electrode potentials listed in Table 15.2 suggest that all of them, except copper, should react with dilute strong acids such as $1 \, mol \, dm^{-3}$ hydrochloric acid. In practice, many of the metals react very slowly with dilute acids because the metal is protected by a thin, unreactive layer of oxide. Chromium provides a very good example of this. Despite the predictions from its standard electrode potential, it is used as a protective, non-rusting metal owing to the presence of an unreactive, non-porous layer of chromium(III) oxide, Cr_2O_3.

Table 15.2 Standard electrode potentials of the transition metals from V to Cu.

Element	Standard electrode potential, $E^{\ominus}$ for $M^{2+}(aq) \mid M(s)/V$
Vanadium	−1.20
Chromium	−0.91
Manganese	−1.19
Iron	−0.44
Cobalt	−0.28
Nickel	−0.25
Copper	+0.34

Copper is the least reactive of the transition metals in Period 4. It does not react with dilute non-oxidising acids, such as dilute HCl and dilute H_2SO_4, and it oxidises only very slowly in moist air. Copper is also a good conductor of electricity, which leads to its use in electricity cables, and for domestic water pipes. Copper's mechanical properties are enhanced by making alloys such as brass and bronze (Figure 15.6).

Figure 15.6 Saxophones are made of brass. Brass is an alloy of 60-80% copper and 20-40% zinc. It is easily worked, has an attractive gold colour and does not corrode.

Test yourself

3 Why can scandium and zinc be described as d-block elements, but not as transition metals?

4 Suggest a reason why zinc only forms compounds in the +2 oxidation state.

5 a) What is the general trend in standard electrode potentials of the $M^{2+}(aq) \mid M(s)$ systems for the transition metals in Table 15.2?

  b) What does this suggest about the reactivity of transition metals across Period 4 in the periodic table?

6 Explain why the atomic radius falls from 0.15 nm in titanium to 0.14 nm in vanadium and then 0.13 nm in chromium.

15.4 Variable oxidation numbers

Most of the d-block elements in Period 4 form a range of compounds in which they are present in different oxidation states. The main reason for this is that the transition metals from titanium to copper have electrons of similar energy in both the 3d and 4s levels. This means that each of these elements can form ions of roughly the same stability in aqueous solution or in crystalline solids by losing different numbers of electrons. This contrasts with the metals in Groups 1 and 2 of the periodic table for which there is a large jump in ionisation energy after the electrons in the outer shell have been removed.

The formulae of the common oxides of the elements from scandium to zinc are shown in Figure 15.7 along with the stable oxidation states of each element in its compounds. The main oxidation states of the elements are shown in bold blue print.

	Sc	Ti	V	Cr	Mn	Fe	Co	Ni	Cu	Zn
Common oxides	Sc_2O_3	Ti_2O_3	V_2O_3	Cr_2O_3	MnO	FeO	CoO	NiO	Cu_2O	ZnO
		TiO_2	V_2O_5	CrO_3	MnO_2	Fe_2O_3	Co_2O_3		CuO	
					Mn_2O_7					

	Sc	Ti	V	Cr	Mn	Fe	Co	Ni	Cu	Zn
					+7					
				+6	+6					
			+5							
		+4	+4		+4					
	+3	+3	+3	+3	+3	+3	+3			
		+2	+2	+2	+2	+2	+2	+2	+2	+2
									+1	

Figure 15.7 Oxidation states and common oxides of the elements scandium to zinc with the main oxidation states in bold blue print.

The elements at each end of the series in Figure 15.7 give rise to only one oxidation state. The elements near the middle of the series have the greatest range of oxidation states. Most of the elements form compounds in the +2 state corresponding to the use of both 4s electrons in bonding.

The +2 state is a main oxidation state for all elements in the second half of the series, whereas +3 is a main oxidation state for all elements in the first part. Across the series, the +2 state becomes more stable relative to the +3 state.

From scandium to manganese, the highest oxidation state corresponds to the total number of electrons in the 3d and 4s energy levels. However, these higher oxidation states never exist as simple ions. Typically, they occur in compounds in which the metal is covalently bonded to an electronegative atom, usually oxygen, as in the dichromate(VI) ion, $Cr_2O_7^{2-}$, and the manganate(VII) ion, MnO_4^-.

One of the most attractive and effective demonstrations of the range of oxidation states in a transition element can be shown by shaking a solution of ammonium vanadate(V), NH_4VO_3, in dilute sulfuric acid with zinc. Before adding zinc, H^+ ions in the sulfuric acid react with VO_3^- ions to form dioxovanadium(V) ions and the solution is yellow.

$$VO_3^-(aq) + 2H^+(aq) \rightarrow VO_2^+(aq) + H_2O(l)$$

When the yellow solution, containing dioxovanadium(V) ions, is shaken with zinc, it is reduced first to blue oxovanadium(IV) ions, $VO^{2+}(aq)$, then to green vanadium(III) ions, $V^{3+}(aq)$, and finally to violet vanadium(II) ions, $V^{2+}(aq)$ (Figure 15.8).

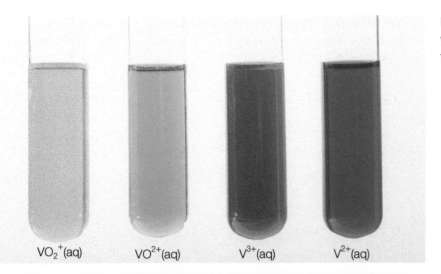

Figure 15.8 The oxidation states of vanadium showing the colours of its ions in the +5, +4, +3 and +2 oxidation states.

VO_2^+(aq) VO^{2+}(aq) V^{3+}(aq) V^{2+}(aq)

Test yourself

For Questions 10 and 11, refer to the data sheet for Chapter 15 headed 'Standard electrode potentials', which you can access via the QR code for this chapter on page 321.

7 Write down four generalisations about the oxidation states of transition metals based on Figure 15.7 and the text in Section 15.4.

8 Give examples of compounds other than oxides of:

a) chromium in the +3 and +6 states

b) manganese in the +2 and +7 states

c) iron in the +2 and +3 states

d) copper in the +1 and +2 states.

9 a) Show that the oxidation state of vanadium in VO_2^+ ions is +5.

b) Refer to Figure 15.8. Predict the colour of the solution when a solution containing VO_2^+ ions is half reduced to VO^{2+} ions.

10 a) Write half-equations for:

   i) the reduction of dioxovanadium(v) ions, VO_2^+, to oxovanadium(iv) ions in acid solution

   ii) the oxidation of zinc to zinc(ii) ions.

b) Use the data sheet 'Standard electrode potentials' to show that:

   i) iodide ions reduce VO_2^+ ions to VO^{2+} ions in acid solution

   ii) tin reduces VO_2^+ ions to V^{3+} ions in acid solution

   iii) zinc reduces VO_2^+ ions to V^{2+} ions in acid solution.

11 a) Write a half-equation involving electrons for:

   i) the oxidation of Cu^+(aq) to Cu^{2+}(aq)

   ii) the reduction of Cu^+(aq) to Cu(s).

b) Use the data sheet 'Standard electrode potentials' to find the standard electrode potentials for the two half-equations in part (a).

c) Using your data in part (b), explain why Cu^+(aq) ions disproportionate in aqueous solution.

Figure 15.9 Solutions containing ions in two of the oxidation states of chromium. On the left, a solution of dichromate(VI) ions, $Cr_2O_7{}^{2-}$; on the right, a solution of chromium(III) ions, Cr^{3+}.

Chromium forms compounds in three oxidation states, +2, +3 and +6. In the +3 state, chromium exists as Cr^{3+} ions, which can be both oxidised and reduced (Figure 15.9).

Under alkaline conditions, hydrogen peroxide oxidises green chromium(III) ions, Cr^{3+}(aq), to yellow chromium(VI) in chromate ions, $CrO_4{}^{2-}$(aq).

$$H_2O_2(aq) + 2e^- \rightarrow 2OH^-(aq)$$

$$Cr^{3+}(aq) + 8OH^-(aq) \rightarrow Cr_2O_4{}^{2-}(aq) + 4H_2O(l) + 3e^-$$

In contrast to this, zinc reduces green Cr^{3+}(aq) to blue–violet Cr^{2+}(aq) ions.

$$Zn(s) \rightarrow Zn^{2+}(aq) + 2e^-$$

$$Cr^{3+}(aq) + e^- \rightarrow Cr^{2+}(aq)$$

Chromium(II) ions are powerful reducing agents which are rapidly converted to chromium(III) by oxygen in the air. This means that air has to be excluded when zinc and acid are used to reduce chromium(III) to the +2 state.

Note that chromium in the +6 state can be either orange or yellow, depending on the pH, as a result of this equilibrium:

$$2CrO_4{}^{2-}(aq) + 2H^+(aq) \rightleftharpoons Cr_2O_7{}^{2-}(aq) + H_2O(l)$$

    yellow   orange

Test yourself

12 Use the data sheet 'Standard electrode potentials', which you can access via the QR code for Chapter 15 on page 321, to show that zinc reduces Cr^{3+}(aq) to Cr^{2+}(aq) ions.

13 a) Explain why an orange solution of dichromate(VI) ions turns yellow on adding alkali, and then orange again if the solution is acidified.

    b) Is the change of $CrO_4{}^{2-}$(aq) to $Cr_2O_7{}^{2-}$(aq) a redox reaction?

15.5 Coloured ions

Most coloured compounds get their colour by absorbing some of the radiation in the visible region of the electromagnetic spectrum with wavelengths between 400 nm and 700 nm. When light hits a substance, part is absorbed, part is transmitted (if the substance is transparent) and part is usually reflected. If all the light is absorbed, the substance looks black. If all the light is reflected, the substance looks white. If very little light is absorbed, and all the radiations in the visible region of the electromagnetic spectrum are transmitted equally, the substance is colourless like water.

However, many compounds, and particularly those of transition metals, absorb radiations in only certain areas of the visible spectrum. This means that the substances take on the colour of the light that they transmit or reflect. For example, if a material absorbs all radiations in the green-blue-violet region of the spectrum, it appears red-orange in white light (Figure 15.10).

Colour of compound	Wavelength absorbed/nm	Colour of light absorbed
greenish yellow	400–430	violet
yellow to orange	430–490	blue
red	490–510	blue-green
purple	510–530	green
violet	530–560	yellow-green
blue	560–590	yellow
greenish blue	590–610	orange
blue-green to green	610–700	red

Figure 15.10 A chart showing complementary colours in the left and right-hand columns. The colour of a compound is the colour complementary to the light it absorbs.

It is the electrons in coloured compounds that absorb radiation and jump from their normal state to a higher excited state. According to the quantum theory, there is a fixed relationship between the size of the energy 'jump' and the wavelength of the radiation absorbed. In many compounds, the electron 'jumps' between one sub-level and the next are so large that the radiation absorbed is in the ultraviolet region of the spectrum. These compounds are therefore white or colourless because they are not absorbing any of the radiation in the visible region of the electromagnetic spectrum.

> ### Tip
>
> The **quantum theory** states that radiation is emitted or absorbed in tiny, discrete amounts called energy quanta. Quanta have energy, $E = hv$ where h is Planck's constant and v is the frequency of the radiation.

However, the colour of transition metal ions arises from the possibility of transitions between the orbitals within the d sub-shell.

In a free gaseous atom or ion, the five 3d orbitals are all at the same energy level even though they do not all have the same shape. But when the ion of a d-block element is surrounded by other ions in a crystalline solid, or by molecules such as water in aqueous solutions, the differences in shape cause the five orbitals to split into two groups. When there are six molecules or ions around the central metal atom, two of the 3d orbitals move to a slightly higher energy level than the other three. As a result, ions such as $Cu^{2+}(aq)$ appear coloured because light of a particular frequency can be absorbed from visible light as electrons jump from a lower to a higher 3d orbital (Figure 15.11). If all the d orbitals are full, or empty, there is no possibility of electronic transitions between them.

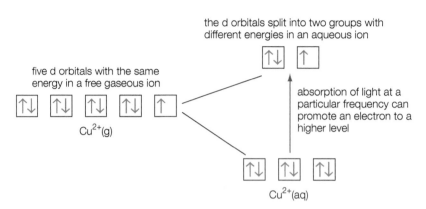

the d orbitals split into two groups with different energies in an aqueous ion

five d orbitals with the same energy in a free gaseous ion

Cu^{2+}(g)

absorption of light at a particular frequency can promote an electron to a higher level

Cu^{2+}(aq)

> **Tip**
>
> The colour of transition metal ions results from the absorption of part of the visible radiation in white light as electrons move from a lower to a higher level. This contrasts with flame colours, which arise from the emission of radiation as electrons fall from a higher to a lower level.

The explanation of the colour of transition metal ions, illustrates the limitations of the simple energy-level model of the electronic structures of atoms. The need for more sophisticated explanations is clear, bearing in mind the existence of sub-shells and the different shapes of orbitals within d sub-shells.

> **Test yourself**
>
> **14 a)** Explain why Zn^{2+}, Cu^+ and Sc^{3+} ions are usually colourless in solution and white in solids by writing out their electronic configurations.
>
> **b)** What colours of light are absorbed most effectively by a Cu^{2+} ion?

> **Tip**
>
> The terms 'hydronium ion' and 'hydroxonium ion' are sometimes used for the oxonium ion, H_3O^+.

15.6 Formation of complex ions

The symbol H^+(aq) does not represent a simple ion in acid solutions. In aqueous solution, H^+ ions are strongly attached to water molecules by dative covalent bonds forming oxonium ions, H_3O^+ (Figure 15.12).

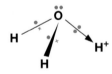

Figure 15.12 An oxygen atom in a water molecule forming a dative covalent bond with a hydrogen ion to form an aqueous H_3O^+ ion. The oxygen atom donates both electrons of a lone pair to form the bond.

In the same way as H^+, other cations can also exist in aqueous solution as hydrated ions. So Cr^{3+}(aq), Cu^{2+}(aq) and Ag^+(aq) can be represented more completely as $[Cr(H_2O)_6]^{3+}$(aq), $[Cu(H_2O)_6]^{2+}$(aq) and $[Ag(H_2O)_2]^+$(aq) in aqueous solution. The larger size of these other cations relative to H^+ enables them to associate with up to six water molecules (Figure 15.13).

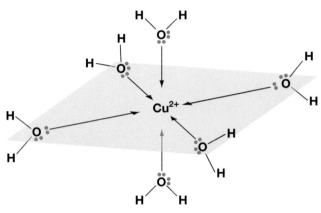

Figure 15.13 Dative covalent bonding in an aqueous Cu^{2+} ion. Each water molecule uses a lone pair of electrons to form a dative covalent bond with the central metal ion.

> **Tip**
>
> In aqueous solution, the copper(ɪɪ) ion is surrounded by six water molecules to form the complex ion $[Cu(H_2O)_6]^{2+}$. In solid hydrated copper(ɪɪ) sulfate ($CuSO_4.5H_2O$), however, there are only four water molecules co-ordinated with each copper(ɪɪ) ion. The fifth water molecule in the solid copper(ɪɪ) sulfate is associated with a sulfate ion, SO_4^{2-}.

Other polar molecules, besides water, can form dative covalent bonds with metal ions. For example, in excess ammonia solution, Cr^{3+} ions form $[Cr(NH_3)_6]^{3+}$, Cu^{2+} ions form $[Cu(NH_3)_4(H_2O)_2]^{2+}$ and Ag^+ ions form $[Ag(NH_3)_2]^+$. In addition to polar molecules, anions can also associate with cations using dative covalent bonds. For example, when anhydrous copper(ɪɪ) sulfate is added to concentrated hydrochloric acid, the solution contains yellow $[CuCl_4]^{2-}$ ions.

Ions such as $[Cu(H_2O)_6]^{2+}$, $[Cu(NH_3)_4(H_2O)_2]^{2+}$ and $[CuCl_4]^{2-}$ in which a metal ion is associated with a number of molecules or anions are called **complex ions**, and the anions and molecules attached to the central metal ion are called **ligands**. Each ligand must have at least one lone pair of electrons which it uses to form a dative covalent bond with the metal ion. The number of ligands in a complex ion is typically two, four or six.

Chemists have an alternative name for dative covalent bonds which they often prefer when describing complex ions. The alternative name is 'co-ordinate bond', which also gives rise to the terms 'co-ordination compound' and 'co-ordination number'. A co-ordination compound is one that contains a complex ion, and the **co-ordination number** of a complex ion is the number of co-ordinate bonds from the ligands to the central metal ion.

Co-ordination compounds contain complexes which may be cations, anions or neutral molecules (Figure 15.14). Examples of co-ordination compounds include:

- $K_3[Fe(CN)_6]$ containing the negatively charged complex ion $[Fe(CN)_6]^{3-}$
- $Fe(NO_3)_3.6H_2O$ containing the positively charged complex ion $[Fe(H_2O)_6]^{3+}$
- $Ni(CO)_4$ containing a neutral complex between nickel atoms and carbon monoxide molecules.

> **Tip**
>
> When anions act as ligands, the overall charge on the complex ion does not equal the oxidation number of the central metal ion.

> **Key terms**
>
> A **complex ion** is an ion in which a number of molecules or anions are bound to a central metal cation by co-ordinate bonds.
>
> A **ligand** is a molecule or anion bound to the central metal ion in a complex ion by co-ordinate bonding.
>
> The **co-ordination number** of a metal ion in a complex is the number of co-ordinate bonds to the metal ion from the surrounding ligands.

Figure 15.14 Crystals of co-ordination compounds. From left to right these are: $NiSO_4.7H_2O$, $FeSO_4.7H_2O$, $CoCl_2.6H_2O$, $CuSO_4.5H_2O$, $Cr_2(SO_4)_3.18H_2O$ and $K_3[Fe(CN)_6]$.

There are two common visible signs that a reaction has occurred during the formation of a new complex ion:

- a colour change
- an insoluble solid dissolving.

A familiar example of a colour change occurs when excess ammonia solution is added to copper(II) sulfate solution. Ammonia molecules displace water molecules from hydrated copper(II) ions forming $[Cu(NH_3)_4(H_2O)_2]^{2+}(aq)$ ions and the colour changes from pale blue to deep blue.

$$[Cu(H_2O)_6]^{2+}(aq) + 4NH_3(aq) \rightarrow [Cu(NH_3)_4(H_2O)_2]^{2+}(aq) + 4H_2O(l)$$

The test for chloride ions, using aqueous silver nitrate followed by ammonia solution, is an example of an insoluble solid dissolving as a complex ion forms. Adding silver nitrate to a solution of chloride ions produces a white precipitate of silver chloride, AgCl. This precipitate dissolves on adding ammonia solution as silver ions form the complex ion, $[Ag(NH_3)_2]^+(aq)$, with ammonia molecules.

$$AgCl(s) + 2NH_3(aq) \rightarrow [Ag(NH_3)_2]^+(aq) + Cl^-(aq)$$

Co-ordination compounds and complex ions are not only important in the inorganic chemistry of transition metals. They are also very important in the natural world. Chlorophyll in the leaves of plants, myoglobin in muscles and haemoglobin in red blood cells are all examples of complexes between metal ions and organic molecules. Zinc is an essential trace element in the human diet because zinc ions are an essential part of important enzymes. Amino acids in the protein chain form dative covalent bonds with the zinc.

15.7 Naming complex ions

There are four rules to follow when naming a complex ion.

1 Identify the number of ligands around the central cation using Greek prefixes: mono-, di-, tri-, tetra-, and so on.

2 Name the ligand using names ending in -o for anions, e.g. chloro- for Cl^-, fluoro- for F^-, cyano- for CN^-, hydroxo- for OH^-. Use aqua for H_2O and ammine for NH_3.

3 Name the central metal ion using the normal name of the metal for positive and neutral complex ions and the Latinised name ending in -ate for negative complex ions, e.g. ferrate for iron, cuprate for copper, argentate for silver.

4 Finally, add the oxidation number of the central metal ion.

The examples in Table 15.3 illustrate how you should use the rules.

Table 15.3 Writing the systematic names of complex ions.

Formula of complex ion	1 Identify the number of ligands	2 Name the ligand	3 Name the central metal ion	4 Add the oxidation number of the central metal ion
$[Ag(NH_3)_2]^+$	di	ammine	silver	(I)
$[Cu(H_2O)_6]^{2+}$	hexa	aqua	copper	(II)
$[CuCl_4]^{2-}$	tetra	chloro	cuprate	(II)
$[Fe(CN)_6]^{3-}$	hexa	cyano	ferrate	(III)

Test yourself

15 What is the co-ordination number of the named ions in the given complex ion?

  a) Cu^{2+} ions in $[CuCl_4]^{2-}$

  b) Cu^{2+} ions in $[Cu(H_2O)_6]^{2+}$

  c) Fe^{3+} ions in $[Fe(CN)_6]^{3-}$

16 Write the systematic name of each complex ion.

  a) $[Co(NH_3)_6]^{3+}$

  b) $[Zn(OH)_4]^{2-}$

  c) $[AlH_4]^-$

  d) $[Ni(H_2O)_6]^{2+}$

17 What is the oxidation state of the metal ion in the following complex ions?

  a) $[NiCl_4]^{2-}$

  b) $[Ag(NH_3)_2]^+$

  c) $[Fe(H_2O)_6]^{3+}$

  d) $[Fe(CN)_6]^{4-}$

18 A solution of thiosulfate ions, $S_2O_3^{2-}$, can dissolve a precipitate of silver bromide. Each silver ion forms a complex ion with two thiosulfate ions as the silver bromide dissolves. Write an equation for this reaction.

Tip

Notice that ammonia, NH_3, in complexes is described as 'ammine', whereas the $-NH_2$ group in organic compounds such as CH_3NH_2 is described as 'amine'.

15.8 The shapes of complex ions

The shapes of complex ions depend on the number of ligands around the central metal ion. There is no simple, definitive rule for predicting the shapes of complexes from their formulae, but:

- in complexes with a co-ordination number of six, the ligands usually occupy octahedral positions so that the six electron pairs around the central atom are repelled as far as possible (Figure 15.15)

- in complexes with a co-ordination number of four, the ligands usually occupy tetrahedral positions although there are a few complexes with four-fold co-ordination, such as $[Pt(NH_3)_2Cl_2]$, that have a square planar structure (Figure 15.15)
- in complexes with a co-ordination number of two, the ligands usually form a linear structure with the central metal ion (Figure 15.15).

Two common tetrahedral complexes are $[CuCl_4]^{2-}$ and $[CoCl_4]^{2-}$. The relatively large size of the chloride ions, compared to oxygen atoms in water molecules, means that it is not possible for more than four ligands to fit round the central metal ion.

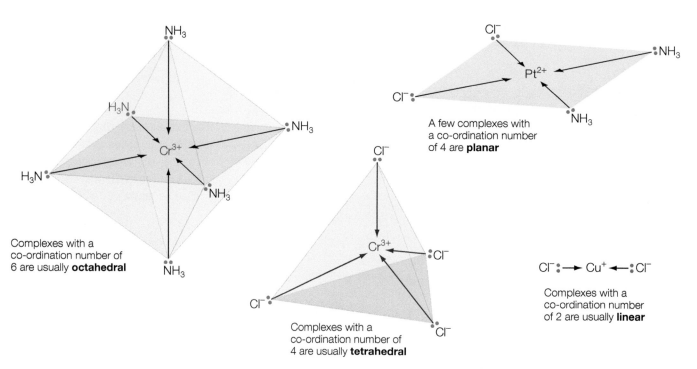

Complexes with a co-ordination number of 6 are usually **octahedral**

A few complexes with a co-ordination number of 4 are **planar**

Complexes with a co-ordination number of 4 are usually **tetrahedral**

Complexes with a co-ordination number of 2 are usually **linear**

Figure 15.15 The shapes of complex ions.

Types of ligand

Most ligands use only one lone pair of electrons to form a co-ordinate bond with the central metal ion. These ligands are described as monodentate because they have only 'one tooth' to hold onto the central cation (*dens* is Latin for tooth). Examples of monodentate ligands include H_2O, NH_3, Cl^-, OH^- and CN^-.

Some ligands have more than one lone pair of electrons that can form co-ordinate bonds with the same metal ion. Bidentate ('two-toothed') ligands, for example, form two dative covalent bonds with metal ions in complexes. Bidentate ligands include 1,2-diaminoethane, $H_2NCH_2CH_2NH_2$, the ethanedioate ion, $C_2O_4^{2-}$, and amino acids (Figure 15.16).

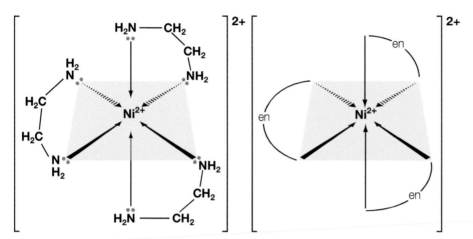

Figure 15.16 Representations of a complex formed by the bidentate ligand 1,2-diaminoethane with nickel(II) ions. Note the use of 'en' as an abbreviation for the ligand.

The hexadentate ligand EDTA^{4-} is particularly impressive because it can form six co-ordinate bonds with the central metal ion in complexes. EDTA^{4-} is the common abbreviation for this ion, which binds so firmly with metal ions that it holds them in solution and makes them chemically inactive. Figure 15.17 shows how the hexadentate ligand can fold itself around metal ions, such as Pb^{2+}, so that four oxygen atoms and two nitrogen atoms form co-ordinate bonds to the metal ion. This is the ion formed when EDTA^{4-} is used to treat lead poisoning. The EDTA^{4-} ion forms such a stable complex with Pb^{2+} ions that they can be excreted through the kidneys.

The disodium salt of EDTA^{4-} is added to commercially produced salad dressings to extend their shelf life. The EDTA^{4-} ion traps traces of metal ions that would otherwise catalyse the oxidation of vegetable oils. The disodium salt of EDTA^{4-} is also an ingredient of bathroom cleaners to help remove scale by dissolving Ca^{2+} ions from the calcium carbonate left by hard water.

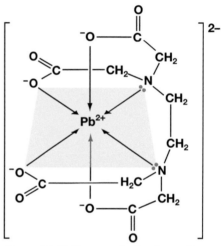

Figure 15.17 The complex ion formed by the EDTA^{4-} ion with a Pb^{2+} ion.

> **Tip**
>
> EDTA crystals consist of the disodium salt of **e**thylene**d**iamine**t**etra**a**cetic **a**cid. Chemists sometimes use the abbreviation Na_2H_2Y for the salt, where Y represents the 4− ion.

Ligands like those in Figures 15.16 and 15.17, which form more than one co-ordinate bond with metal ions, are sometimes called **multidentate ligands**, and the complexes which these ligands form are called **chelates** (pronounced 'keelates'). The term 'chelate' comes from a Greek word for a crab's claw, reflecting the claw-like way in which chelating ligands grip metal ions. Powerful chelating agents trap metal ions and effectively isolate them in solution.

> **Tip**
>
> Multidentate ligands are sometimes called polydentate ligands.

> **Key terms**
>
> **Multidentate ligands** form more than one co-ordinate bond with the same metal ion.
>
> **Chelates** are complex ions involving multidentate ligands.

Cis-platin – an important chemotherapy drug

The neutral complex, $PtCl_2(NH_3)_2$, in which Cl^- ions and NH_3 molecules act as ligands, has two isomers. These isomers have different melting temperatures and different chemical properties. One isomer called *cis*-platin is used as a chemotherapy drug in the treatment of certain cancers, whereas the other isomer is ineffective against cancer. The *trans* isomer is more toxic and not effective as a cancer drug.

Patients are given an intravenous injection of *cis*-platin, which circulates all around the body, including the cancerous area. *Cis*-platin diffuses relatively easily through the tumour cell membrane because it has no overall charge, like the cell membrane.

Once inside the cell, *cis*-platin exchanges one of its chloride ions for a molecule of water to form $[Pt(NH_3)_2(Cl)(H_2O)]^+$, which is the 'active principle' (Figure 15.18). This positively charged ion then enters the cell nucleus where it readily bonds with two sites on the DNA. Binding involves co-ordinate bonding from the nitrogen or oxygen atoms in the bases of DNA to the platinum ion.

The *cis*-platin binding changes the overall structure of the DNA helix, pulling it out of shape and shortening the helical turn. The badly shaped DNA can no longer replicate and divide to form new cells, although the affected cells continue to grow. Eventually, the cells die and, if enough of the cancerous cells absorb *cis*-platin, the tumour is destroyed.

Unfortunately, *cis*-platin is not a miracle cure without risks or drawbacks. It is toxic, resulting in unpleasant side-effects, and can cause kidney failure. Clinical trials have, however, led to the discovery of other platinum complexes that cause fewer problems and are already used as anti-cancer drugs.

1 Why is it possible to conclude that *cis*-platin has a square planar rather than a tetrahedral structure?

2 What type of isomerism do *cis*-platin and its isomer show?

3 a) What is the oxidation number of platinum in *cis*-platin?
  b) Write the systematic name of *cis*-platin.
  c) Draw the structure of *cis*-platin.

4 Why does *cis*-platin diffuse easily through the membrane of cells?

5 What is meant by the term 'active principle' applied to $[Pt(NH_3)_2(Cl)(H_2O)]^+$?

6 When $[Pt(NH_3)_2(Cl)(H_2O)]^+$ has formed inside the cell, it cannot diffuse out through the cell membrane. Why is this?

7 Why is a cell with *cis*-platin binding to its DNA unable to replicate?

8 Why is the binding to *cis*-platin from nitrogen and oxygen atoms rather than from carbon and hydrogen atoms in the bases of DNA?

9 Why is *cis*-platin more likely to affect cancerous cells than normal cells?

10 Why is any anti-cancer chemotherapy drug that acts like *cis*-platin likely to have undesirable side-effects?

11 Why is it important to test samples of *cis*-platin to make sure that they are free of the *trans* isomer?

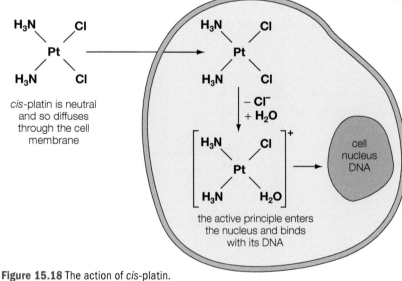

cis-platin is neutral and so diffuses through the cell membrane

− Cl⁻
+ H_2O

the active principle enters the nucleus and binds with its DNA

cell nucleus DNA

Figure 15.18 The action of *cis*-platin.

19 Predict the likely shape of the following complex ions:

a) $[Ag(CN)_2]^-$
b) $[Fe(CN)_6]^{3-}$
c) $[NiCl_4]^{2-}$
d) $[CrCl_2(H_2O)_4]^+$

20 Explain how the amino acid glycine (H_2NCH_2COOH) can act as a bidentate ligand.

21 a) Why is $EDTA^{4-}$ described as a hexadentate ligand?

b) What is the overall shape of the $EDTA^{4-}$ complex in Figure 15.17?

22 a) Draw a diagram to represent the complex ion formed between a Cr^{3+} ion and three ethanedioate ions $^-O_2C-CO_2^-$.

b) What is the overall shape of this complex ion?

23 a) Predict an order of stability for the complex ions $[Ni(NH_3)_6]^{2+}$, $[Ni(en)_3]^{2+}$ and $[Ni(EDTA)]^{2-}$.

b) Explain your prediction.

24 Refer to Figure 15.19. Write the formula of:

a) the hexacyanoferrate(II) ion
b) iron(III) hexacyanoferrate(II).

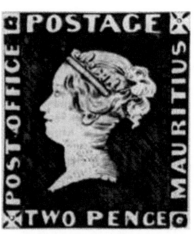

Figure 15.19 The blue pigment in this rare Mauritius two pence stamp from 1847 is called Prussian blue. Its correct chemical name is iron(III) hexacyanoferrate(II).

15.9 Ligand exchange reactions

Complex ions often react by exchanging one ligand for another. These ligand exchange reactions are often reversible and the changes of ligand are sometimes accompanied by colour changes. For example, when excess concentrated ammonia solution is added to pale blue copper(II) sulfate solution, ammonia molecules are exchanged for water molecules around the central Cu^{2+} ion and the colour changes to a deep blue. The reaction takes place in two stages. At first the alkaline solution of ammonia removes protons from the hydrated copper(II) ions to give a pale blue precipitate of the hydrated hydroxide.

$$[Cu(H_2O)_6]^{2+}(aq) + 2OH^-(aq) \rightleftharpoons [Cu(H_2O)_4(OH)_2](s) + 2H_2O(l)$$
pale blue solution   pale blue precipitate

Then the ligand exchange takes place as the precipitate redissolves to give the deep blue solution.

$$[Cu(H_2O)_4(OH)_2](s) + 4NH_3(aq)$$
pale blue precipitate

$$\rightleftharpoons [Cu(NH_3)_4(H_2O)_2]^{2+}(aq) + 2H_2O(l) + 2OH^-(aq)$$
deep blue solution

The ligands NH_3 and H_2O are both uncharged and similar in size. This allows exchange reactions between these ligands without a change in co-ordination number of the metal ion.

A ligand exchange reaction also occurs when concentrated hydrochloric acid is added to copper(II) sulfate solution. This time, the colour changes from pale blue to yellow as Cl^- ions replace water molecules around the Cu^{2+} ion.

$$[Cu(H_2O)_6]^{2+}(aq) + 4Cl^-(aq) \rightleftharpoons [CuCl_4]^{2-}(aq) + 6H_2O(l)$$
pale blue   yellow

In this case, however, the ligand exchange involves a change in co-ordination number. Chloride ions are larger than water molecules, so fewer chloride ions can fit round the central Cu^{2+} ion.

> **Tip**
>
> A solution of a copper(II) salt may turn green and not yellow if the concentration of added chloride ions is not high enough to convert all the blue hydrated ions to the yellow chloro complex.

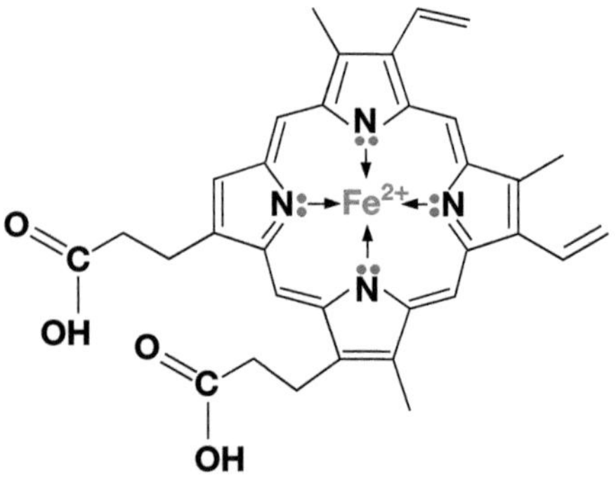

Figure 15.20 A haem group with its Fe^{2+} ion. There are four co-ordinate bonds between N atoms in the haem group and the metal ion. In haemoglobin there is a fifth co-ordinate bond between an N atom in one of the polypeptide chains and the Fe^{2+} ion. This leaves one site on the Fe^{2+} ion that can accept a pair of electrons from an oxygen molecule.

Another example of a colour change associated with a change of ligand and a change of co-ordination number is observed on adding concentrated hydrochloric acid to an aqueous solution of a cobalt(II) salt. The octahedral hexaaquocobalt(II) ion is pink. The solutions turns to a deep blue colour as the tetrahedral tetrachlorocobaltate(II) ion forms.

$$[Co(H_2O)_6]^{2+}(aq) + 4Cl^-(aq) \rightleftharpoons [CoCl_4]^{2-}(aq) + 6H_2O(l)$$

Ligand exchange reactions also take place in living organisms. Haemoglobin is the red protein in blood that carries oxygen from the lungs to the cells in body tissues. A haemoglobin molecule consists of four polypeptide chains, each with a nitrogen atom forming a dative bond to an Fe^{2+} ion in a haem group (Figure 15.20). So there are four haem groups in each haemoglobin molecule.

Nitrogen atoms in the porphyrin ring form four more dative bonds to the Fe^{2+} ion. So the porphyrin ring acts as a multidentate ligand. This leaves one remaining site on the metal ion which can accept a pair of electrons from an oxygen molecule (in oxyhaemoglobin which is bright red) or a water molecule (in deoxyhaemoglobin which is dull red).

The reactions between haem groups and oxygen or water are reversible ligand substitution reactions, allowing haemoglobin to pick up and release oxygen. The reaction with carbon monoxide is irreversible, which explains why the gas is dangerously toxic.

> ## Test yourself
>
> 25 Explain why breathing in carbon monoxide leads to death.
> 26 Write equations for the ligand exchange reactions that occur when:
>   a) hexaaquacobalt(II) ions react with ammonia molecules to form hexaamminecobalt(II) ions
>   b) hexaamminecobalt(II) ions react with chloride ions to from tetrachlorocobaltate(II) ions
>   c) hexaaquairon(II) ions react with cyanide ions to form hexacyanoferrate(II) ions.
> 27 A dilute solution of cobalt(II) chloride is pink because it contains hydrated cobalt(II) ions. The solution turns blue on adding concentrated hydrochloric acid with the formation of tetrachlorocobaltate(II) ions.
>   a) Write an equation for the reaction that occurs when concentrated HCl is added to dilute cobalt(II) chloride solution and indicate the colour of all species.
>   b) Explain the chemical basis for the test illustrated in Figure 15.21.

Figure 15.21 Filter paper soaked in pink cobalt(II) chloride solution and dried in an oven until it is blue, can be used to test for the presence of water.

Relative stability of complex ions

In aqueous solution, the simple compounds of most transition metals contain complex ions with formulae such as $[Cu(H_2O)_6]^{2+}$, $[Cr(H_2O)_6]^{3+}$ and $[Co(H_2O)_6]^{2+}$.

When solutions containing other ligands, such as Cl^-, are added to aqueous solutions of these hydrated cations, the mixture comes to an equilibrium in which the water molecules of some complexes have been replaced by the added ligands. For example, the equilibrium which results when concentrated sodium chloride solution is added to aqueous copper(II) ions is:

$$[Cu(H_2O)_6]^{2+}(aq) + 4Cl^-(aq) \rightleftharpoons [CuCl_4]^{2-}(aq) + 6H_2O(l)$$

The equilibrium constant, K_c, for this reaction is:

$$K_c = \frac{[CuCl_4^{2-}(aq)]}{[Cu(H_2O)_6^{2+}(aq)][Cl^-(aq)]^4}$$

$[H_2O(l)]$ is constant and therefore it is not included in the equation for K_c.

Equilibrium constants like this for the formation of complex ions in aqueous solution are called stability constants and the symbol K_{stab} is sometimes used in place of K_c.

Stability constants enable chemists to compare the stabilities of the complex ions of a cation with different ligands. The larger the stability constant, the more stable is the complex ion compared with that containing water.

Table 15.4 shows the stability constants of three complexes of the copper(II) ion. These show that the relative stabilities of the three copper(II) complexes are:

$$[Cu(EDTA)]^{2-} > [Cu(NH_3)_4(H_2O)_2]^{2+} > [CuCl_4]^{2-}$$

Ligand	Complex ion	K
Cl^-	$[CuCl_4]^{2-}$	4.0×10^5
NH_3	$[Cu(NH_3)_4(H_2O)_2]^{2+}$	1.3×10^{13}
$EDTA^{4-}$	$[Cu(EDTA)]^{2-}$	6.3×10^{18}

Table 15.4 The stability constants of three copper(II) complexes.

Complex ions and entropy

When a bidentate ligand, such as 1,2-diaminoethane, replaces a monodentate ligand, such as water, there is an increase in entropy of the system. One molecule of the bidentate ligand replaces two molecules of the monodentate ligand and this results in an increase in the number of product particles. For example, in the reaction:

$$[Cu(H_2O)_6]^{2+}(aq) + 3H_2NCH_2CH_2NH_2(aq)$$
$$\rightarrow [Cu(H_2NCH_2CH_2NH_2)_3]^{2+}(aq) + 6H_2O(l)$$

there are seven product particles but only four reactant particles. In this reaction the enthalpy change is very small and so the entropy change in the surroundings is close to zero.

As the entropy of any system depends on the number of particles present, the entropy of the system increases when this reaction occurs. In other words, ΔS_{system} is positive.

When a polydentate ligand, such as $EDTA^{4-}$, replaces a monodentate ligand, an even larger increase occurs in the entropy of the system.

$$[Cu(H_2O)_6]^{2+}(aq) + EDTA^{4-}(aq) \rightarrow [Cu(EDTA)]^{2-}(aq) + 6H_2O(l)$$

Because of this increase in the entropy of the system, complexes with polydentate ligands are usually more stable than those with bidentate ligands, which in turn are more stable than complexes with monodentate ligands.

Tip

For a spontaneous change to occur, ΔS_{total} must be positive.

$$\Delta S_{total} = \Delta S_{system} + \Delta S_{surroundings}$$

If ΔS_{total} is positive, the products of the system are more thermodynamically stable than the reactants.

Table 15.5

Complex	Stability constant	Colour
$[Co(H_2O)_6]^{2+}$	1.0	Pink
$[Co(NH_3)_6]^{2+}$	3×10^4	Green
$[Co(EDTA)]^{2-}$	2×10^{16}	Pink

Test yourself

28 The stability constants and colour of some cobalt(II) complexes are shown in Table 15.5.

What would you expect to see when:

a) ammonia solution is added to an aqueous solution of cobalt(II) chloride

b) EDTA solution is added to a solution of cobalt(II) chloride in aqueous ammonia?

29 Suggest two reasons why the stability constant of $[Co(EDTA)]^{2-}$ is so much larger than those of $[Co(H_2O)_6]^{2+}$ and $[Co(NH_3)_6]^{2+}$.

Core practical 12

The preparation of a transition metal complex

This is a summary of the procedure for preparing and purifying the complex salt called hexamminecobalt(III) chloride.

Preparation of the salt

A Add 8 g ammonium chloride and 12 g hydrated cobalt(II) chloride, $[Co(H_2O)_6]Cl_3$, to a measured volume of water in a flask. Then add a measured amount of powdered charcoal and bring to the boil.

B Cool the mixture from A and then add 25 cm^3 of concentrated ammonia solution, and cool again.

C Add a total of 25 cm^3 of 20-volume hydrogen peroxide a small amount at a time, shaking after each addition. Then heat the mixture to about 60 °C and keep the solution at this temperature for 30 minutes.

D Cool the flask in iced water to precipitate the impure product.

Purification of the salt

E Filter off the impure crystals with the charcoal catalyst.

F Add the solid from the filter paper to boiling water acidified with concentrated hydrochloric acid. Stir and the filter again, retaining the filtrate.

G Cool the filtrate in iced water, then filter off the crystals that separate.

H Rinse the crystals on the filter paper first with a very little cold water and then with ethanol.

I Leave the crystals to dry, then measure the mass of the product.

Questions

1 Identify the particularly hazardous chemicals used in the preparation and state the special precautions needed when handling them.

2 There is a ligand exchange reaction in step B to change the hexaqua complex into a hexamminecobalt(II) ion. Write an ionic equation for the reaction.

3 Suggest a reason for adding ammonium chloride to the reaction mixture as well as concentrated ammonia.

4 Step C involves the oxidation of the cobalt(II) complex to the required cobalt(III) complex.

a) Use these standard electrode potentials to explain why the oxidation is possible with the ammine complex but not the aqua complex.

$$[Co(NH_3)_6]^{3+}(aq) + e^- \rightleftharpoons [Co(NH_3)_6]^{2+}(aq)$$
$$E^\ominus = +0.10\,V$$

$$H_2O_2(aq) + 2H^+(aq) + 2e^- \rightleftharpoons 2H_2O(l)$$
$$E^\ominus = +1.77\,V$$

$$[Co(H_2O)_6]^{3+}(aq) + e^- \rightleftharpoons [Co(H_2O)_6]^{2+}(aq)$$
$$E^\ominus = +1.82\,V$$

b) What do the electrode potentials show about the relative stability of the cobalt(II) and cobalt(III) states when complexed with water or with ammonia?

5 Write the overall balanced ionic equation for the oxidation reaction.

6 The charcoal is added as a catalyst for the formation of the required product. What is the evidence that this is a heterogeneous catalyst?

7 What is the name of the procedure in steps F and G and why does it purify the product?

8 Suggest a reason for adding hydrochloric acid in step F to increase the yield of the product.

9 Explain the purpose of the small amount of cold water and then the ethanol used in step H.

10 Calculate the theoretical yield of product and the mass of the golden-brown crystals obtained if the procedure gives a 70% yield.

15.10 Reactions of transition metal ions with aqueous alkalis

The acid–base and ligand-exchange reactions of aqueous metal ions can be used in qualitative analysis to identify the positive ions in salts. Adding aqueous sodium hydroxide produces a precipitate if the metal hydroxide is insoluble. The precipitate dissolves in excess of the alkali if the hydroxide is amphoteric.

Adding ammonia solution also precipitates insoluble hydroxides. These redissolve in excess if the metal ion forms stable complex ions with ammonia molecules (Table 15.6).

Table 15.6 Results of adding aqueous sodium hydroxide and aqueous ammonia solutions to samples of transition metal ions.

Positive ion in solution	Observations on adding sodium hydroxide solution drop by drop and then in excess	Observations on adding ammonia solution drop by drop and then in excess
Chromium(III), Cr^{3+} green	Green precipitate which dissolves in excess reagent to form a dark green solution	Grey-green precipitate slightly soluble in excess reagent to give a purple solution
Iron(II), Fe^{2+} pale green	Dirty green precipitate insoluble in excess reagent	Green precipitate insoluble in excess reagent
Iron(III), Fe^{3+} yellow	Browny-red precipitate insoluble in excess reagent	Browny-red precipitate insoluble in excess reagent
Cobalt(II), Co^{2+} pink	Blue precipitate from a pink solution, insoluble in excess reagent	Blue precipitate from a pink solution dissolving in excess reagent to give a pale brown solution that gradually darkens.
Copper(II), Cu^{2+} blue	Pale blue precipitate insoluble in excess reagent	Pale blue precipitate dissolving in excess reagent to form a dark blue solution

The behaviour of aqueous chromium(III) ions shown in Table 15.6 is typical of hydrated metals ions in the 3+ state. The behaviour of aqueous iron(III) is the exception. The chromium(III) ions are hydrated. In the complex, the electrons in the water molecules are pulled towards the highly polarising Cr^{3+} ion, making it easier for the water molecules linked to the chromium ion to give away protons. The hydrated ion is partially ionised in solution. It is a strong enough acid to form carbon dioxide when added to a solution of sodium carbonate.

$$[Cr(H_2O)_6]^{3+} \rightleftharpoons [Cr(H_2O)_5OH]^{2+}(aq) + H^+(aq)$$

$$[Cr(H_2O)_5OH]^{2+}(aq) \rightleftharpoons [Cr(H_2O)_4(OH)_2]^+(aq) + H^+(aq)$$

Adding hydroxide ions, a base, to a solution of chromium(III) ions removes a third proton, producing an uncharged complex. The uncharged complex is much less soluble in water and precipitates as a white jelly-like precipitate of hydrated chromium(III) hydroxide.

$$[Cr(H_2O)_4(OH)_2]^+(aq) + OH^-(aq) \rightleftharpoons [Cr(H_2O)_3(OH)_3](s) + H_2O(l)$$

Key term

An **amphoteric** hydroxide is one that can dissolve in either aqueous acid or aqueous alkali.

Analysis of an inorganic unknown

A student carried out a series of tests on a compound X as described in the first column of Table 15.7. The results are shown in column 2 of the table. Column 3 shows some, but not all, of the student's interpretation of the observations.

Table 15.7

Test		Observations	Inferences
A	Describe the appearance of X.	Pale green crystals	X might be a salt of a transition metal.
B	Heat a sample of X in a test tube, first gently and then more strongly.	At first condensation appears on the cooler parts of the test tube. On strong heating gas is given off and the solid turns reddish-brown. The gas turns blue litmus red and reacts with a solution potassium dichromate(VI) turning it from orange to green.	X is a hydrated salt. It decomposes on strong heating. The gas is a reducing agent.
C	Add a few drops of NaOH(aq) to a solution of X, then add an excess of the alkali.	A green precipitate forms. The precipitate does not dissolve in excess alkali.	X could be an iron(II) salt. It cannot be a chromium(III) salt.
D	Add a few drops of NH_3(aq) to a solution of X, then add an excess of the ammonia solution.	A green precipitate forms. The precipitate does not dissolve in excess ammonia solution.	This is consistent with the result for test C.
E	Add aqueous chlorine to a solution of X, then add a few drops of NaOH(aq) to the solution followed by an excess of the alkali.	The very pale green solution turns yellow on adding chlorine. Then adding alkali produces a browny-red precipitate insoluble in excess reagent.	
F	Acidify a solution of X with dilute nitric acid, then add a few drops of $AgNO_3$(aq) to the solution.	No precipitate forms.	
G	Acidify a solution of X with dilute nitric acid, then add a few drops of $Ba(NO_3)_2$(aq) to the solution.	A white precipitate forms.	

1 Test A: Which of the first row of transition metals, in which oxidation states, form salts that are green?

2 Test B:
- Why is the conclusion that the gas is a reducing agent justified?
- Given an example of an acidic gas that is a reducing agent.

3 Test C: Why do the results of the test show that X cannot be a chromium(III) salt?

4 Test E:
- What can be inferred from the results of this test?
- Write an ionic equation for the reaction between X and chlorine.

5 Tests F and G:
- What can be inferred from the results of tests F and G?
- Write an ionic equation for the reaction in test G.

6 What is the evidence that a redox reaction takes place on heating crystals of X? Suggest an equation for the decomposition reaction.

7 The traditional name for iron(II) sulfate is green vitriol. This is because it produces oil of vitriol (sulfuric acid) when the gases given off on heating the salt are condensed. Use your answer to Question 6 to suggest an explanation for the formation of sulfuric acid in this way.

Tip

Analysis of an organic unknown is covered in Core practical 15 (part 2), in Section 18.3.3.

Refer to Practical skills sheet 9, 'Analysing inorganic unknowns', which you can access via the QR code for Chapter 15 on page 321.

Adding excess alkali removes yet another proton, now producing a negatively charged ion which is soluble in water so that the precipitate redissolves. This explains the amphoteric properties of chromium(III) hydroxide.

$$[Cr(H_2O)_3(OH)_3](s) + OH^-(aq) \rightleftharpoons [Cr(H_2O)_2(OH)_4]^-(aq) + H_2O(l)$$

All these changes are reversed by adding a solution of a strong acid such as hydrochloric acid, which reacts with the hydroxide ions in the complex to form water molecules.

Test yourself

30 Explain the following changes with the help of ionic equations.

   a) Adding a small amount of ammonia solution to a pale blue solution of hydrated copper(II) ions produces a pale blue precipitate of the hydrated hydroxide.

   b) On adding more ammonia solution, the precipitate dissolves to give a deep blue solution.

31 Explain the following observations with the help of ionic equations.

   a) A solution of iron(III) chloride is acidic. A browny-red precipitate forms on adding aqueous sodium hydroxide but the precipitate is not soluble in excess alkali.

   b) Adding aqueous sodium hydroxide to a solution of cobalt(II) chloride produces a blue precipitate which does not dissolve in excess alkali.

15.11 Catalysts based on transition elements and their compounds

Transition elements and their compounds play a crucial role as catalysts in industry. Table 15.8 lists some important examples of transition metals and their compounds as catalysts.

Catalysts can be divided into two types – heterogeneous and homogeneous.

Heterogeneous catalysis

Heterogeneous catalysis involves a catalyst in a different state from the reactants it is catalysing. It is used in almost every large-scale manufacturing process, for example the manufacture of ammonia in the Haber process (Table 15.8), in which nitrogen and hydrogen gas flow through a reactor containing lumps of iron or iron(III) oxide.

Table 15.8 Some important examples of transition metals and their compounds as heterogeneous catalysts.

Transition element/compound used as catalyst	Reaction catalysed
Vanadium(V) oxide, V_2O_5	Contact process in the manufacture of sulfuric acid: $2SO_2(g) + O_2(g) \rightleftharpoons 2SO_3(g)$
Iron or iron(III) oxide	Haber process to manufacture ammonia: $N_2(g) + 3H_2(g) \rightleftharpoons 2NH_3(g)$
Nickel, platinum or palladium	Hydrogenation of unsaturated vegetable oils: $RCH{=}CH_2(l) + H_2(g) \rightarrow RCH_2CH_3(l)$
Platinum or platinum–rhodium alloys	Conversion of NO and CO to CO_2 and N_2 in catalytic converters in vehicles: $2CO(g) + 2NO(g) \rightarrow 2CO_2(g) + N_2(g)$
Platinum	Reforming straight-chain alkanes as cyclic alkanes and arenes: $CH_3(CH_2)_5CH_3 \rightarrow CH_3{-}C_6H_5 + 4H_2$ heptane methylbenzene

Heterogeneous catalysts work by adsorbing reactants at active sites on their surface. Nickel, for example, acts as a catalyst for the addition of hydrogen to unsaturated compounds with carbon–carbon double bonds (Table 15.8). Hydrogen molecules are **adsorbed** on the catalyst surface, where they are thought to split into single atoms (free radicals). These highly reactive hydrogen atoms undergo addition with molecules of unsaturated compounds, like ethene, when the unsaturated compounds approach the catalyst surface (Figure 15.22).

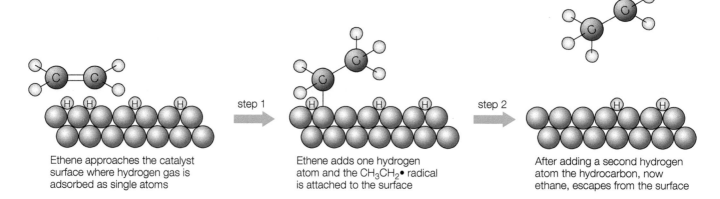

Ethene approaches the catalyst surface where hydrogen gas is adsorbed as single atoms	Ethene adds one hydrogen atom and the $CH_3CH_2\bullet$ radical is attached to the surface	After adding a second hydrogen atom the hydrocarbon, now ethane, escapes from the surface

Figure 15.22 A possible mechanism for the hydrogenation of an alkene using a nickel catalyst. The reaction takes place on the surface of the catalyst, which adsorbs hydrogen molecules and then splits them into atoms.

If a metal is to be a good catalyst for the addition of hydrogen, it must not adsorb the hydrogen so strongly that the hydrogen atoms become unreactive. This happens with tungsten. Equally, if adsorption is too weak there are insufficient adsorbed atoms for the reaction to occur at a useful rate, and this is the case with silver. The strength of adsorption must have a suitable intermediate value, which is the case with nickel, platinum and palladium.

The Contact process for making sulfuric acid gets its names from the 'contact' between the reacting gases and the surface of the heterogeneous catalyst. The vanadium(v) oxide catalyst is effective because the metal can change its oxidation state reversibly. First the vanadium(v) oxide is reduced to vanadium(IV) oxide as it oxidises sulfur dioxide to sulfur trioxide. Then the vanadium(IV) oxide is reoxidised to vanadium(v) oxide by oxygen in the mixture of reacting gases.

Test yourself

32 a) Write equations to describe the catalytic action of vanadium(v) oxide in the Contact process.

   b) Why is the vanadium(v) oxide in the Contact process described as a catalyst given that it reacts with sulfur dioxide?

Catalytic converters

Catalytic converters have done a great deal to improve air quality in our towns and cities. The catalyst speeds up reactions that remove pollutants from motor vehicle exhausts. The reactions convert oxides of nitrogen to nitrogen and oxygen. They also convert carbon monoxide to carbon dioxide, and unburnt hydrocarbons to carbon dioxide and steam (Figure 15.23).

The catalyst in a catalytic converter is made from a combinations of platinum, palladium and rhodium. The pollutants are **adsorbed** onto the surface of the catalyst, where they react (Figure 15.24). Then the products are desorbed into the stream of exhaust gases.

The catalyst must not adsorb molecules so strongly that the reactive sites on the surface of the metal are inactivated. However, the interaction between pollutant molecules and the metal surface has to be strong enough to weaken bonds and provide a reaction mechanism that is fast enough under the conditions in the exhaust system. Then the reaction products have to be so weakly attracted that they are quickly released into the gas stream.

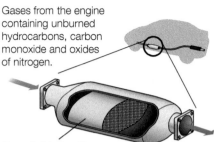

Gases from the engine containing unburned hydrocarbons, carbon monoxide and oxides of nitrogen.

Ceramic block with a structure like a honey-comb. The channels have a very large surface area which is coated with the catalyst.

Exhaust gas containing carbon dioxide, nitrogen and steam.

Figure 15.23 Function of a catalytic converter.

Figure 15.24 The surface of the metal catalyst in a catalytic converter adsorbs the pollutants NO and CO, where they react to form N_2 and CO_2. In this computer graphic, oxygen atoms are coloured red, nitrogen atoms are coloured blue and carbon atoms are coloured green.

1 Why do you think the catalyst in a catalytic converter is present as a very thin layer on the surface of many fine holes running through a block of inert ceramic?

2 Suggest reasons why the catalyst in a catalytic converter is only fully effective:
   a) after the engine has been running for some time
   b) if the engine is properly maintained so that it runs with the right mixture of air and fuel.

3 Write equations for the reactions catalysed by a catalytic converter that remove carbon monoxide and nitrogen monoxide from exhaust gases.

4 How does the catalyst speed up the reactions that destroy pollutants?

5 Why would the catalyst be ineffective if the bonding between the catalyst surface:
   a) and the reactants is too weak
   b) and the products is too strong?

6 Suggest a reason why using petrol containing lead additives in a car engine rapidly stops the catalytic converter being effective.

7 Identify two ways, other than fitting catalytic converters, to reduce air pollution from motor vehicles in cities.

8 What contribution, if any, do catalytic converters make to solving the problem of climate change?

Homogeneous catalysis

Homogeneous catalysis involves a catalyst in the same state as the reactants it is catalysing. Homogeneous catalysts are very important in biological systems because enzymes (proteins) act as catalysts in the metabolic processes of all organisms. Very often transition metal ions act as co-enzymes in these processes, enhancing the catalytic activity of the associated enzyme. Cytochrome oxidase is an important enzyme containing copper. This enzyme is involved when energy is released from the oxidation of food. In the absence of copper, cytochrome oxidase is totally ineffective and the animal or plant is unable to metabolise successfully.

Transition metal ions can also be effective as homogeneous catalysts because they can gain and lose electrons, changing from one oxidation state to another. The oxidation of iodide ions by peroxodisulfate(VI) ions using iron(III) ions as a catalyst is a good example of this.

$$2I^-(aq) + S_2O_8^{2-}(aq) \rightarrow I_2(aq) + 2SO_4^{2-}(aq)$$

In the absence of Fe^{3+} ions the reaction is very slow, but with Fe^{3+} ions in the mixture the reaction is many times faster. A possible mechanism is that Fe^{3+} ions are reduced to Fe^{2+} as they oxidise iodide ions to iodine. Then the $S_2O_8^{2-}$ ions oxidise Fe^{2+} ions back to Fe^{3+}, ready to oxidise more of the iodide ions, and so on.

Sometimes one of the products of a reaction can act as a catalyst for the process. This is called autocatalysis. An autocatalytic reaction starts slowly, but then speeds up as the catalytic product is formed. Mn^{2+} ions act as an autocatalyst in the oxidation of ethanedioate ions, $C_2O_4^{2-}$, by manganate(VII) ions in acid solution. This is also an example of homogeneous catalysis.

$$2MnO_4^-(aq) + 16H^+(aq) + 5C_2O_4^{2-}(aq)$$
$$\rightarrow 2Mn^{2+}(aq) + 8H_2O(l) + 10CO_2(g)$$

Test yourself

33 What is the advantage of using a solid heterogeneous catalyst in:

a) a continuous industrial process

b) an industrial batch process?

34 a) Suggest a reason why the reaction of between iodide ions and peroxodisulfate ions is slow in the absence of a catalyst.

b) Write half-equations to explain the mechanism by which iron(III) ions catalyse the reaction between iodide ions and peroxodisulfate ions.

c) Do you think Fe^{2+} ions can also catalyse this reaction? Explain your answer.

35 a) Suggest two methods of speeding up the reaction between $MnO_4^-(aq)$ and $C_2O_4^{2-}(aq)$ from the start of the reaction.

b) What would you expect to see when a solution of potassium manganate(VII) is added to an acidified solution of potassium ethanedioate:

i) at the start of the reaction

ii) as the reaction gets underway?

The development of new catalysts

One of the real challenges and priority areas in chemical research today is the development of new and improved catalysts. Catalysts are particularly important in the drive for a 'greener world' and a 'greener future'.

- Catalysts speed up reactions and products are obtained faster.
- Catalysts allow processes to operate at lower temperatures with savings on energy and fuel.
- Catalysts make possible processes that have high atom economies and produce less waste.
- Catalysts can be highly selective so that only the desired product is formed without the obvious waste from side-reactions.

The manufacture of ethanoic acid illustrates the advantages of developing new catalysts. Until the 1970s, the main method of manufacturing ethanoic acid was to oxidise hydrocarbons from crude oil in the presence of a cobalt(II) ethanoate catalyst. The process operated at 180–200 °C and 40–50 times atmospheric pressure. Only 35% by mass of the products was ethanoic acid.

Today, ethanoic acid is manufactured as the only product in the direct combination of methanol and carbon monoxide. The catalyst of iridium metal is mixed with ruthenium compounds that act as catalyst promoters and triple the rate of reaction. In addition, the process operates at lower temperature and lower pressure, producing only ethanoic acid.

As scientists develop new techniques, such as the improved catalytic process for ethanoic acid, and propose new ideas, it is important that their work is reported, checked and validated. This reporting and validating is carried out in three ways.

- *Through reports, journals and conferences* at which scientists discuss their work with others.
- *By peer review* in which others working in the same area look closely at the validity of the experimental methods used, the accuracy of the results obtained and the appropriateness of the conclusions drawn.
- *By replicating experiments*, which is perhaps the ultimate test of reliability and validity for any innovation. If other scientists repeat the work and achieve the same results, then its integrity is not questioned.

Test yourself

36 a) Write an equation for the production of ethanoic acid from methanol and carbon monoxide.

   b) What is the atom economy of the process?

37 You hear, from one of the popular media, about a newly discovered catalyst that promises great economic and environmental benefits. Suggest some of the questions that you should ask before deciding whether or not to take the claims seriously.

Exam practice questions

1 a) Write the electronic structure of:
    i)   a scandium atom   *(1)*
    ii)   a copper atom   *(1)*
    iii) a Cu^{2+} ion.   *(1)*

b) Both scandium and copper are *d-block elements*, but only copper is a *transition element*. Explain the meaning of the terms in italics. *(3)*

c) Aqueous Cu^{2+} ions react with excess ammonia solution to form $[Cu(NH_3)_4(H_2O)_2]^{2+}$ ions.
    i)   Write the name of the $[Cu(NH_3)_4(H_2O)_2]^{2+}$ ion. *(1)*
    ii)   What is the overall shape of the $[Cu(NH_3)_4(H_2O)_2]^{2+}$ ion? *(1)*

d) Explain why the complexes of copper(II) ions are usually coloured. *(4)*

2 This question concerns the chemistry of transition metals.
a) Define and explain the terms giving examples:
    i)   transition metal *(3)*
    ii)   oxidation number *(3)*
    iii) complex ion. *(3)*

b) Discuss, with examples, equations and observations, the typical reactions of the ions of transition metals. *(6)*

3 Chromium shows its highest oxidation state in the oxoanion, CrO_4^{2-}.
a) State the oxidation number of chromium in CrO_4^{2-}. *(1)*

b) The mixture changes colour when dilute acid is added to a solution of CrO_4^{2-} ions. State the changes in colour and write an equation for the reaction that occurs. *(3)*

c) When sulfur dioxide is bubbled into a solution of CrO_4^{2-} ions, the colour changes and chromium is reduced to a simple ion. State the new colour and the formula of the new simple ion. *(2)*

d) A 0.012 mol sample of an oxochloride of vanadium, $VOCl_x$, required $20.0\,cm^3$ of $0.100\,mol\,dm^{-3}$ potassium dichromate(VI) solution for oxidation of the vanadium to its +5 oxidation state.
    i)   Copy and balance the half-equation below for the action of dichromate(VI) as an oxidising agent.
$$Cr_2O_7^{2-}(aq) + \ldots..H^+(aq) + \ldots..e^-$$
$$\rightarrow \ldots..Cr^{3+}(aq) + \ldots H_2O(l) \; (1)$$
    ii)   How many moles of dichromate(VI) reacted with the oxochloride of vanadium? *(1)*
    iii) How many moles of electrons were gained by the $Cr_2O_7^{2-}$ ions? *(1)*
    iv) Calculate the change in oxidation state of the vanadium during the reaction. *(1)*
    v)   Write the formula of the oxochloride of vanadium showing the correct value of *x*. *(1)*

4 a) Classify these catalysts as homogeneous or heterogeneous:
    i)   the platinum alloy in a catalytic converter *(1)*
    ii)   the sodium hydroxide used to hydrolyse a halogenoalkane *(1)*
    iii) the zeolite use to crack oil fractions *(1)*
    iv) the nickel used to hydrogenate unsaturated fats. *(1)*

b) The activation energy for the decomposition of ammonia into nitrogen and hydrogen is $335\,kJ\,mol^{-1}$ in the absence of a catalyst but $162\,kJ\,mol^{-1}$ in the presence of a tungsten catalyst. Explain the significance of these values in terms of transition state theory. *(4)*

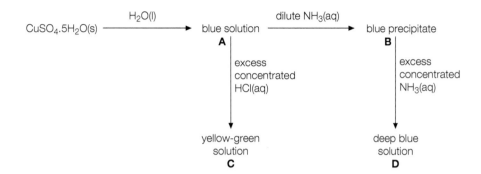

$$CuSO_4.5H_2O(s) \xrightarrow{\quad H_2O(l) \quad} \underset{\mathbf{A}}{\text{blue solution}} \xrightarrow{\quad \text{dilute } NH_3(aq) \quad} \underset{\mathbf{B}}{\text{blue precipitate}}$$

A → (excess concentrated HCl(aq)) → yellow-green solution **C**

B → (excess concentrated NH₃(aq)) → deep blue solution **D**

5 The reaction scheme above involves various compounds of copper.
  a) Write the formulae of the species responsible for the colour in each of the products A to D. *(4)*
  b) Describe and explain, with an equation, what you would see when solution C is diluted with excess water. *(6)*
  c) When aqueous sodium hydroxide is added to copper(II) sulfate solution, a blue precipitate is formed. If, however, excess EDTA^{4-} solution is first added to the copper(II) sulfate solution before the aqueous sodium hydroxide, no precipitate forms.

  Write an equation for the formation of the blue precipitate with aqueous sodium hydroxide, and explain why no precipitate forms if excess EDTA^{4-} is added to the copper(II) sulfate solution before the sodium hydroxide. *(5)*

6 There are two main types of catalyst: homogeneous and heterogeneous.
  a) Explain the term 'homogeneous catalysis' and state the most important feature of transition metal ions that allows them to act as homogeneous catalysts. *(2)*
  b) In aqueous solution, I$^-$ ions slowly reduce $S_2O_8^{2-}$ ions to SO_4^{2-} ions.
    i) Write an equation (or two half-equations) for the reaction. *(2)*
    ii) Suggest why the activation energy of the reaction is high, resulting in a slow reaction in the absence of a catalyst. *(1)*
    iii) Write two equations (or two pairs of half-equations) to show the role of iron salts in catalysing the reaction. *(2)*
  c) In Periods 5 and 6, the catalytic efficiency of transition metals as heterogeneous catalysts tends to be poor at the ends of the transition series, but high in the middle of the series.
    i) Suggest two reasons why the catalytic efficiency of metals is poor at the ends of the transition series. *(2)*
    ii) Suggest why the catalytic efficiency is high in the middle of the transition series. *(2)*
  d) In catalytic converters used to 'clean' the exhaust gases from petrol engines, a catalyst reduces nitrogen oxides using another pollutant gas as the reducing agent. State a suitable catalyst for catalytic converters, identify the reducing agent and write an equation for a possible reaction that results. *(3)*

7 Hydrazine, H_2NNH_2, is a powerful reducing agent in alkaline solution. It is oxidised to nitrogen gas and water.

Vanadium exists in several oxidation states and two of its electrode (reduction) potentials are shown below.

$$VO^{2+}(aq) + H_2O(l) + e^-$$
$$\rightleftharpoons V^{3+}(aq) + 2OH^-(aq)$$
$$E^\ominus = -1.32\,V$$

$$VO_2^+(aq) + H_2O(l) + e^-$$
$$\rightleftharpoons VO^{2+}(aq) + 2OH^-(aq)$$
$$E^\ominus = -0.66\,V$$

  a) Deduce the ionic half-equation for the oxidation of hydrazine in alkaline solution. *(2)*
  b) Hydrazine reduces vanadium(V) but not vanadium(IV) in alkaline solution. What does this tell you about the value of the electrode potential for the reduction half-equation that is the reverse of your answer to part (a)? *(2)*

c) Write the overall ionic equation for the reduction of vanadium(v) by hydrazine in alkaline solution. *(2)*

d) i) Explain, with an example, what is meant by the term 'disproportionation'. *(2)*

ii) Why must an element have at least three oxidation states if it is to undergo disproportionation? *(1)*

iii) Derive an equation for the disproportionation of vanadium(IV) into vanadium(III) and vanadium(V) in alkaline solution. *(1)*

iv) Say whether this disproportionation will occur and explain your answer. *(3)*

8 The complexes $[Ni(NH_3)_2Cl_2]$ and $[Co(NH_3)_4Cl_2]^+$ both form *cis* and *trans* stereoisomers.

a) Define the terms 'complex' and 'stereoisomer'. *(4)*

b) Write down the oxidation number of:
i) Ni in $[Ni(NH_3)_2Cl_2]$ *(1)*
ii) Co in $[Co(NH_3)_4Cl_2]^+$. *(1)*

c) Write down the co-ordination number of:
i) Ni in $[Ni(NH_3)_2Cl_2]$ *(1)*
ii) Co in $[Co(NH_3)_4Cl_2]^+$. *(1)*

d) Write the systematic name for:
i) $[Ni(NH_3)_2Cl_2]$ *(1)*
ii) $[Co(NH_3)_4Cl_2]^+$. *(1)*

e) Draw and describe the shape of the *cis* and *trans* isomers of $[Ni(NH_3)_2Cl_2]$. *(4)*

f) Draw and describe the shape of the *cis* and *trans* isomers of $[Co(NH_3)_4Cl_2]^+$. *(4)*

Kinetics II

16

16.1 Factors that affect reaction rates

Reaction kinetics is the study of the rates of chemical reactions. Several factors influence the rate of chemical change including the concentration of the reactants, the surface area of solids, the temperature of the reaction mixture and the presence of a catalyst. Chemists have found that they can learn much more about reactions by studying these effects quantitatively.

> **Tip**
>
> The first section of this chapter is a very brief summary of ideas introduced in Chapter 9 in Student Book 1. The rest of this chapter shows that there is much that chemists can learn about reactions from the quantitative study of kinetics. In particular, the chapter ends by showing how chemists' understanding of the mechanisms of organic reactions depends on key evidence from kinetics experiments.

Figure 16.1 Understanding the factors which determine the rate of chemical change is essential in the design of processes to manufacture drugs for the pharmaceutical industry.

> **Key term**
>
> The **activation energy** of a reaction is the minimum energy needed in a collision between molecules if they are to react. The activation energy is the height of the energy barrier separating reactants and products during the progress of a reaction.

Chemists use a collision model to explain the effects of the factors that alter reaction rates. This model is based on kinetic theory and the Maxwell–Boltzmann distribution of energies in a collection of molecules. The idea is that a chemical reaction happens when the molecules or ions of reactant collide, making some bonds break and allowing new bonds to form. However, it is not enough for the molecules to collide. In soft collisions the molecules simply bounce off each other. Molecules are in rapid random motion and if every collision led to reaction all reactions would be explosively fast. Only pairs of molecules that collide with enough energy to stretch and break chemical bonds can lead to new products. Reactant molecules have to overcome the **activation energy**.

Chemists apply the theory of chemical kinetics to drug design and to the formulation of medicines to make sure that patients receive treatments

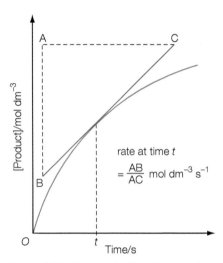

rate at time t

$$= \frac{AB}{AC} \text{ mol dm}^{-3} \text{ s}^{-1}$$

Figure 16.2 A concentration–time graph for the formation of a product. The rate of formation of product at time t is the gradient (or slope) of the curve at this point.

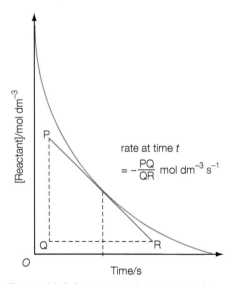

rate at time t

$$= -\frac{PQ}{QR} \text{ mol dm}^{-3} \text{ s}^{-1}$$

Figure 16.3 A concentration–time graph for the disappearance of a reactant. Here the gradient is negative and so the rate of reaction at time t is minus the gradient (or slope) of the curve at this point.

that are effective for some time without causing harmful side-effects. The theory can also help to account for the damage arising from pollutants in the atmosphere and to explore ways for reducing or preventing the problems.

16.2 Measuring reaction rates

Balanced chemical equations say nothing about how quickly the reactions occur. In order to get this information, chemists have to do experiments to measure the rates of reactions under various conditions.

The amounts of the reactants and products change during any chemical reaction. Products form as reactants disappear. The rates at which these changes happen give a measure of the **rate of reaction**.

Chemists define the rate of reaction as the change in concentration of a product, or a reactant, divided by the time for the change. Usually the rate is not constant but varies as the reaction proceeds. Normally the rate decreases with time as the concentrations of reactants fall. However, a reaction may get faster and faster if it is exothermic and the temperature rises (Section 16.5), or if the reaction gives a product that can act as a catalyst for the reaction (Section 15.11).

The first step in analysing the results of an experiment is to plot a concentration–time graph. The gradient (or slope) of the graph at any point gives a measure of the rate of reaction at that time (Figures 16.2 and 16.3).

Practical methods

Ideally, chemists look for methods for measuring reaction rates that do not interfere with the reaction mixture, as shown in Figures 16.4–16.7. Sometimes, however, it is necessary to withdraw samples of the reaction mixture at regular intervals and analyse the concentration of a reactant or product by titration, as illustrated in Figure 16.8.

Tip

The units for rate of reaction are always $mol\,dm^{-3}\,s^{-1}$. Do not confuse this with the units for rate constants, k, which vary according to the order of the reaction. See Section 4 in 'Mathematics in A Level chemistry', which you can access via the QR code for Chapter 16 on page 321, for guidance on how to work out and interpret the gradient of a tangent to a graph.

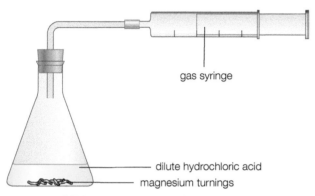

gas syringe

dilute hydrochloric acid
magnesium turnings

Figure 16.4 Following the course of a reaction with time by collecting and measuring the volume of a gas formed.

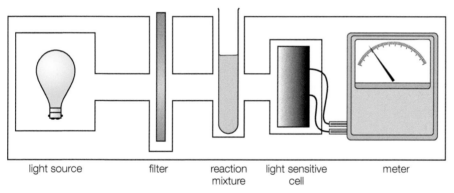

light source   filter   reaction   light sensitive   meter
                         mixture   cell

Figure 16.5 Using a colorimeter to follow the formation of a coloured product or the removal of a coloured reactant.

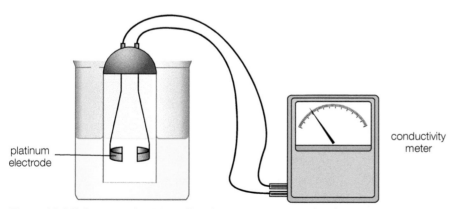

platinum
electrode

conductivity
meter

Figure 16.6 Using a conductivity cell and meter to measure the changes in electrical conductivity of the reaction mixture as the number or nature of the ions changes.

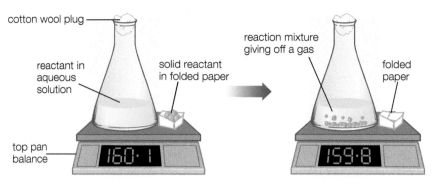

Figure 16.7 Following the course of a reaction by measuring the change of mass as the reaction gives off a dense gas that is lost from the system.

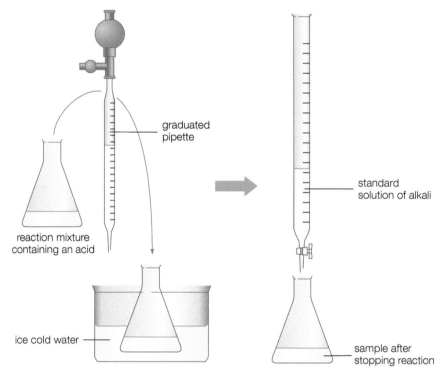

Figure 16.8 Following the course of a reaction during which there is a change in concentration of acids present by removing measured samples of the mixture at intervals, stopping the reaction by running the sample into an alkali, and then determining the concentration of one reactant or product by titration. Further samples are taken at regular intervals.

Test yourself

5 Suggest a suitable method for measuring the rate of each of these reactions:

a) $Br_2(aq) + HCOOH(aq) \rightarrow 2HBr(aq) + CO_2(g)$

b) $CH_3COOCH_3(l) + H_2O(l) \rightarrow CH_3COOH(aq) + CH_3OH(aq)$

c) $C_4H_9Br(l) + H_2O(l) \rightarrow C_4H_9OH(l) + H^+(aq) + Br^-(aq)$

d) $MgCO_3(s) + 2HCl(aq) \rightarrow MgCl_2(aq) + CO_2(g) + H_2O(l)$

Investigating the effect of concentration on the rate of a reaction

Bromine oxidises methanoic acid in aqueous solution to carbon dioxide. The reaction is catalysed by hydrogen ions.

$$Br_2(aq) + HCOOH(aq) \rightarrow 2Br^-(aq) + 2H^+(aq) + CO_2(g)$$

The reaction can be followed using a colorimeter.

Table 16.1 shows some typical results. The concentration of methanoic acid was kept constant throughout the experiment by having it present in large excess.

Table 16.1 Results of an experiment to investigate the rate of reaction of bromine with methanoic acid. Note that the bromine concentrations are multiplied by 1000. The actual bromine concentration at 90 seconds, for example, was $7.3 \times 10^{-3}\,mol\,dm^{-3}$ = $0.0073\,mol\,dm^{-3}$.

Time/s	Concentration of bromine/$10^{-3}\,mol\,dm^{-3}$
0	10.0
10	9.0
30	8.1
90	7.3
120	6.6
180	5.3
240	4.4
360	2.8
480	2.0
600	1.3

1 Explain why it is possible to follow the rate of this reaction using a colorimeter.
2 Suggest a suitable chemical to use as the catalyst for the reaction.
3 Explain the purpose of adding a large excess of methanoic acid.

4 Plot a graph of bromine concentration against time using the results in Table 16.1.
5 Draw tangents to the graph and measure the gradient to obtain values for the rate of reaction at two points during the experiment. Take values at 100 s and 500 s. (Remember when calculating the gradients that the bromine concentrations are 1000 times less than the numbers in the table.)
6 Table 16.2 shows values for the reaction rate obtained by drawing gradients at other times on the concentration–time graph. Plot a graph of rate against concentration using the two values you have found in answering Question 5 and the values in Table 16.2.

Table 16.2 Rate values obtained by finding gradients of tangents to the concentration–time graph. Note that the rate of reaction values are multiplied by 100 000. The actual rate at 300 seconds, for example, was $1.2 \times 10^{-5}\,mol\,dm^{-3}\,s^{-1}$.

Time/s	Concentration of bromine/ $10^{-3}\,mol\,dm^{-3}$	Rate of reaction from gradients to the concentration–time graph/ $10^{-5}\,mol\,dm^{-3}\,s^{-1}$
50	8.3	2.9
200	5.0	1.7
300	3.5	1.2
400	2.5	0.8

7 How does the bromine concentration change with time?
8 How does the rate of reaction change with time?
9 How does the rate of reaction depend on the bromine concentration?

16.3 Rate equations

Chemists have found that they can summarise the results of investigating the rate of reaction in the form of a rate equation. A **rate equation** shows how changes in the concentrations of reactants affect the rate of a reaction.

Take the example of a general reaction for which $x\,mol$ A react with $y\,mol$ B to form products:

$$xA + yB \rightarrow products$$

The equation that describes how the rate varies with the concentrations of the reactants takes this form:

$$rate = k[A]^m[B]^n$$

The powers *m* and *n* are the **reaction orders** with respect to the reactants A and B that appear in this equation.

The **overall order of the reaction** is sum of the orders for all the substances that appear in the rate equation; here $(m + n)$.

Table 16.3 The units of the rate constant for different reaction orders. These units can be worked out from the rate equations.

Overall order	Units of the rate constant
Zero	$mol\,dm^{-3}\,s^{-1}$
First	s^{-1}
Second	$dm^3\,mol^{-1}\,s^{-1}$

Tip

A rate equation cannot be deduced from the balanced equation: it has to be found by experiment. In the general example, rate = $k[A]^m[B]^n$, the values of *m* and *n* in the rate equation may or may not be the same as the values of *x* and *y* in the balanced equation for the reaction.

where [A] and [B] represent the concentrations of the reactants in moles per cubic decimetre.

The powers *m* and *n* are the **reaction orders**. The reaction above is order *m* with respect to A and order *n* with respect to B. The **overall order of the reaction** is $(m + n)$.

The rate constant, *k*, is only constant for a particular temperature. In other words, the value of *k* varies with temperature (see Section 16.5). The units of the rate constant depend on the overall order of the reaction (Table 16.3).

Example

The decomposition of ethanal to methane and carbon monoxide is second order with respect to ethanal. When the concentration of ethanal in the gas phase is $0.20\,mol\,dm^{-3}$, the rate of reaction is $0.080\,mol\,dm^{-3}\,s^{-1}$ at a certain temperature. What is the value of the rate constant at this temperature? (Give the units with your answer.)

Notes on the method

Start by writing out the rate equation based on the information given. There is no need to write the equation for the reaction because the rate equation cannot be deduced from the balanced chemical equation.

Substitute values in the rate equation, including the units as well as the values. Then rearrange the equation to find the value of *k*. Check that the units are as expected for a second order reaction.

Answer

The rate equation: rate = $k[\text{ethanal}]^2$

Substituting: $0.080\,mol\,dm^{-3}\,s^{-1} = k \times (0.20\,mol\,dm^{-3})^2$

Rearranging: $k = \dfrac{0.080\ mol\ dm^{-3}\ s^{-1}}{(0.20\,mol\ dm^{-3})^2}$

Hence: $k = 2.0\,dm^3\,mol^{-1}\,s^{-1}$

Test yourself

6 The rate of decomposition of an organic peroxide is first order with respect to the peroxide. Calculate the rate constant for the reaction at 107 °C if the rate of decomposition of the peroxide at this temperature is $7.4 \times 10^{-6}\,mol\,dm^{-3}\,s^{-1}$ when the concentration of peroxide is $0.02\,mol\,dm^{-3}$. Show that the unit of the rate constant is s^{-1}.

7 The hydrolysis of the ester methyl ethanoate in alkali is first order with respect to both the ester and hydroxide ions. The rate of reaction is $0.00069\,mol\,dm^{-3}\,s^{-1}$, at a given temperature, when the ester concentration is $0.05\,mol\,dm^{-3}$ and the hydroxide ion concentration is $0.10\,mol\,dm^{-3}$. Write the rate equation for the reaction and calculate the rate constant. Show that the unit of the rate constant is $dm^3\,mol^{-1}\,s^{-1}$.

First order reactions

A reaction is first order with respect to a reactant if the rate of reaction is proportional to the concentration of that reactant. The concentration term for this reactant is raised to the power one in the rate equation.

$$\text{rate} = k[X]^1 = k[X]$$

This means that doubling the concentration of the chemical X leads to a doubling of the rate of reaction.

The fact that the rate of reaction is directly proportional to the concentration of the reactant means that a plot of rate against concentration gives a straight line passing through the origin (Figure 16.9).

One of the easier ways to spot a first order reaction is to plot a concentration–time graph and then study the time taken for the concentration to fall by half (Figure 16.10). This is the **half-life**, $t_{\frac{1}{2}}$. It can be shown mathematically from the rate equation that for a first order reaction:

$$t_{\frac{1}{2}} = \frac{\ln 2}{k} = \frac{0.69}{k}$$

where k is the rate constant.

This shows that, at a constant temperature, the half-life of a first order reaction is the same wherever it is measured on a concentration–time graph. It is independent of the initial concentration.

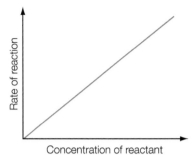

Concentration of reactant

Figure 16.9 The variation of reaction rate with concentration for a first order reaction.

Key term

The **half-life** of a reaction is the time for the concentration of one of the reactants to fall by half.

Test yourself

8 Refer to your answers to the activity in Section 16.2.

   a) From your rate–concentration graph, what is the order of the reaction of the reaction of bromine with methanoic acid with respect to bromine?

   b) i) Determine three values for half-lives for the reaction from your concentration–time graph.

      ii) Are your values consistent with your answer to part (a)?

9 Explain how the rate constant can be found from a rate–concentration graph such as in Figure 16.9.

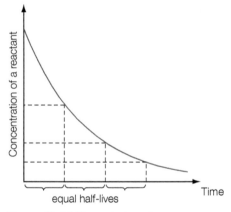

equal half-lives   Time

Figure 16.10 The variation of concentration of a reactant plotted against time for a first order reaction. The half-life for a first order reaction is a constant, so it is the same wherever it is read off the curve. It is independent of the initial concentration.

Second order reactions

A reaction is second order with respect to a reactant if the rate of reaction is proportional to the concentration of that reactant squared. This means that the concentration term for this reactant is raised to the power two in the rate equation. At its simplest, the rate equation for a second order reaction takes this form:

$$\text{rate} = k[\text{reactant}]^2$$

This means that doubling the concentration of X increases the rate by a factor of four.

Tip

Logarithms, including natural logarithms (ln), are explained in Section 3 of 'Mathematics in A Level chemistry', which you can access via the QR code for Chapter 16 on page 321. There is also more information about half-lives in Section 4.

A concentration–time graph of a second order reaction has unequal half-lives (Figure 16.11). The time for the concentration to fall from its initial value c to $\frac{c}{2}$ is half the time for the concentration to fall from $\frac{c}{2}$ to $\frac{c}{4}$. The half-life is inversely proportional to the starting concentration.

The variation of rate with concentration for a second order reaction can be found, as before, by drawing tangents to the curve of the concentration–time graph. However, a rate–concentration graph is not a straight line for a second order reaction; instead it is a curve, as shown in Figure 16.12.

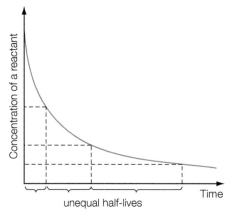

Figure 16.11 The variation of concentration of a reactant plotted against time for a second order reaction.

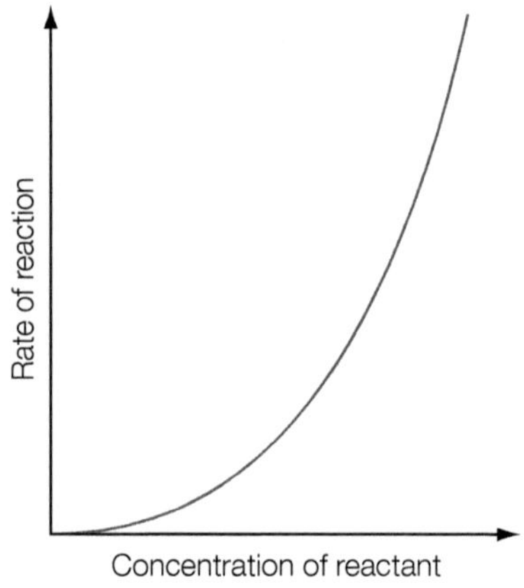

Figure 16.12 The variation of reaction rate with concentration for a second order reaction.

Test yourself

10 The rate of reaction of 1-bromopropane with hydroxide ions is first order with respect to the halogenoalkane and first order with respect to hydroxide ions.

   a) Write the rate equation for the reaction.

   b) What is the overall order of reaction?

   c) What are the units of the rate constant?

Zero order reactions

At first sight is seems odd that there can be zero order reactions. However, there are examples of reactions for which the gradient of a concentration–time graph does not change with time. The constant gradient in Figure 16.13 shows that the rate of the reaction stays the same, even though the concentration of the reactant is falling.

A reaction of this kind is zero order with respect to a reactant because the rate of reaction is unaffected by changes in the concentration of that reactant (Figure 16.14). Chemists have found a way to account for zero order reactions in terms of the mechanisms of these reactions (see Section 16.4).

In a rate equation for a zero order reaction, the concentration term for the reactant is raised to the power zero and so the rate is equal to the rate constant.

$$\text{rate} = k[\text{reactant}]^0 = k$$

Tip

Any term raised to the power zero equals 1 (see Section 1 of 'Mathematics in A Level chemistry', which you can access via the QR code for Chapter 16 on page 321). So $[\text{reactant}]^0 = 1$.

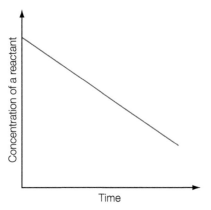

Figure 16.13 The variation of concentration of a reactant plotted against time for a zero order reaction.

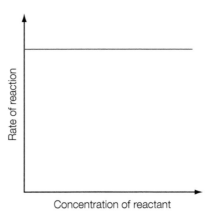

Figure 16.14 The variation of reaction rate with concentration for a zero order reaction.

Test yourself

11 Ammonia gas decomposes to nitrogen and hydrogen in the presence of a hot platinum wire. Experiments show that the reaction continues at a constant rate until all the ammonia has disappeared.

a) Sketch a concentration–time graph for the reaction.

b) Write both the balanced chemical equation and the rate equation for this reaction.

The nitration of methylbenzene is an example where the conditions can be such that the reaction is zero order with respect to the aromatic compound. The methylbenzene reacts with nitronium ions (NO_2^+) formed from nitric acid (see Section 18.1.6). The reaction creating the NO_2^+ ions is relatively slow, but as soon as the ions form they react with methylbenzene. As a result the rate is not affected by the methylbenzene concentration.

The initial-rate method

The most general method for determining reaction orders is the initial-rate method. The method is based on finding the rate immediately after the start of a reaction. This is the one point when all the concentrations are known.

The investigator makes up a series of mixtures in which all the initial concentrations are the same except one. A suitable method is used to measure the change of concentration with time for each mixture (see Section 16.2). The results are used to plot concentration–time graphs. The initial rate for each mixture is then found by the drawing tangent to the curve at the start and calculating the gradient.

The initial-rate method was used to study the reaction:

$$BrO_3^-(aq) + 5Br^-(aq) + 6H^+(aq) \rightarrow 3Br_2(aq) + 3H_2O(l)$$

The initial rate was calculated from four graphs plotted to show how the concentration of $BrO_3^-(aq)$ varied with time for different initial concentrations of reactants. The results are shown in Table 16.4.

What is:

a) the rate equation for the reaction

b) the value of the rate constant?

Table 16.4

Experiment	Initial concentration of BrO_3^-/ $mol\,dm^{-3}$	Initial concentration of Br^-/ $mol\,dm^{-3}$	Initial concentration of H^+/ $mol\,dm^{-3}$	Initial rate of reaction/ $mol\,dm^{-3}\,s^{-1}$
1	0.10	0.10	0.10	1.2×10^{-3}
2	0.20	0.10	0.10	2.4×10^{-3}
3	0.10	0.30	0.10	3.6×10^{-3}
4	0.20	0.10	0.20	9.6×10^{-3}

Notes on the method

Recall that the rate equation cannot be worked out from the balanced equation for the reaction.

First, study the experiments in which the concentration of BrO_3^- varies but the concentration of the other two reactants stays the same. How does doubling the concentration of BrO_3^- affect the rate?

Then study, in turn, the experiments in which the concentrations of first Br^- and then H^+ vary, while the concentrations of the other two reactants stay the same. How does doubling or tripling the concentration of a reactant affect the rate?

Substitute values for any one experiment in the rate equation to find the value of the rate constant, k. Take care with the units.

Answer

From experiments 1 and 2:

doubling $[BrO_3^-]_{initial}$ increases the rate by a factor of 2. So rate $\propto [BrO_3^-]^1$.

From experiments 1 and 3:

tripling $[Br^-]_{initial}$ triples the rate. So rate $\propto [Br^-]^1$

From experiments 2 and 4:

doubling $[H^+]_{initial}$ increases the rate by a factor of 4 (2^2). So rate $\propto [H^+]^2$.

The reaction is first order with respect to BrO_3^- and Br^- but second order with respect to H^+.

The rate equation is: $\quad\quad$ rate $= k\,[BrO_3^-][Br^-][H^+]^2$

Rearranging this equation, and substituting values from experiment 4:

$$k = \frac{\text{rate}}{[BrO_3^-][Br^-][H^+]^2}$$

$$= \frac{9.6 \times 10^{-3}\,\text{mol dm}^{-3}\,\text{s}^{-1}}{0.2\,\text{mol dm}^{-3} \times 0.1\,\text{mol dm}^{-3} \times (0.2\,\text{mol dm}^{-3})^2}$$

$$k = 12.0\,\text{dm}^9\,\text{mol}^{-3}\,\text{s}^{-1}$$

Test yourself

12 This data refers to the reaction of the halogenoalkane 1-bromobutane (here represented as RBr) with hydroxide ions. The results are shown in Table 16.5.

a) Deduce the rate equation for the reaction.

b) Calculate the value of the rate constant.

Experiment	[RBr]/mol dm^{-3}	[OH$^-$]/mol dm^{-3}	Rate of reaction/mol dm^{-3} s^{-1}
1	0.020	0.020	1.36
2	0.010	0.020	0.68
3	0.010	0.005	0.17

Table 16.5

13 This data refers to the reaction of halogenoalkane 2-bromo-2-methylbutane (here represented as R'Br) with hydroxide ions. The results are shown in Table 16.6.

a) Deduce the rate equation for the reaction.

b) Calculate the value of the rate constant.

Experiment	[R'Br]/mol dm^{-3}	[OH$^-$]/mol dm^{-3}	Rate of reaction/mol dm^{-3} s^{-1}
1	0.020	0.020	40.40
2	0.010	0.020	20.19
3	0.010	0.005	20.20

Table 16.6

14 Hydrogen gas reacts with nitrogen monoxide gas to form steam and nitrogen. Doubling the concentration of hydrogen doubles the rate of reaction. Tripling the concentration of NO gas increases the rate by a factor of nine.

a) Write the balanced equation for the reaction.

b) Write the rate equation for the reaction.

Clock reactions

A variant on the initial-rate method is to use a 'clock reaction', so called because the reaction is set up to produce a sudden colour change after a certain time when it has produced a fixed amount of one reactant.

The reaction of hydrogen peroxide with iodide ions in acid solution can be set up as a clock reaction:

$$H_2O_2(aq) + 2H^+(aq) + 2I^-(aq) \rightarrow I_2(aq) + 2H_2O(l)$$

A small, known amount of sodium thiosulfate ions is added to the reaction mixture, which also contains starch indicator. At first the thiosulfate ions react with any iodine, I_2, as soon as it is formed, turning it back to iodide ions, so there is no colour change. At the instant when all the thiosulfate ions have been used up, free iodine is produced and this immediately gives a deep blue–black colour with the starch.

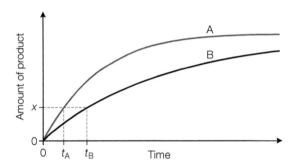

Figure 16.15 Two plots showing the formation of a product with time under different conditions. In a clock reaction the reaction mixture includes an indicator that gives a sudden colour change when the amount x of product has formed.

The experiment can be repeated with different conditions but each time with the same amount of sodium thiosulfate added. This means that the sudden colour change always happens when the same amount of iodine has been formed (represented by amount x in Figure 16.15).

Line A shows the formation of a product under one set of conditions. An amount of product x forms in time t_A. Line B shows the formation of the same product under a different set of conditions. The same amount of product x forms in the longer time t_B.

The average rate of formation of product on line $A = \dfrac{x}{t_A}$

The average rate of formation of product on line $B = \dfrac{x}{t_B}$

If x is kept the same, it follows that the average rate near the start $\propto \dfrac{1}{t}$.

This means that it is possible to use $\dfrac{1}{t}$ as a measure of the initial rate of a reaction by determining how long the reaction takes to produce the small, fixed amount of product needed for the colour change in the clock reaction.

Test yourself

15 Explain why the estimate of the initial rate of a reaction determined by a clock reaction is close to, but not equal to, the true initial rate.

Investigating the reaction of hydrogen peroxide with iodide ions by a clock reaction

The equation for the reaction of hydrogen peroxide with iodide ions in acid solution is:

$$H_2O_2(aq) + 2H^+(aq) + 2I^-(aq) \longrightarrow I_2(aq) + 2H_2O(l)$$

The rate equation for the reaction takes the form:

$$Rate = k[H_2O_2]^m[I^-]^n[H^+]^p$$

A student followed these instructions to determine the order of the reaction with respect to iodide ions.

A Use a pipette to add $10\,cm^3$ of $0.10\,mol\,dm^{-3}$ hydrogen peroxide to a clean beaker. Then use a measuring cylinder to add $25\,cm^3$ of $0.25\,mol\,dm^{-3}$ sulfuric acid to the same beaker.

B Use burettes to add to a second clean beaker: $5.0\,cm^3$ of $0.10\,mol\,dm^{-3}$ potassium iodide solution; $20.0\,cm^3$ of distilled water, $2.0\,cm^3$ of $0.050\,mol\,dm^{-3}$ sodium thiosulfate solution and $1\,cm^3$ of starch solution.

C Pour the contents of the first beaker into the second beaker, start timing and swirl the contents to mix thoroughly. Record the time (to an appropriate accuracy) taken for the blue colour of the starch–iodine complex to appear.

D Repeat steps A and C using the same quantities in step A but with different volumes of the potassium iodide solution and water in step B (making sure that the total volume of potassium iodide solution and water is $25\,cm^3$ each time).

The student's results are shown in Table 16.7.

Table 16.7

Run	Volume of 0.01 mol dm⁻³ KI(aq)/ cm³	Volume of water/ cm³	Time, t, taken for the blue colour to appear/s
1	5.0	20.0	258
2	10.0	15.0	136
3	15.0	10.0	98
4	20.0	5.0	72
5	25.0	0.0	58

Questions

1 Explain why the instructions require that the total volume of potassium iodide solution and water is $25\,cm^3$ for each run of the experiment.

2 What is the relationship between the concentration of iodide ions in each run and the volume of KI(aq)?

3 Give the equation for the reaction of sodium thiosulfate with iodine and explain why $\frac{1}{t}$ is a measure of the initial rate of reaction where t is the time for the blue colour to appear.

4 Show that under the conditions of this experiment the rate equation takes the form: rate = constant × $[I^-]^n$

5 Explain why a graph of log(rate) against $\log[I^-]$ gives a straight line with gradient n (see Section 4 in 'Mathematics in A Level chemistry', which you can access via the QR code for Chapter 16 on page 321).

6 Draw up a table with the headings shown and enter the values for each run.

Volume of 0.01 mol dm⁻³ KI(aq)/cm³	log (volume of aqueous KI)	$\frac{1}{t}$/s⁻¹	$\log\left(\frac{1}{t}\right)$

7 Plot a graph of $\log\left(\frac{1}{t}\right)$ against log (volume of aqueous KI) and hence determine the order of the reaction with respect to iodide ions.

8 What changes to the procedure would be needed to determine the order of the reaction with respect to hydrogen peroxide?

Tip

For practical guidance, refer to Practical skills sheet 8, 'Investigating reaction orders and activation energies', which you can access via the QR code for Chapter 16 on page 321.

16.4 Rate equations and reaction mechanisms

Rate equations were some of the first pieces of evidence to set chemists thinking about the mechanism of reactions. They wanted to understand why a rate equation cannot be predicted from the balanced equation for the reaction. They were puzzled that similar reactions turned out to have different rate equations.

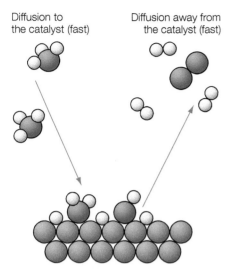

Diffusion to the catalyst (fast)　　　Diffusion away from the catalyst (fast)

Bonds breaking and new bonds forming. Rate determined by the surface area of the catalyst, which is a constant (rate-determining)

Figure 16.16 Three steps in the decomposition of ammonia gas in the presence of a platinum catalyst.

Key terms

The **mechanism** of a reaction describes how the reaction takes, place showing step by step the bonds that break and the new bonds that form.

Adsorption is a process in which atoms, molecules or ions are held on the surface of a solid.

The **rate-determining step** in a multi-step reaction is the slowest step: the one with the highest activation energy.

Figure 16.17 A one-step mechanism for the hydrolysis of 1-bromobutane.

Key term

An **S_N2 reaction** is a nucleophilic substitution reaction with a mechanism that involves two molecules or ions in the rate-determining step.

Multi-step reactions

The key to understanding the **mechanism** of a reaction was the realisation that most reactions do not take place in one step, as suggested by the balanced equation, but in a series of steps.

It is unexpected that the decomposition of ammonia gas in the presence of a hot platinum catalyst is a zero order reaction. How can it be that the concentration of the only reactant does not affect the rate? A possible explanation is illustrated in Figure 16.16.

Ammonia rapidly diffuses to the surface of the metal and is **adsorbed** onto the surface. This happens fast. Bonds break and atoms rearrange to make new molecules on the surface of the metal. This is the slowest process. Once formed, the nitrogen and hydrogen rapidly break away from the metal into the gas phase.

So there is a **rate-determining step** which can only happen on the surface of the platinum. The rate of reaction is determined by the surface area of the platinum, which is a constant. This means that the rate of reaction is a constant as long as there is enough ammonia to be adsorbed all over the metal surface. The rate is independent of the ammonia concentration.

Test yourself

16 Give an analogy from the everyday world to explain the idea of a rate-determining step. You could base your example on people getting their meals in a busy self-service canteen, or heavy traffic on a motorway affected by lane closures.

Hydrolysis of halogenoalkanes

Another puzzle for chemists was the discovery that there are different rate equations for the reactions between hydroxide ions and two isomers with the formula C_4H_9Br (see 'Test yourself' Questions 12 and 13 in Section 16.3).

Hydrolysis of a primary halogenoalkane, such as 1-bromobutane, is overall second order. The rate equation has the form: rate = $k[C_4H_9Br][OH^-]$.

To account for this chemists have suggested a mechanism showing the C−Br bond breaking at the same time as the nucleophile, OH^-, forms a C−OH bond. In this mechanism both reactants are involved in the single, rate-determining step (Figure 16.17).

| 1-bromobutane | transition state | butan-1-ol |

In this example of a substitution reaction the nucleophile is the hydroxide ion. Chemists label this mechanism S_N2, where the '2' shows that there are two molecules or ions involved in the rate-determining step.

Hydrolysis of tertiary halogenoalkanes such as 2-bromo-2-methylpropane, however, is overall first order. The rate equation has the form:

rate = $k[C_4H_9Br]$

The suggested mechanism shows the C–Br bond breaking first in a slow step to form an ionic intermediate. This is the rate-determining step that forms a tertiary carbocation. Then the nucleophile, OH⁻, rapidly forms a new bond with the positively charge carbon atom (Figure 16.18).

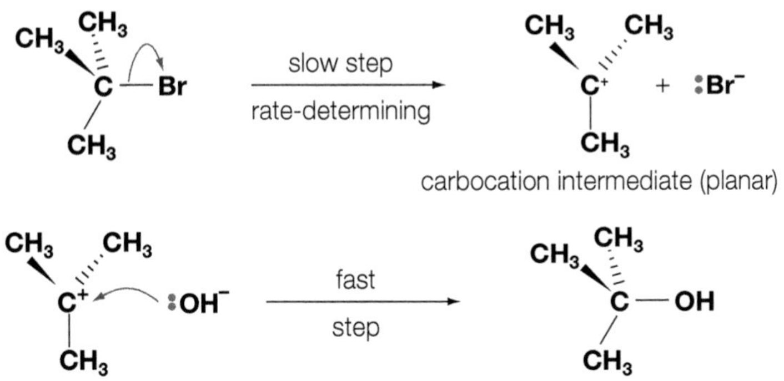

Figure 16.18 A two-step mechanism for the hydrolysis of 2-bromo-2-methylpropane.

In this example of a substitution reaction the nucleophile is also the hydroxide ion. Chemists label the mechanism S_N1, where the '1' shows that there is just one molecule or ion involved in the rate-determining step. The concentration of the hydroxide ions does not affect the rate of reaction because hydroxide ions are not involved in the rate-determining step.

What these examples show is that it is generally the molecules or ions involved (directly or indirectly) in the rate-determining step that appear in the rate equation for the reaction.

> **Key term**
>
> An S_N1 reaction is a nucleophilic substitution reaction with a mechanism that involves only one molecule or ion in the rate-determining step.

Test yourself

17 Explain, in terms of bonding, why the first step in the S_N1 mechanism is slow while the second step is fast.

18 In the proposed two-step mechanism for the reaction of nitrogen dioxide gas with carbon monoxide gas, the first step is slow while the second step is fast:

$2NO_2(g) \rightarrow NO_3(g) + NO(g)$ slow

$NO_3(g) + CO(g) \rightarrow NO_2(g) + CO_2(g)$ fast

a) What is the overall equation for the reaction?

b) Suggest a rate equation that is consistent with this mechanism.

c) What, according to your suggested rate equation, is the order of reaction with respect to carbon monoxide?

> **Tip**
>
> More evidence to suggest that the S_N1 and S_N2 mechanisms provide a correct description of the reactions comes from the study of the shapes of molecules (see Section 17.1.3).

The reaction of iodine with propanone

The reaction of iodine with propanone is another example which shows that it is not possible to deduce the rate equation from the balanced equation for the reaction.

The results of experiments, such as those outlined in Core practical 13a, suggest that the rate-determining step, for the reaction between iodine and propanone, involves propanone and hydrogen ions but not iodine.

Tip

Refer to Practical skills sheet 8, 'Investigating reaction orders and activation energies', which you can access via the QR code for Chapter 16 on page 321.

Contrast the reaction of propanone with iodine in acid conditions in Core practical 13a with the reaction forming triiodomethane under alkaline conditions (Section 17.2.6).

Core practical 13a

Investigating the rate equation for the reaction of iodine with propanone

Iodine reacts with propanone in the presence of an acid catalyst.

$$I_2(aq) + CH_3COCH_3(aq) \rightarrow CH_2ICOCH_3(aq) + H^+(aq) + I^-(aq)$$

A student first investigated the order of reaction with respect to iodine by using a titration method.

She mixed $50\,cm^3$ of $0.020\,mol\,dm^{-3}$ aqueous iodine with $50\,cm^3$ of an acidified $0.25\,mol\,dm^{-3}$ solution of propanone in a flask. Every five minutes after the start, she used a pipette to remove $10.0\,cm^3$ of the reaction mixture which she ran into an excess of sodium hydrogencarbonate solution. She then titrated the iodine remaining against a standard solution of sodium thiosulfate solution. She plotted a graph of her results, as shown in Figure 16.19.

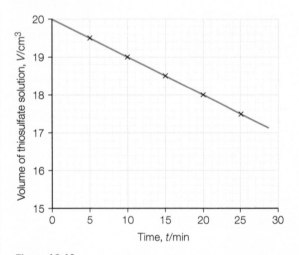

Figure 16.19

1 Why did the student add the $10\,cm^3$ samples of the reaction mixture to excess sodium hydrogencarbonate immediately after removing them from the flask?

2 How did the rate of change of iodine concentration change during the experiment?

3 What is the order of the reaction with respect to iodine?

The student carried out a second experiment, with a series of runs using the initial rate method, to check the result of the first experiment and to find the order with respect to propanone and hydrogen ions. She made up a series of mixtures of acid, propanone and water as shown in Table 16.8. Then she added a measured volume of iodine solution to each mixture and recorded the time taken for the iodine colour to disappear. Her results are shown in Table 16.8.

Table 16.8

	Run 1	Run 2	Run 3	Run 4
Volume of $2.0\,mol\,dm^{-3}$ HCl(aq)/cm^3	20.0	10.0	20.0	20.0
Volume of $2.0\,mol\,dm^{-3}$ CH_3COCH_3(aq)/cm^3	8.0	8.0	4.0	8.0
Volume of water/cm^3	0	10.0	4.0	2.0
Volume of $0.001\,mol\,dm^{-3}$ I_2(aq)/cm^3	4.0	4.0	4.0	2.0
Time, t, for iodine colour to disappear/s	115	264	243	58
Rate of reaction/cm^3 iodine solution per second	0.035	0.015	0.016	0.034

4 Explain why the shape of the line in Figure 16.19 justifies the method used in the second experiment.

5 Why was it important to measure the volumes of solutions with pipettes or burettes?

6 Show how the student arrived at the values of the rate of reaction in runs 3 and 4.

7 Show that her results confirm the value for the order of reaction with respect to iodine found in the first experiment.

8 What are the orders of reaction with respect to propanone and hydrogen ions?

9 Give the rate equation for the reaction.

10 What does the rate equation suggest about the mechanism of the reaction of propanone with iodine?

The suggested mechanism for the acid-catalysed reaction is that propanone molecules react relatively slowly to form an intermediate molecule with a double bond and an −OH group. This is followed by fast steps in which the intermediate reacts with iodine, as shown in Figure 16.20. Only the concentrations of chemical species involved the rate-determining step appear in the rate equation.

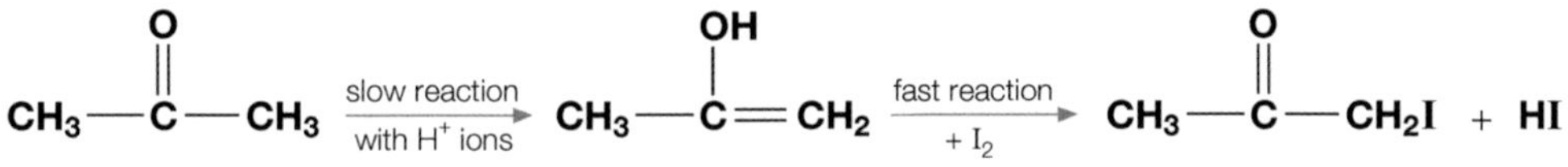

Figure 16.20 Outline of the suggested mechanism to account for the rate equation for the reaction of propanone with iodine.

This reaction shows that the form of the rate equation, the order of reaction and the value of the rate constant are all likely to be different when a catalyst is added to speed up a reaction. The reaction of iodine with propanone is an example of the way that a catalyst can change the mechanism of a reaction by combining with one of the reactants to form an intermediate. The intermediate then reacts to give the products and the catalyst is released so that it is freed up to interact with further reactant molecules and continue the reaction.

Test yourself

19 Why do chemists use the term 'enol' to describe the intermediate formed during the reaction of iodine with propanone?

20 Why does the formula for hydrogen ions appear in the rate equation for the iodination of propanone but not as a reactant in the balanced equation for the reaction?

21 Bromine reacts with propanone in a similar way to iodine. The mechanism for the reaction is the same. Explain why bromine reacts with propanone at the same rate as iodine under similar conditions.

16.5 The effect of temperature on reaction rates

Raising the temperature often has a dramatic effect on the rate of a reaction, especially reactions that involve the breaking of strong covalent bonds. This explains why the practical procedure for most organic reactions involves heating the reaction mixtures. With the help of collision theory it is possible to make predictions about the effect of temperature changes on rates.

The constant k in a rate equation is only a constant at a specified temperature. Generally, the value of the rate constant increases as the temperature rises and this means that the rate of reaction increases.

Collision theory accounts for the effect of temperature on reaction rates by supposing that chemical changes pass through a **transition state**. The transition state is at a higher energy than the reactants so there is an energy

Key term

A **transition state** is the state of the reacting atoms, molecules or ions when they are at the top of the activation energy barrier for a reaction step.

barrier or activation energy as shown by the **reaction profile** in Figure 16.21. Reactant molecules must collide with enough energy to overcome the activation energy barrier. This means that the only collisions which lead to reaction are those with enough energy to break existing bonds and allow the atoms to rearrange to form new bonds in the product molecules.

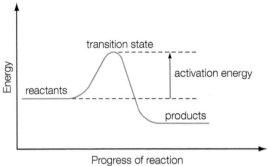

Figure 16.21 Reaction profile showing the activation energy for a reaction.

Activation energies account for the fact that reactions go much more slowly than would be expected if every collision in a mixture of chemicals led to reaction. Only a very small proportion of collisions bring about chemical change because molecules can only react if they collide with enough energy to overcome the energy barrier. For many reactions, at around room temperature only about 1 in 10^{10} molecules have enough energy to react.

The Maxwell–Boltzmann curve shows the distribution of the kinetic energies of molecules. As Figure 16.22 shows, the proportion of molecules which can collide with energies greater than the activation energy is small at around 300 K.

Figure 16.22 The Maxwell-Boltzmann distribution of molecular kinetic energies in a gas at two temperatures. The modal energy gets higher as the temperature rises. The area under the curve gives the total number of molecules. This does not change as the temperature rises so the peak height falls as the curve widens.

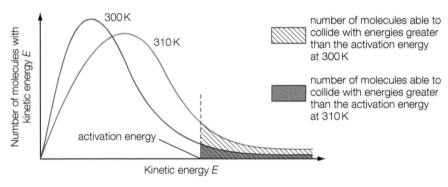

The shaded areas in Figure 16.22 are a measure of the proportions of molecules able to collide with enough energy for a reaction at two temperatures. The area is bigger at a higher temperature. So at a higher temperature there are more molecules with enough energy to react when they collide, and the reaction goes faster.

The effect of temperature changes on rate constants

The Swedish physical chemist Svante Arrhenius (1859–1927) found that he obtained a straight line if he plotted the natural logarithm of the rate constant for a reaction against $1/T$ (the inverse of the absolute temperature). His equation to describe the relationship between rate constant and temperature is:

$$k = A\,\mathrm{e}^{-E_a/RT}$$

where k is the rate constant, R is the gas constant, T the absolute temperature, E_a is the activation energy for the reaction and A is another constant.

After taking natural logarithms of both sides, the Arrhenius equation takes this form:

$$\ln k = \frac{-E_a}{R} \times \frac{1}{T} + \text{constant}$$

A useful rough guide, based on the Arrhenius equation, is that at about room temperature the value of the rate constant doubles for each 10 degree rise in temperature if the activation energy for the reaction is about $50\,kJ\,mol^{-1}$.

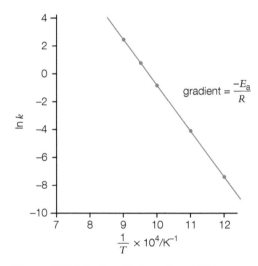

Figure 16.23 A plot of $\ln k$ against $1/T$ for a reaction. The activation energy can be calculated from the gradient. The general equation for a straight line is $y = mx + c$, where m is the gradient and c is the intercept on the y-axis. Here c is the constant in the Arrhenius equation and the gradient m is $-E_a/R$.

Test yourself

23 Table 16.9 shows the value of the rate constant for the reaction of a diazonium salt with water at four temperatures.

Temperature/K	Rate constant/$10^{-5}\,s^{-1}$
278	0.15
298	4.1
308	20
323	140

Table 16.9

a) What do the units of the rate constant tell you about the form of the rate equation?

b) What do the values tell you about the effect of temperature on the rate of the reaction?

c) What is the effect of a 10 degree rise in temperature on the rate of the reaction?

24 Show that the Arrhenius equation signifies that:

a) the higher the temperature, the greater the value of k and hence the faster the reaction

b) a reaction with a relatively high activation energy has a relatively small rate constant.

25 The rate constant for the decomposition of hydrogen peroxide is $4.93 \times 10^{-4}\,s^{-1}$ at $295\,K$. It increases to $1.40 \times 10^{-3}\,s^{-1}$ at $305\,K$. Estimate the activation energy for the reaction.

Finding the activation energy of a reaction

A student used the clock method for determining initial rates to find the activation for the oxidation of iodide ions by peroxodisulfate(VI) ions:

$$S_2O_8^{2-}(aq) + 2I^-_2(aq) \rightarrow 2SO_4^{2-}(aq) + I_2(aq)$$

In a series of experiments he kept the concentrations of the reactants constant while varying temperature of the reaction mixture over a range of values.

The reaction mixture included a small, measured amount of sodium thiosulfate and some starch solution, as shown in Figure 16.24.

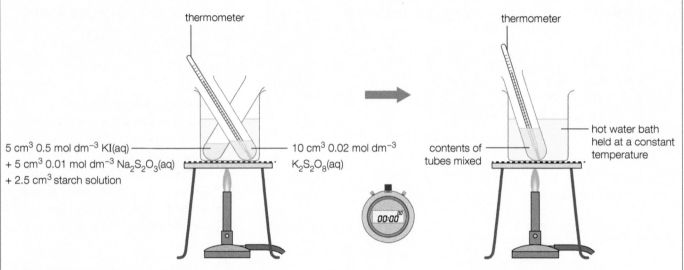

5 cm³ 0.5 mol dm⁻³ KI(aq) + 5 cm³ 0.01 mol dm⁻³ $Na_2S_2O_3$(aq) + 2.5 cm³ starch solution

10 cm³ 0.02 mol dm⁻³ $K_2S_2O_8$(aq)

contents of tubes mixed

hot water bath held at a constant temperature

Figure 16.24 Outline of the experimental procedure.

Table 16.10 shows the student's results from a series of runs with temperatures in the range 30–51 °C.

Table 16.10

Temperature/°C	30	36	39	45	51
Temperature/K	303	309			
Time, t, for the blue colour to appear/s	204	138	115	75	55
ln(1/t)		−5.32	−4.93		
$\frac{1000}{T}$/K⁻¹		3.30	3.24		

1 Why did the tube containing potassium iodide solution also include sodium thiosulfate and starch?

2 Why was the solution of potassium peroxodisulfate(VI) measured into a separate tube at the start, and when should the solutions be mixed and timing started?

3 The table includes some calculated values for ln(1/t) and 1000/TK⁻¹ corresponding to the variables in the Arrhenius equation. Copy and complete the table by calculating the values to include in the empty cells.

4 What is the advantage of calculating 1000/T rather than 1/T?

5 Plot a graph of ln(1/t) against 1000/T.

6 Use the graph to calculate the activation energy for the reaction.

Tip

Refer to Practical skills sheet 8, 'Investigating reaction orders and activation energies', which you can access via the QR code for this chapter on page 321.

Exam practice questions

1 Hydrogen peroxide oxidises iodide ions to iodine in the presence of hydrogen ions. The other product is water. The reaction is first order with respect to hydrogen peroxide, first order with respect to iodide ions, but zero order with respect to hydrogen ions.
   a) Write a balanced equation for the reaction. *(1)*
   b) Write a rate equation for the reaction. *(2)*
   c) What is the overall order of the reaction? *(1)*
   d) A proposed mechanism for the reaction involves three steps:
   $$H_2O_2 + I^- \rightarrow H_2O + IO^-$$
   $$H^+ + IO^- \rightarrow HIO$$
   $$HIO + H^+ + I^- \rightarrow I_2 + H_2O$$
   Which step is likely to be the rate-determining step and why? *(2)*

2 The data in the table refers to the decomposition of hydrogen peroxide, H_2O_2.

Time/10^3 s	[H_2O_2]/10^{-3} mol dm^{-3}
0	20.0
12	16.0
24	13.1
36	10.6
48	8.6
60	6.9
72	5.6
96	3.7
120	2.4

   a) Plot a concentration–time graph for the decomposition reaction. *(3)*
   b) Read off three half-lives from the graph and show that this is a first-order reaction. *(3)*
   c) Draw tangents to the curve in your graph at four different concentrations and hence find the gradient of the curve at each point. *(4)*
   d) Plot a graph of rate against concentration using your results from (c) and hence find the value for the rate constant at the temperature of the experiment. *(4)*

3 The results in the table come from a study of the rate of reaction of iodine with a large excess of hex-1-ene dissolved in ethanoic acid.

Time/10^3 s	[I_2]/10^{-3} mol dm^{-3}
0	20.0
1	15.6
2	12.8
3	11.0
4	9.4
5	8.3
6	7.5
7	6.8
8	6.2

   a) Plot a concentration–time graph and show that the half-life is not constant. *(5)*
   b) From your graph, find the rate of reaction at four concentrations. *(3)*
   c) Use your results from part (b) to plot a graph of log (rate) against log (concentration) to find the order of the reaction with respect to iodine. *(6)*

4 Two gases X and Y react according to this equation:

   $$X(g) + 2Y(g) \rightarrow XY_2(g)$$

   This reaction was studied at 400 K, giving the results shown in the table below.

Experiment number	Initial concentration of X/mol dm^{-3}	Initial concentration of Y/mol dm^{-3}	Initial rate of formation of XY$_2$/ mol dm^{-3} s^{-1}
1	0.10	0.10	0.0001
2	0.10	0.20	0.0004
3	0.10	0.30	0.0009
4	0.20	0.10	0.0001
5	0.30	0.10	0.0001

   a) What is the order of the reaction with respect to:
   i) X
   ii) Y?
   Explain your answers. *(4)*
   b) Write a rate equation for the reaction of X with Y. *(2)*
   c) Use the results of the first experiment to calculate a value of the rate constant and give its units. *(2)*

d) Suggest a possible mechanism for the reaction. *(3)*

e) Explain why chemists are interested in determining rate equations and measuring rate constants. *(3)*

5 The table below shows data obtained for the reaction between nitrogen monoxide and hydrogen at 750 °C.

$$2NO(g) + 2H_2(g) \rightarrow N_2(g) + 2H_2O(g)$$

Experiment	1	2	3
Initial concentration of hydrogen/mol dm^{-3}	0.012	0.012	0.024
Initial concentration of NO/ mol dm^{-3}	0.002	0.004	0.002
Initial rate/mol dm^{-3} s^{-1}	1.20	2.40	4.80

a) Explain the advantage of investigating reaction kinetics by measuring initial rates. *(2)*

b) What is the order of the reaction with respect to:
  i) hydrogen *(1)*
  ii) nitrogen monoxide? *(1)*

c) Give the rate equation for the reaction. *(1)*

d) Calculate the value of the rate constant and give the units. *(2)*

e) Why is it not possible to deduce the reaction orders from the balanced chemical equation? *(2)*

f) i) What happens to the value of the rate constant for this reaction as the temperature rises above 750 °C? *(1)*
  ii) Explain why the rate constant changes as it does when the temperature increases. *(4)*

6 The table below shows the results of a series of experiments to determine the activation energy for the oxidation of iodide ions by peroxodisulfate(VI) ions in the presence of iron(III) ions.

$$S_2O_8{}^{2-}(aq) + 2I^-(aq) \rightarrow 2SO_4{}^{2-}(aq) + I_2(aq)$$

The results were obtained by the 'clock' method. Each reaction mixture included a small, measured amount of aqueous sodium thiosulfate and a few drops of starch solution.

Temperature, T/K	288	292	299	308	315
Time, t, for the blue colour to appear/s	10.0	7.0	5.0	3.5	2.5

a) Explain why each reaction mixture included a small amount sodium thiosulfate solution and starch solution. *(3)*

b) Analyse the results and plot a graph to find a value for the activation energy in the presence of iron(III) ions. (The gas constant $R = 8.31\,\text{J K}^{-1}\,\text{mol}^{-1}$.) *(8)*

c) In the absence of iron(III) ions the activation energy for the reaction is $52.9\,\text{kJ mol}^{-1}$. Suggest an explanation for the effect of adding iron(III) ions. *(4)*

7 Chlorate(I) ions disproportionate on heating an aqueous solution of sodium chlorate(I). The rate equation is:
$$\text{Rate} = k[\text{ClO}^-]^2$$

a) Write an equation for the disproportionation of chlorate(I) ions and show changes in the oxidation states of chlorine. *(3)*

b) What is the effect on the rate of reaction of halving the concentration of chlorate(I) ions if all other conditions are kept the same? *(1)*

c) Suggest a mechanism for the reaction that is consistent with the rate equation. Give your reasoning. *(5)*

d) Suggest what evidence you might look for to test whether your proposed mechanism is correct. *(2)*

8 a) Explain, with the help of sketch graphs, how the Maxwell–Boltzmann distribution can be used to explain the effect of temperature on reaction rates. *(4)*

b) Calculations based on theory show that the rates of chemical reactions increase with a rise in temperature far more than would be predicted just from the increase in the rate of collisions between molecules. How can this be explained? *(3)*

c) The rate constants (k) for the decomposition of hydrogen iodide at different temperatures are given in the table. Plot a graph and use it to determine the activation energy for the reaction. (The gas constant $R = 8.31\,\text{J K}^{-1}\,\text{mol}^{-1}$.) *(7)*

Rate constant, k/dm^{-3} mol^{-1}s^{-1}	Temperature, T/K
3.74×10^{-9}	500
6.65×10^{-6}	600
1.15×10^{-3}	700
7.75×10^{-2}	800

17.1.1 Isomerism

> **Tip**
>
> The first section of this chapter revisits ideas about isomerism first introduced in Chapters 6.1 and 6.2 of Student Book 1. Structural isomerism and the *E/Z* form of stereoisomerism were introduced in the first year of the A Level course. This chapter expands on what you already know about stereoisomerism and introduces chirality and optical isomerism. Some of the 'Test yourself' questions are designed to help you revise ideas from the first year of the A Level course.

Key terms

Structural isomerism occurs where compounds have the same molecular formula but different structural formulas.

Stereoisomerism occurs where molecules have the same structural formula but the atoms are arranged differently in space.

Figure 17.1.1 Caraway seeds (left) and spearmint leaves (right). The compound carvone is largely responsible for the different tastes and smells of caraway and spearmint. There are two forms of carvone molecules (see Figure 17.1.15). These forms have the same formula and structure but subtly different shapes and so different smells and tastes. Chemists describe them as optical isomers (see Section 17.1.3).

If two molecules have the same molecular formula, but a different arrangement of their atoms, they are isomers. These isomers are distinct compounds with different physical properties and, in most cases, different chemical properties. Isomers and isomerism occur most commonly with carbon compounds because of the way in which carbon atoms can form chains and rings.

There are two ways in which the atoms can be rearranged to give isomers.

- The atoms are joined together in a different order forming different structures. This is called **structural isomerism** (Figure 17.1.2).
- The atoms are joined together in the same order, but they occupy different positions in space. This is called **stereoisomerism**.

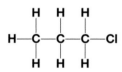

1-chloropropane

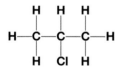

2-chloropropane

Figure 17.1.2 The two structural isomers of C_3H_7Cl.

Figure 17.1.3 shows how the two different types of isomerism are further divided. Structural isomerism can be divided into three different types: **chain isomerism**, **position isomerism** and **functional group isomerism**.

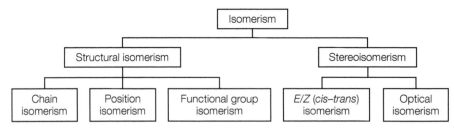

Figure 17.1.3 A 'family tree' showing the relationship between different forms of isomerism.

There are two different types of stereoisomerism: *E/Z* (or *cis–trans*) isomerism and optical isomerism. In both these forms of stereoisomerism, the stereoisomers have the same molecular formula and the same structural formula, but different three-dimensional shapes in which their atoms occupy different positions in space.

Test yourself

1 Draw the structures and name the chain isomers of 2,2-dimethylbutane.

2 Draw the structures and name the position isomers of 1-bromopentane.

3 Draw the structures and name two functional group isomers with the molecular formula $C_4H_{10}O$.

4 There are three isomers with the formula C_5H_{12}. Their boiling temperatures are: 10 °C, 28 °C and 36 °C. Draw the structures of the three compounds. Match the structures with the boiling temperatures and justify your answer.

17.1.2 *E/Z* isomerism

The traditional system for naming the isomers of alkenes, in which the same groups are arranged differently, is to name them as *cis* or *trans*. This is illustrated by the sex attractant, bombycol, secreted by female silk moths (Figures 17.1.4 and 17.1.5). This messenger molecule strongly attracts male moths of the same species. Analysis shows that two double bonds help to determine the shape of the compound. Chemists are interested in pheromones because they offer an alternative to pesticides for controlling damaging insects. By baiting insect traps with sex attractants it is possible to capture large numbers of insects before they mate.

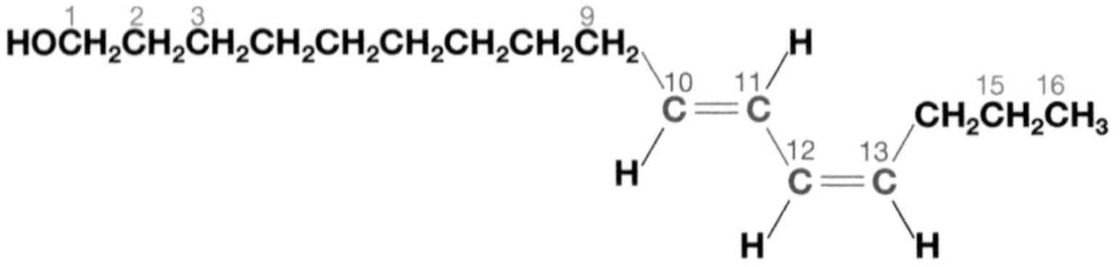

Figure 17.1.4 Silk moths and cocoons.

$$\overset{1}{HOCH_2}\overset{2}{CH_2}\overset{3}{CH_2}CH_2CH_2CH_2CH_2\overset{9}{CH_2}$$

Figure 17.1.5 The sex attractant bombycol, which is hexadeca-*trans*,10-*cis*,12-dien-1-ol.

However, there are many examples where the *cis–trans* system is not easily applied and the *E/Z* system was developed to name these more complex molecules. The advantage of the *E/Z* system is that it always works, whereas the *cis–trans* system can break down in some cases.

Follow these steps in applying the *E/Z* naming system:

1 Look at the atoms bonded to the first carbon atom in the C=C bond. The atom with the higher atomic number has the higher priority.

2 If two atoms with the same atomic number, but in different groups, are attached to the first carbon atom, then the next bonded atom is taken into account. Thus, CH_3CH_2- has precedence over CH_3-.

3 Similarly, identify the group with the higher priority of the two attached to the second carbon atom in the C=C bond.

4 If the two groups of higher priority are on the same side of the double bond, the isomer is designated *Z*-, but if the two groups of highest priority are on opposite sides of the double bond, the isomer is designated *E*- (Figure 17.1.6).

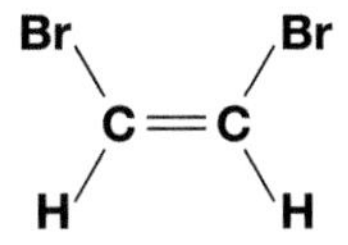

Tip

Z comes from the German word *zusammen*, meaning together.
E comes from the German word *entgegen*, meaning opposite.

Z-1,2-dibromoethene
melting temperature −53°C
boiling temperature 110°C

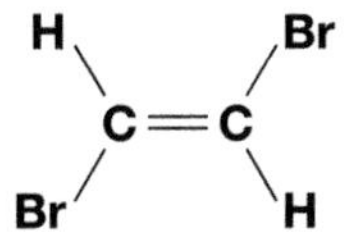

E-1,2-dibromoethene
melting temperature −9°C
boiling temperature 108°C

Figure 17.1.6 The *E* and *Z* isomers of 1,2-dibromoethene are distinct compounds with different melting temperatures and different boiling temperatures. The relative atomic masses of bromine atoms are higher than the relative atomic masses of hydrogen atoms so they have the higher priority.

Tip

In many cases a compound classed as *trans* in one system is *E* in the other and a compound classed as *cis* is *Z* in the other, but there are examples where this is **not** the case.

Test yourself

5 Draw the *E/Z* isomers of the following compounds and name them using the *E/Z* system. State also whether each structure is a *cis* or a *trans* isomer.

  a) pent-2-ene

  b) 2-bromobut-2-ene

  c) 1-chloro-2-methylbut-1-ene

6 Draw the skeletal formula of:

  a) (1*E*,4*Z*)-1,5-dichlorohexa-1,4-diene

  b) (*E*)-3-methyl-4-propyloct-3-ene.

17.1.3 Chirality and optical isomerism

Mirror-image molecules

Every molecule has a mirror image. Generally, the mirror image of a molecule can be turned around to show that it is identical to the original molecule. Sometimes, however, a molecule and its mirror image are not quite the same. The molecule and its mirror image cannot be superimposed.

Figure 17.1.7 shows left and right hands. If a mirror is placed vertically between the two hands, the reflection of the left hand in the mirror is the same as the right hand. Each hand is the mirror image of the other. But the right hand does not match the left when placed on top of it (without turning either over) so the right hand (the mirror image of the left) cannot be superimposed on the left. The hands are non-superimposable.

A **chiral molecule**, like a hand, cannot be superimposed on its mirror image. The word 'chiral' (pronounced 'kiral') comes from the Greek for 'hand'.

Many everyday objects are chiral. Figure 17.1.8 shows some objects which are chiral and some which are not.

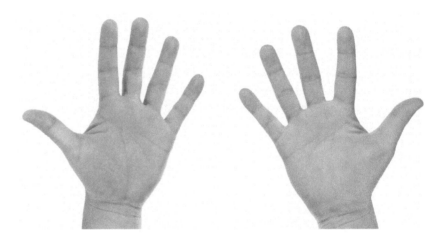

Figure 17.1.7 Left and right hands are mirror images of each other.

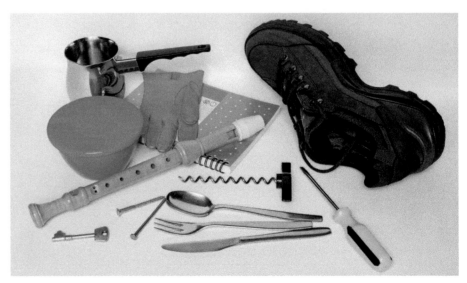

Figure 17.1.8 Everyday objects – chiral or not?

The most common chiral compounds are organic molecules in which there is a single chiral centre due to a carbon atom attached to four different atoms or groups. These carbon atoms are said to be **asymmetric**.

Figure 17.1.9 shows the two structures of lactic acid, which form when milk turns sour. The two molecules each have the same four atoms or groups attached to their central carbon atom: a CH_3- group, an $-OH$ group, a $-COOH$ group and an $H-$ atom. However, it is impossible to superimpose the mirror images of lactic acid. No matter how the molecules are rotated, it is not possible to get the two to look identical with groups and atoms in the same positions in space.

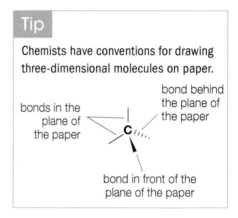
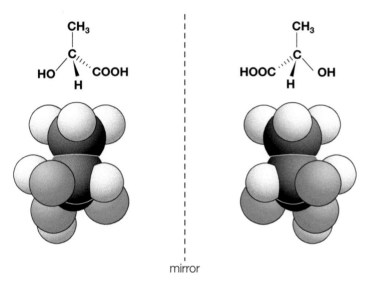

mirror

Figure 17.1.9 Molecules of lactic acid (2-hydroxypropanoic acid) are chiral. It is not possible to superimpose the two mirror-image molecules.

The two forms of lactic acid behave identically in all their chemical reactions and all their physical properties except for their effect on polarised light. This optical property is the only way of telling the two forms of lactic acid apart. So, chemists call them **optical isomers**. The word 'enantiomers' is also used to describe mirror–image molecules that are optical isomers. The word 'enantiomer' comes from a Greek word meaning 'opposite'.

Key terms

An **asymmetric** carbon atom is joined to four different groups or atoms.

Asymmetric molecules are molecules with no centre, axis or plane of symmetry. Asymmetric molecules are chiral and exist in mirror-image forms. Any carbon atom with four different groups or atoms attached to it is asymmetric and chiral.

Optical isomers or **enantiomers** occur in pairs made up of a chiral molecule and its non-superimposable mirror image. One mirror-image form rotates the plane of plane-polarised light clockwise. The other form rotates the plane of plane-polarised light anticlockwise.

Optical isomerism and polarised light

A light beam becomes polarised after passing through a sheet of Polaroid®, the material used to make some sunglasses. The Polaroid prevents vibrations of the light waves in all but one plane. So, in polarised light, all the waves are vibrating in the same plane (Figure 17.1.10).

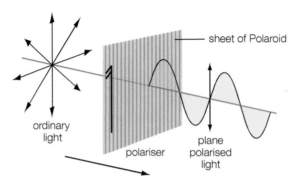

Figure 17.1.10 Ordinary light and polarised light.

Light is said to be plane polarised after passing through a sheet of Polaroid. If the polarised light is then directed at a second sheet of Polaroid, all the polarised beam passes through if the second sheet of Polaroid is aligned in the same way as the first (Figure 17.1.11a). However, no light gets through if the second sheet is rotated through 90° relative to the first (Figure 17.1.11b).

Figure 17.1.11 The effect of a second sheet of Polaroid on polarised light.

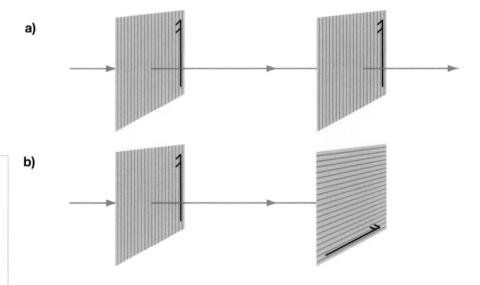

When polarised light passes through a solution of just one of a pair of optical isomers, it rotates the plane of polarisation. One isomer rotates the plane of plane-polarised light clockwise. This is named the + isomer. The other isomer rotates the plane of plane-polarised light anticlockwise and this is the − isomer.

When polarised light passes through a solution containing equal amounts of both optical isomers, the effects cancel so this solution is optically inactive. Solutions of this type are described as containing a **racemic mixture**.

For accurate results, chemists measure the rotations with monochromatic light (light of one colour or frequency) in an instrument called a polarimeter (Figure 17.1.12).

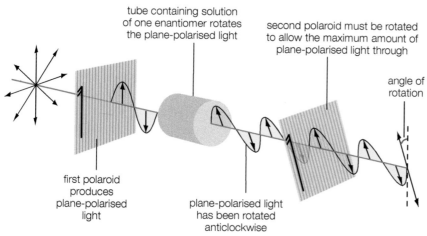

tube containing solution of one enantiomer rotates the plane-polarised light

second polaroid must be rotated to allow the maximum amount of plane-polarised light through

angle of rotation

first polaroid produces plane-polarised light

plane-polarised light has been rotated anticlockwise

Figure 17.1.12 The effect of passing plane-polarised light through a solution of a chiral compound.

Test yourself

7 Identify the chiral objects in Figure 17.1.8.

8 With the help of molecular models, decide which of the following molecules are chiral:

NH_3, CH_2Cl_2, CH_2ClBr, $CH_3CHClBr$, $CH_3CH(NH_2)COOH$.

9 Which of these alcohols can exist as optical isomers?

butan-1-ol, butan-2-ol, pentan-1-ol, pentan-2-ol, pentan-3-ol

10 Look at the upper representations of lactic acid in Figure 17.1.9. Using the same convention, draw three-dimensional representations of the two enantiomers of 2-chlorobutane, showing that they are mirror images.

17.1.4 Optical isomerism and reaction mechanisms

The optical activity of the reactants and products of organic reactions can help chemists to determine the mechanisms of reactions. This is illustrated by the outcomes of the S_N1 and S_N2 mechanisms in nucleophilic substitution reactions (see Section 16.4). This information provides additional evidence for the mechanism.

During the one-step S_N2 mechanism, the three groups that remain attached to the central carbon atom are turned inside out. The molecule is inverted like an umbrella in a high wind (Figure 17.1.13).

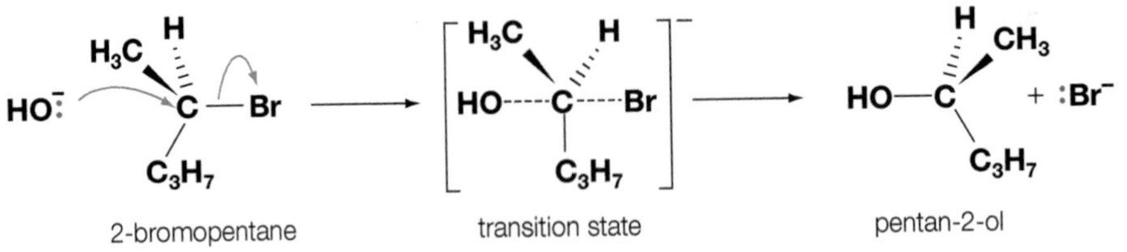

Figure 17.1.13 Nucleophilic substitution by the S_N2 mechanism with a molecule that is chiral.

2-bromopentane   transition state   pentan-2-ol

This means that an optically active halogenoalkane gives rise to an optically active alcohol if substitution takes place by the S_N2 mechanism.

> ### Tip
>
> There no simple relationship between the three-dimensional shape of a chiral compound and the direction that it rotates polarised light. What this means is that molecular inversion of the + isomer of a halogenoalkane during an S_N2 reaction does give an optical isomer of the alcohol, but it is not possible to predict whether it will be the + or the − isomer.

In the two-step S_N1 mechanism, a planar intermediate is formed after the first step. However, attack by the nucleophile during the second step can happen with equal probability from either above or below the planar intermediate. The result is that starting with one optical isomer of a halogenoalkane leads to a product that is a racemic mixture of the two forms of the chiral alcohol. This means that the product is optically inactive (Figure 17.1.14).

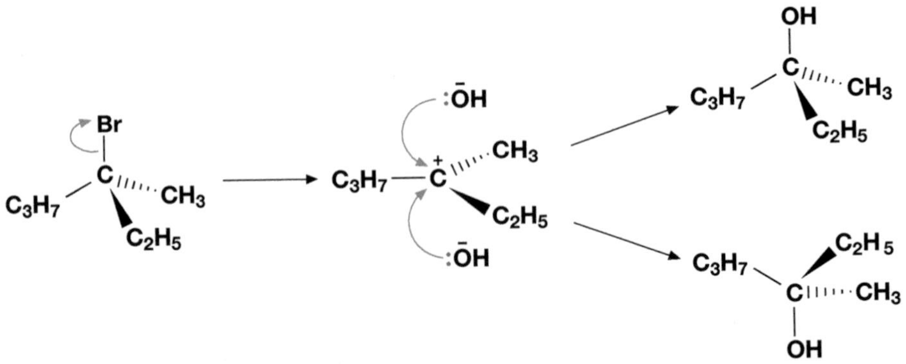

Figure 17.1.14 Hydrolysis of an optically active halogenoalkane by the S_N1 mechanism. The product is a mixture of equal amounts of the two optical isomers of the alcohol. This is a racemic mixture.

Test yourself

11 Suggest explanations to account for the fact that the reaction of iodide ions with an optically active isomer of 3-chloro-3-methylhexane gives:

   a) a mixture of the two optical isomers of 3-iodo-3-methylhexane

   b) a product mixture which is slightly optically active.

12 Account for the fact that the reaction of 2-bromooctane with sodium hydroxide in an aqueous solvent is stereospecific: (+) 2-bromooctane reacts to form (−) octan-2-ol while (−) 2-bromooctane gives (+) octan-2-ol.

Chirality and living things

Human senses are sensitive to molecular shape. The optical isomers of some molecules have different tastes and smells (Figure 17.1.15).

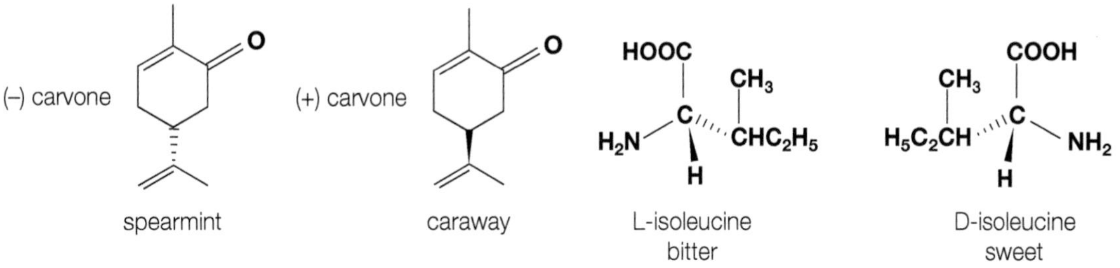

Figure 17.1.15 Optical isomers with differing tastes and smells.

What is true of the sensitive cells in the nose and on the tongue is also true of most of the rest of the human body. Living cells are full of messenger and carrier molecules that interact selectively with the active sites and receptors in other molecules such as enzymes. These messenger and carrier molecules are all chiral and the body works with only one of the mirror-image forms. This is particularly true of amino acids and proteins (see Chapter 18.2).

The chemists who synthesise and test new drugs have to pay close attention to chirality, as molecular shape can also subtly alter the physiological effects of drugs. Perhaps the most tragic example of this is thalidomide (Figure 17.1.16). This drug was first used in 1957 as a mild sedative that reduced morning sickness in early pregnancy. It was banned in 1961 after children were born with stunted and distorted limbs and evidence showed that thalidomide was responsible. The drug had been produced and sold as a racemic mixture. It is now known that one of the optical isomers was beneficial and harmless, while the other enantiomer was toxic. This tragedy caused many countries to introduce much stricter rules for the testing of new drugs before licences are granted.

1 Identify the functional groups in:
   a) carvone
   b) isoleucine.
2 Identify the chiral centres in:
   a) isoleucine
   b) carvone
   c) thalidomide.
3 Explain why the amino acid alanine, CH_3CHNH_2COOH, has optical isomers while the amino acid glycine, CH_2NH_2COOH, does not.
4 Suggest reasons why chemists working on new drugs need to develop effective methods to separate optical isomers of new compounds or methods to synthesise selectively each of the pairs of isomers.
5 In the last few years thalidomide has been discovered to be effective in the treatment of leprosy, some AIDS/HIV-related conditions and also of certain cancers. Should the use of this drug be allowed despite the risk?

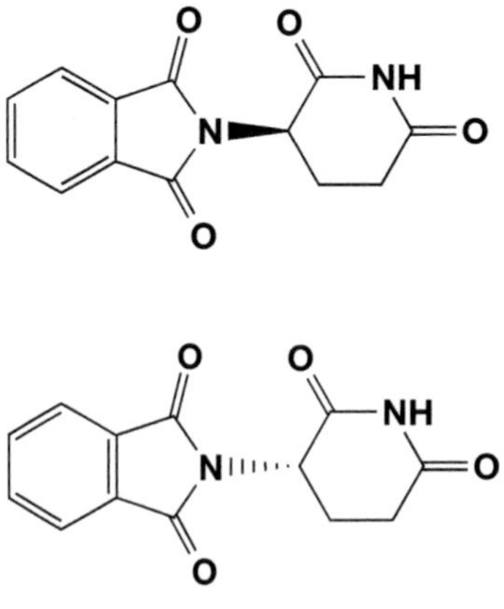

Figure 17.1.16 The two enantiomers of thalidomide.

Exam practice questions

1 Draw the displayed formulae of all the isomers of the following, stating which types of isomerism are involved:
   a) C_3H_7Cl (2)
   b) $C_2HFClBr$ (6)
   c) C_2H_4ClBr. (3)

2 a) State what is meant by a 'chiral centre' in a molecule. (1)
   b) Explain why a chiral centre in a molecule gives rise to optical isomers. (2)
   c) State which of the following compounds can exist as optical isomers. (1)
   $CH_2(NH_2)COOH$ CH_2OHCH_2COOH
   $CH_3CHOHCOOH$
   d) Explain why a racemic mixture of optical isomers has no effect on the plane of plane-polarised light. (2)
   e) Salbutamol is a drug used to relieve the symptoms of asthma.

   i)   Deduce the molecular formula of salbutamol. (1)
   ii) Identify any chiral centres in the salbutamol molecule. (1)

3 Explain the term 'stereoisomerism' and describe the different types of stereoisomerism. Illustrate your answer with suitable examples. (15)

4 The halogenoalkane 2-bromobutane has optical isomers. It reacts with aqueous sodium hydroxide.
   a) State why 2-bromobutane has optical isomers. (1)
   b) Explain why the reaction of an optical isomer of 2-bromobutane with aqueous sodium hydroxide at 25 °C gives a product that is also optically active. (3)

c) Suggest an explanation for the fact that the reaction of an optical isomer of 2-bromobutane with aqueous sodium hydroxide at 80 °C gives a product showing very little optical activity. (5)
d) If the sodium hydroxide reacts in ethanolic solution with 2-bromobutane, a mixture of three isomeric alkenes is formed. Draw and name these three isomers. (3)

5 a) Name and draw a mechanism for the reaction of pent-1-ene with HBr to form the major product. (5)
   b) Explain why major and minor products are formed. (3)
   c) Explain why the major product obtained in the reaction shows no optical activity. (4)

6 a) Draw four non-cyclic functional group isomers of C_3H_6O. (4)
   b) Describe how you could distinguish between them using:
      i)   test-tube reactions (give the reagents used and the observations you would make) (9)
      ii) infrared spectroscopy, excluding the fingerprint region (refer to the infrared spectroscopy data in the Edexcel Data booklet). (4)
   c) Draw two cyclic isomers of C_3H_6O and comment on their relative stabilities. (4)

7 P is a neutral compound with the formula $C_{10}H_{18}O_2$. P slowly dissolves on boiling with aqueous sodium hydroxide to produce ethanol and a solution of the sodium salt of Q. Q is obtained by acidifying this solution. The formula of Q is $C_6H_{10}O_4$. The solid Q decomposes at its melting temperature to give a racemic mixture of isomers of R with the formula $C_5H_{10}O_2$.

Suggest an explanation for these observations and propose structures for P, Q and R. (8)

Carbonyl compounds

17.2

17.2.1 The carbonyl group

The carbonyl group consists of the C=O bond. In aldehydes and ketones, which are known as the carbonyl compounds, only carbon or hydrogen atoms are attached to the carbon atom of the carbonyl group. However, the carbonyl group is also present in carboxylic acids (RCOOH) and their derivatives, acyl chlorides (RCOCl), esters (RCOOR) and amides ($RCONH_2$). In these compounds the carbon of the carbonyl group is also attached to another electronegative atom. These electronegative atoms modify the properties of the carbonyl group and for this reason carboxylic acids and their derivatives are treated as separate functional groups (see Chapter 17.3 and Chapter 18.2).

As expected for compounds containing a double bond, the characteristic reactions of carbonyl compounds are addition reactions. The C=O bond is polar because oxygen is highly electronegative. As a result, the mechanism of addition to carbonyl compounds (Section 17.2.5) is different from the mechanism of addition to alkenes.

17.2.2 Aldehydes

Names and structures

Aldehydes are carbonyl compounds in which a carbonyl group (C=O) is attached to a hydrogen atom and a hydrocarbon group, or in the case of the first aldehyde, methanal, to two hydrogen atoms. So the carbonyl group in aldehydes is at the end of a carbon chain. The names are based on the alkane with the same carbon skeleton with the ending changed from **-ane** to **-anal**.

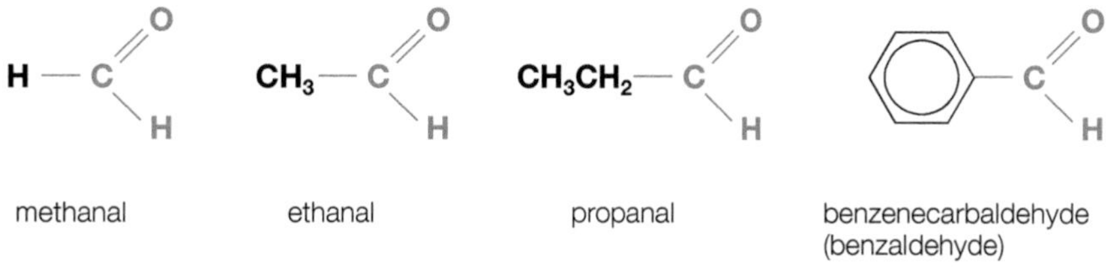

methanal   ethanal   propanal   benzenecarbaldehyde (benzaldehyde)

Figure 17.2.1 Structures and names of aldehydes. The —CHO group is the functional group that gives aldehydes their characteristic reactions.

> **Tip**
>
> Because the aldehyde functional group is always on the end carbon in a chain, no number is needed in the name of an aldehyde to show where this functional group is - it always includes carbon number 1. Always write the aldehyde group as —CHO. Writing —COH is unconventional and easily leads to confusion with alcohols.

> **Tip**
>
> Sections 17.2.2–17.2.4 of this chapter revisit and build on ideas about alcohols and their oxidation first introduced in Chapter 6.3 of Student Book 1. Some of the 'Test yourself' questions are also designed to help revise ideas from the first year of the A Level course.

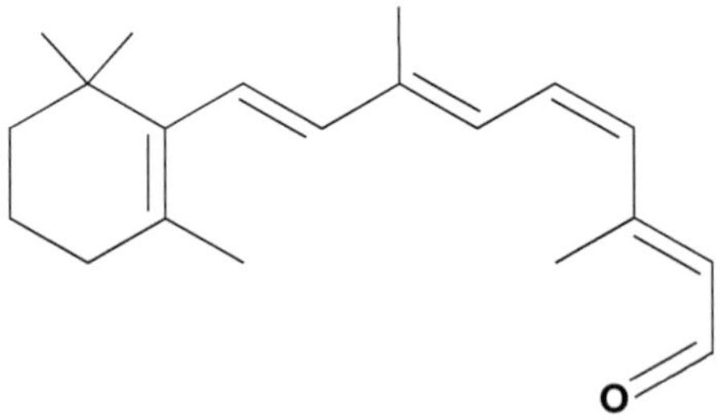

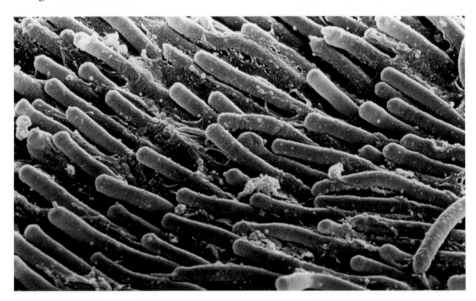

Figure 17.2.2 The skeletal formula of retinal.

Figure 17.2.3 A coloured scanning electron micrograph of rod cells in the retina of an eye. The cells are magnified about 3000 times. Rod cells contain a visual pigment that can respond to dim light but cannot distinguish colours.

Occurrence and uses

Biologists traditionally used a solution of methanal to preserve specimens. Such use is now restricted because of the toxicity of methanal. It has also been the main ingredient of the fluids used by embalmers. Methanal is still an important industrial chemical because it is a raw material for the manufacture for a range of thermosetting plastics.

Retinal is a naturally occurring aldehyde (Figure 17.2.2). Combined with a protein, it forms the light-sensitive part of the visual pigment in the rod cells of the retina (Figure 17.2.3). When light falls on a rod cell, the retinal molecule changes shape. This sets off a series of changes that lead to a signal being sent to the brain.

Formation

Aldehydes are formed by the oxidation of primary alcohols using a mixture of potassium or sodium dichromate(VI) and dilute sulfuric acid. An excess of the alcohol is heated with the oxidising agent and the aldehyde distilled off as it forms (see Figure 17.2.4). Unlike ketones, aldehydes can easily be oxidised further to carboxylic acids by longer heating with an excess of the oxidising agent and using a reflux condenser to prevent escape of the aldehyde (see Section 17.2.5).

Figure 17.2.4 Apparatus used to oxidise a primary alcohol to an aldehyde. The aldehyde distils off as it forms.

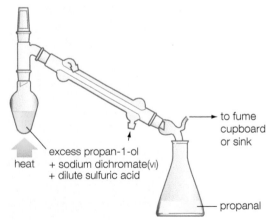

to fume cupboard or sink

heat

excess propan-1-ol
+ sodium dichromate(VI)
+ dilute sulfuric acid

propanal

Oxidation of a primary alcohol can be represented by a simplified equation, where [O] represents the oxidising agent (see also Section 6.3.8 in Student Book 1):

$$\underset{\text{propan-1-ol}}{CH_3CH_2CH_2OH} + [O] \rightarrow \underset{\text{propanal}}{CH_3CH_2CHO} + H_2O$$

Test yourself

1 a) Draw the displayed formula of ethanal.
b) Draw the structural formula of 2-methylbutanal.
c) Draw the skeletal formula of 3-methylpentanedial.

2 An aldehyde can be made by adding butan-1-ol a few drops at a time to a mixture of sodium dichromate(VI) and dilute sulfuric acid. The product is then distilled from the reaction mixture.
a) Write an equation for the reaction. (Represent the oxygen from the oxidising agent as [O].)
b) Suggest a reason for adding the butan-1-ol a few drops at a time.
c) Explain why the reaction mixture is not heated in a flask fitted with a reflux condenser before distilling off the product.

17.2.3 Ketones

Names and structures

In ketones the carbonyl group is attached to two hydrocarbon groups. Chemists name ketones after the alkane with the same carbon skeleton by changing the ending –**ane** to –**anone**. Where necessary a number in the name shows the position of the carbonyl group.

Tip

In ketones, the carbonyl group is in the middle of a chain of carbon atoms. This means that the simplest ketone is propanone, which has the minimum number of three carbon atoms.

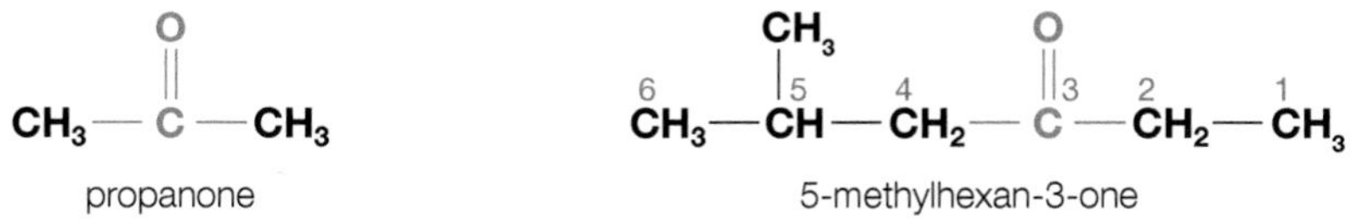

Figure 17.2.5 Structures and names of two ketones.

Tip

The number of the principal functional group in a name is always as low as possible. So the correct name of the second ketone in Figure 17.2.5 uses numbering from the right to give 5-methylhexan-3-one, rather than the alternative name of 2-methylhexan-4-one found by numbering from the left.

Occurrence and uses

The most widely used ketone is propanone, which is a common solvent and is familiar as nail polish remover. It has a low boiling temperature and evaporates quickly, making it suitable for cleaning and drying parts of precision equipment. Propanone is also the starting point for producing the monomer of the glass-like addition polymer in display signs, plastic baths and the cover of car lights. Propanone, and other ketones, form during normal metabolism, especially at night and during fasting when the levels of propanone and other ketones in the blood rise. People with diabetes produce larger amounts of propanone than normal. Other ketones, such as menthone (Figures 17.2.6 and 17.2.7), are used in oils and perfumes.

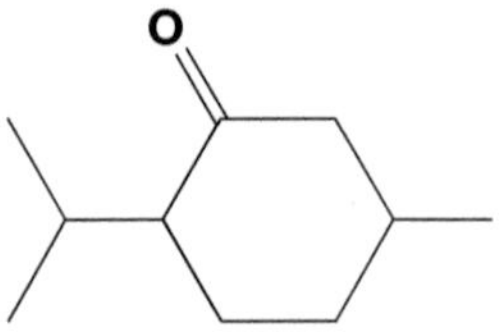

Figure 17.2.7 The skeletal formula of the naturally occurring ketone called menthone, which is found in the oils extracted from some plants.

Formation

Oxidation of secondary alcohols with hot, acidified potassium dichromate(VI) produces ketones which, unlike aldehydes, are not easily oxidised further.

Oxidation of a secondary alcohol can be represented as a simplified equation, where [O] represents the oxidising agent (see also Section 6.3.8 in Student Book 1):

$$\underset{\text{propan-2-ol}}{CH_3CH(OH)CH_3} + [O] \rightarrow \underset{\text{propanone}}{CH_3COCH_3} + H_2O$$

Figure 17.2.6 A bottle of peppermint oil and leaves of the peppermint plant. The oil is used in aromatherapy. Peppermint oil contains menthone, together with a range of chemicals which include menthol, methyl ethanoate and volatile oils.

Test yourself

3 a) Draw the displayed formula of propanone.

  b) Draw the structural formula of 4,4-dimethylpentan-2-one.

  c) Draw the skeletal formula of 2,4-dimethylcyclohexanone.

4 What is the molecular formula of menthone?

5 Show that propanone and propanal are functional group isomers.

6 Write an equation for the oxidation of butan-2-ol to butanone. (Represent the oxygen from the oxidising agent as [O].)

7 The simplest ketone that contains a benzene ring is $C_6H_5COCH_3$, which is called phenylethanone. Why might the ethanone part of the name for this ketone be considered unusual?

17.2.4 Physical properties of carbonyl compounds

The C=O bond in carbonyl compounds is polar (Figure 17.2.8). As a result, the intermolecular forces include dipole–dipole attractions as well as London forces. There are no hydrogen atoms bonded to oxygen in carbonyl compounds so hydrogen bonding cannot occur between the molecules of aldehydes or ketones. However, hydrogen bonding is possible between oxygen atoms in carbonyl groups and the –OH group in water.

Figure 17.2.8 Representations of the carbonyl group in aldehydes and ketones. The bond is polar because the oxygen atom is more electronegative than carbon.

Methanal is a gas at room temperature. Ethanal boils at 21 °C, so it may be a liquid or gas at room temperature, depending on the conditions. Other common aldehydes are also liquids. Similarly, the common ketones are liquids with boiling temperatures similar to those of the corresponding aldehydes.

The simpler aldehydes such, as methanal and ethanal, are freely soluble in water. The simplest ketone, propanone, mixes freely with water.

Test yourself

8 Refer to the data sheet headed 'Properties of alkanes, alcohols, aldehydes and ketones', which you can access via the QR code for Chapter 17.2 on page 321.

   a) Show that the boiling temperatures of aldehydes are higher than those of alkanes with similar relative molecular masses, but lower than those of the corresponding alcohols.

   b) Account for the values of the boiling temperatures of aldehydes relative to those of alkanes and alcohols in terms of intermolecular forces.

9 Explain why propanone is freely soluble in water.

10 Why does an aldehyde such as ethanal mix freely with water whereas hexanal is much less soluble?

17.2.5 Reactions of aldehydes and ketones

Oxidation

Oxidising agents easily convert aldehydes to carboxylic acids (Figure 17.2.9). It is much harder to oxidise ketones. Oxidation of ketones is only possible with powerful oxidising agents which break up the molecules. Chemical tests to distinguish aldehydes and ketones are based on the difference in the ease of oxidation (Section 17.2.6).

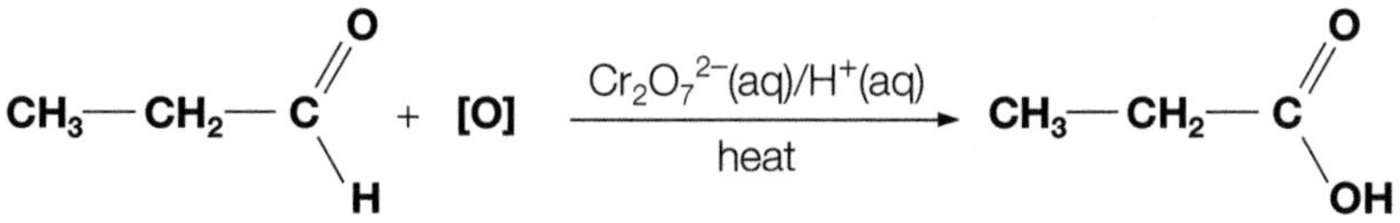

$$CH_3-CH_2-\overset{\displaystyle O}{\underset{\displaystyle H}{C}} \ + \ [O] \ \xrightarrow[\text{heat}]{Cr_2O_7^{2-}(aq)/H^+(aq)} \ CH_3-CH_2-\overset{\displaystyle O}{\underset{\displaystyle OH}{C}}$$

Figure 17.2.9 Oxidation of propanal to propanoic acid.

Acidified potassium dichromate(VI) is orange and contains $Cr_2O_7^{2-}$ ions. After oxidising an aldehyde to a carboxylic acid, a green solution is formed containing green Cr^{3+} ions.

Chemicals in perfumes

The perfume 'Chanel N° 5' was innovative when produced for the first time in 1921. As well as natural extracts from flowers, the scent includes a high proportion of synthetic aldehydes, such as dodecanal. This produces a highly original perfume.

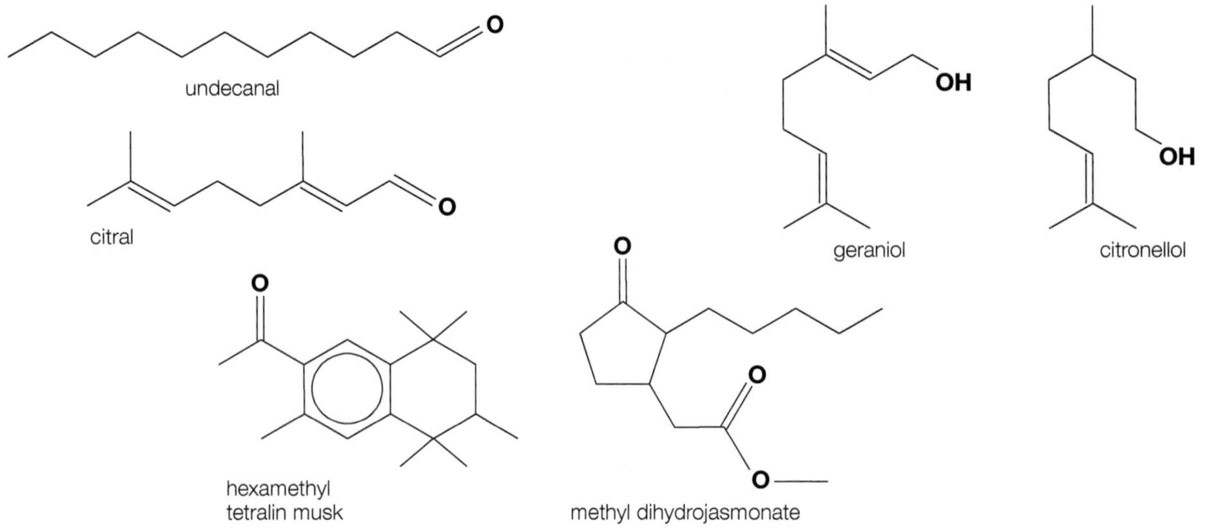

Figure 17.2.10 Skeletal formulae of some perfume chemicals.

The people who devise new perfumes think of the mixture as a sequence of 'notes'. You first smell the 'top notes', but the main effect depends on the 'middle notes', while the more lasting elements of the perfume are the 'end notes'. The overall balance of the three is critical. This means that the volatility of perfume chemicals is of great importance to the perfumer.

Table 17.2.1 Natural and synthetic chemicals used to make perfumes.

Note	Natural chemicals	Synthetic chemicals	Boiling temperature or melting temperature
Top	Citrus oils	Octanal (citrus)	(boils) 168 °C
	Lavender	Undecanal (green)	(boils) 117 °C
Middle	Rose	Geraniol (floral)	(boils) 146 °C
	Violet	Citronellol (rosy)	(boils) 224 °C
End	Balsam	Indane (musk)	(melts) 53 °C
	Musk	Hexamethyl tetralin (musk)	(melts) 55 °C

1 Draw the skeletal formula of octanal.

2 Suggest two advantages for the perfumer of using synthetic chemicals instead of chemicals extracted from living things.

3 Identify the carbonyl compounds among the compounds shown in Figure 17.2.10. In each case state whether the compound includes the functional group of an aldehyde or of a ketone.

4 Like many perfume chemicals, geraniol is a terpene. Terpene molecules are built from units derived from 2-methylbuta-1,3-diene.

a) Draw the structure of 2-methylbuta-1,3-diene.

b) How many 2-methylbuta-1,3-diene units are needed to make up the hydrocarbon skeleton of geraniol?

5 Use your knowledge of intermolecular forces to explain why:

a) aldehydes are useful as top notes while alcohols are more commonly used as middle notes

b) the musks used as 'end notes' also help to 'fix', or retain, the more volatile components of a perfume

c) geraniol is more soluble in water than undecanal.

Test yourself

11 Identify the oxidation states of chromium before and after the oxidation of an aldehyde by a solution of acidified potassium dichromate(VI).

12 **a)** Outline the reagents and conditions for converting butanal to butanoic acid.

b) What apparatus is used **i)** to carry out the reaction and **ii)** to separate the product from the reaction mixture?

c) Write a half-equation for the oxidation of butanal to butanoic acid.

13 Write equations for the oxidation of:

a) ethanediol to ethanedial and

b) ethanedial to ethanedioic acid.

(In both cases, represent the oxygen from the oxidising agent as [O].)

Reduction

Metal hydrides can reduce carbonyl compounds to alcohols. Lithium tetrahydridoaluminate(III), $LiAlH_4$, is a powerful reducing agent that converts aldehydes to primary alcohols, and ketones to secondary alcohols. $LiAlH_4$ is easily hydrolysed so the reagent is dissolved in dry ether (ethoxyethane).

The reaction involves two steps: reduction involving $LiAlH_4$ followed by addition of dilute acid to complete the reaction. Simplified equations, as in Figure 17.2.11, are written for the overall reaction; these use [H] to represent the reducing agent.

Tip

$LiAlH_4$ is also known as lithium aluminium hydride, which is sometimes abbreviated to lithal.

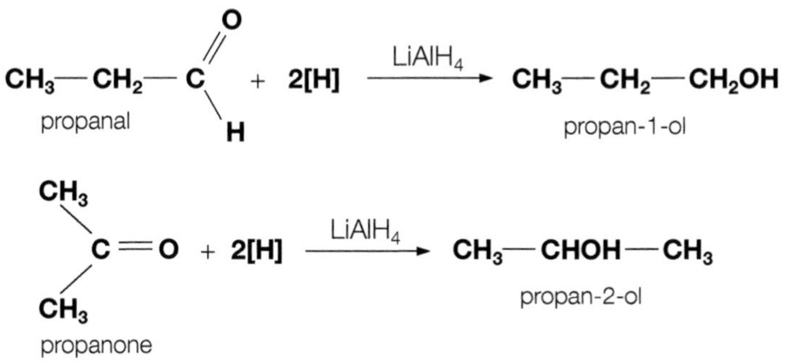

Figure 17.2.11 Reduction of propanal and propanone. The 2[H] comes from the reducing agent. This is a shorthand way of balancing a complex equation involving reduction.

Tip

The reduction step in the reaction of carbonyl compounds with $LiAlH_4$ can be represented by a nucleophilic addition mechanism (see next section).

Test yourself

14 Name the products of reducing butanal and butanone and state which is a primary alcohol and which a secondary alcohol.

15 Show that reduction of an aldehyde or ketone with $LiAlH_4$ has the effect of adding hydrogen to the double bond.

16 Write equations for the reduction, using $LiAlH_4$, of:

a) 2-methylbutanal

b) cyclohexanone.

Use [H] to represent the reducing agent.

Reaction with hydrogen cyanide – a nucleophilic addition

Hydrogen cyanide rapidly adds to carbonyl compounds at room temperature (Figure 17.2.12). Hydrogen cyanide is a highly toxic gas that is formed in the reaction mixture by adding potassium cyanide and dilute sulfuric acid. The potassium cyanide must be in excess to ensure that there are free cyanide ions ready to start the reaction. The product of the addition is a hydroxynitrile. The **nitrile** group, –CN, can be hydrolysed to a carboxylic acid (see Section 17.3.2). The reaction with the cyanide ion adds a carbon atom to the carbon skeleton of the original molecule and is therefore a useful step in synthetic routes to valuable compounds.

Tip

Note that the carbon atom in the –CN group counts as part of the carbon chain when naming nitriles.

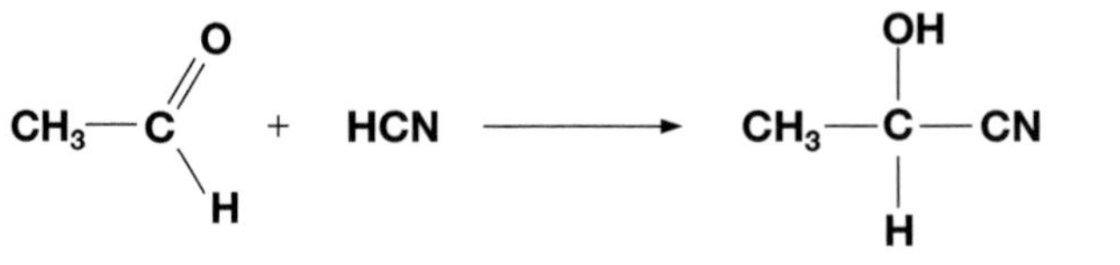

Figure 17.2.12 Addition of hydrogen cyanide to ethanal.

Nucleophilic addition

The reaction of a carbonyl compound with hydrogen cyanide and also its reduction with $LiAlH_4$ are both examples of **nucleophilic addition**. The carbon atom in a carbonyl group is electron deficient because the electronegative oxygen draws electrons away from it. This leaves it open to attack by a **nucleophile**.

The incoming nucleophile uses its lone pair to form a new bond with the δ+ carbon atom. This displaces one pair of electrons from the double bond onto oxygen. Oxygen has gained one electron from carbon and now has a single negative charge (Figure 17.2.13).

Tip

It is the lone pair on carbon, not nitrogen, that is used for nucleophilic attack.

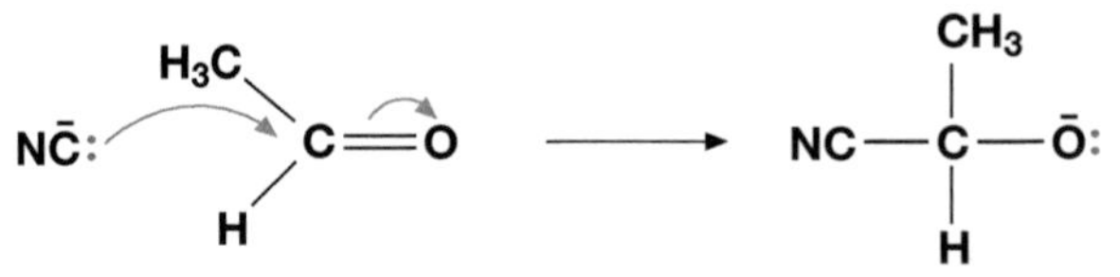

Figure 17.2.13 The first step of the nucleophilic addition of hydrogen cyanide to ethanal.

To complete the reaction, the negatively charged oxygen acts as a base and gains a proton from a hydrogen cyanide molecule (Figure 17.2.14) or from a water molecule.

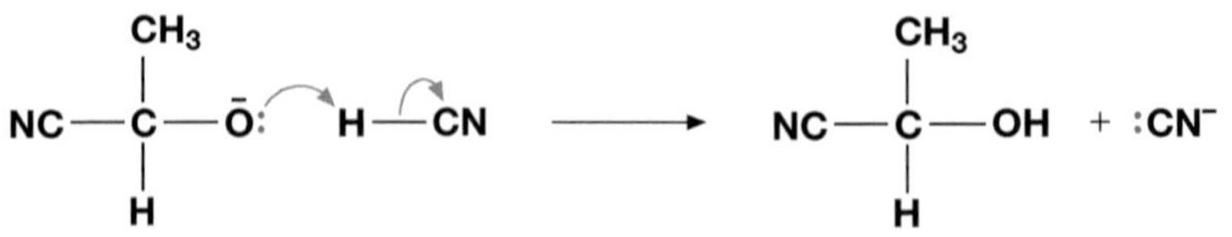

Figure 17.2.14 The second step of the nucleophilic addition of hydrogen cyanide to ethanal. Note that taking a proton from HCN produces another cyanide ion.

Key terms

A **nitrile** is a compound with the functional group –C≡N.

Nucleophilic addition to a carbonyl compound occurs when an electron-rich species, a nucleophile, attacks the electron-deficient carbon atom of the unsaturated C=O bond, leading to the formation of a carbon–oxygen single bond.

A **nucleophile** is an electron-pair donor.

This mechanism helps to account for the fact that if the product of addition is chiral, the outcome is a racemic mixture of the two optical isomers. This is illustrated by Figure 17.2.15, which shows the addition of hydrogen cyanide to ethanal in three dimensions. The atoms around the carbon atom of a carbonyl group lie in a plane. The attacking nucleophile has an equal chance of bonding to the carbon atom from either side of this plane.

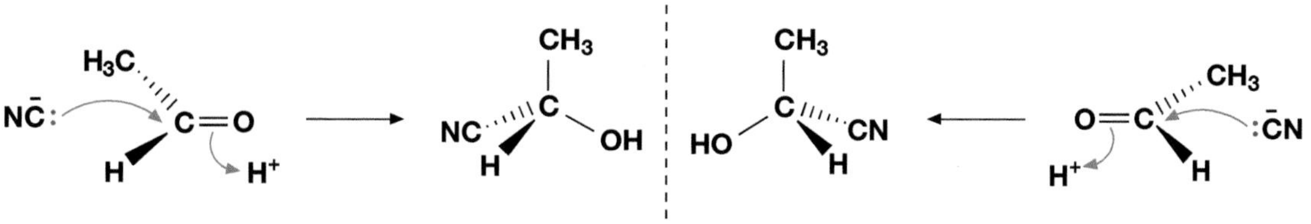

Figure 17.2.15 Formation of two optical isomers by addition of HCN to ethanal. Nucleophiles attack with equal probability from either side of the ethanal molecules, giving rise to equal numbers of the two isomer molecules.

Test yourself

17 These questions are about the nucleophilic addition of hydrogen cyanide to a carbonyl compound.

 a) What features of the cyanide ion means that it is a nucleophile?

 b) What type of bond breaking takes place in each step?

 c) Explain why there is a negative charge on the oxygen atom at the end of step 1.

 d) Which molecule acts as an acid in step 2?

18 a) Show that the hydroxynitrile formed from ethanal and HCN is chiral, but that formed from propanone is not.

 b) Name the product of each of these reactions.

19 a) Write equations to show a nucleophilic addition mechanism for the reduction of ethanal by $LiAlH_4$. You may assume that the nucleophile is the hydride ion, H^-, and that water is involved in the second step of the process.

 b) Explain why $LiAlH_4$ reduces the double bond in carbonyl compounds but not the double bond in alkenes.

17.2.6 Tests for aldehydes and ketones

Recognising carbonyl compounds

Today chemists can identify aldehydes and ketones with the help of instrumental techniques such as mass spectrometry and infrared spectroscopy (Chapter 19). Traditionally chemists characterised these compounds by combining them with a reagent that could convert them to a solid product. The solid, a so-called crystalline **derivative**, is a chemical which can be purified by recrystallisation and then identified by measuring its melting temperature.

Key term

Chemists used to use a **derivative** to identify an unknown organic compound. Converting a compound to a crystalline derivative produces a compound that can be purified by recrystallisation and then identified by measuring its melting temperature.

Figure 17.2.16 A bright orange 2,4-dinitrophenylhydrazone derivative.

The reagent 2,4-dinitrophenylhydrazine reacts with carbonyl compounds to form 2,4-dinitrophenylhydrazone derivatives, which are solid at room temperature and bright yellow or orange (Figure 17.2.16). Figure 17.2.17 shows the equation for the formation of the derivative formed with ethanal. The solid derivative can be filtered off, recrystallised and identified by measuring its melting temperature. This value is then compared to data values, such as those on the data sheet '2,4-Dinitrophenylhydrazine derivatives of carbonyl compounds', which you can access via the QR code for this chapter on page 321. Together with the boiling temperature of the original aldehyde or ketone, this information makes it possible to identify the carbonyl compound.

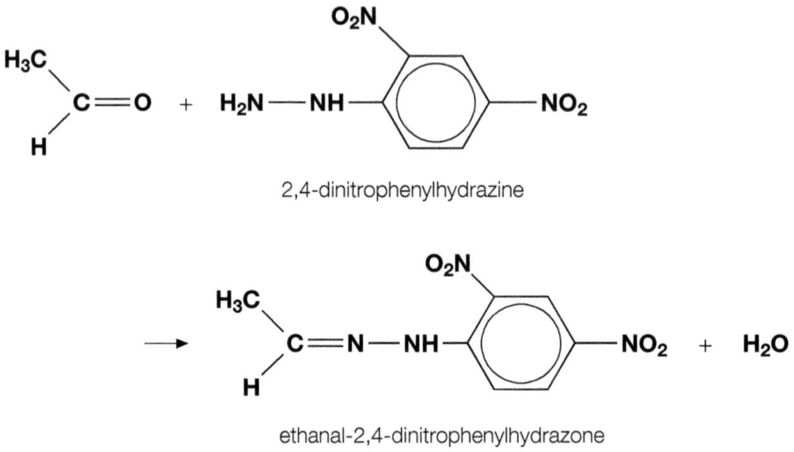

2,4-dinitrophenylhydrazine

ethanal-2,4-dinitrophenylhydrazone

Figure 17.2.17 Equation showing the formation of ethanal-2,4-dinitrophenylhydrazone.

Distinguishing aldehydes and ketones

Aldehydes are easily oxidised. It is more difficult to oxidise ketones, but they can be oxidised by stronger oxidising agents. In order to be absolutely sure that ketones are not affected, three very mild oxidising agents are used to distinguish aldehydes from ketones. These are Fehling's solution, Benedict's solution and Tollens' reagent.

Fehling's reagent does not keep, so it is made when required by mixing two solutions. One solution is copper(II) sulfate in water. The other is a solution of 2,3-dihydroxybutanedioate (tartrate) ions in strong alkali. The 2,3-dihydroxybutanedioate salt forms a complex with copper(II) ions so that they do not precipitate as copper(II) hydroxide with the alkali.

Benedict's solution is similar to Fehling's solution but is more stable. It is less strongly alkaline and does not react so reliably with all aldehydes.

Aldehydes reduce the blue copper(II) ions in Fehling's, or Benedict's, solution to copper(I), which then precipitates in the alkaline conditions to give an orange-brown precipitate of copper(I) oxide, Cu_2O (Figure 17.2.18).

Tollens' reagent (ammoniacal silver nitrate) similarly distinguishes aldehydes from ketones. Tollens' reagent consists of an alkaline solution of diamminesilver(I) ions, $[Ag(NH_3)_2]^+$. Forming a complex ion with ammonia keeps the silver(I) ions in solution under alkaline conditions (see Section 15.6). Aldehydes reduce the silver ions to metallic silver (Figure 17.2.19).

The mild oxidising agents Fehling's, Benedict's and Tollens' oxidise the aldehyde to a carboxylic acid, which is present as a **carboxylate** ion in the alkaline conditions. Ketones do not react with these mild oxidising agents and so the colour remains unchanged.

Key term

A **carboxylate** ion is formed when a carboxylic acid molecule loses a proton.

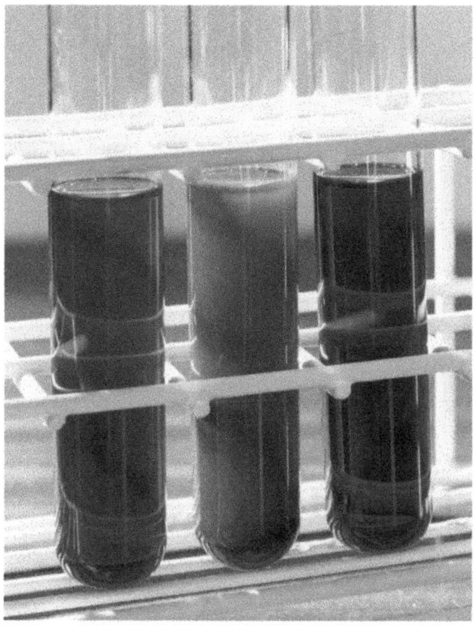

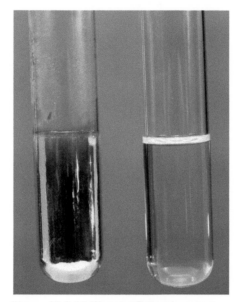

Figure 17.2.18 The test tube in the middle contains Fehling's reagent that has been reduced by an aldehyde to form an orange-brown precipitate of copper(I) oxide. The test tubes on the left and right contain Fehling's reagent and ketones.

Figure 17.2.19 Warming Tollens' reagent with an aldehyde produces a precipitate of silver, which coats clean glass with a shiny layer of silver so that it acts like a mirror (left). There is no reaction with a ketone (right).

Test yourself

20 Write an ionic equation for the reaction of copper(II) ions with an alkali in the absence of 2,3-dihydroxybutanedioate ions.

21 **a)** Write an equation for the reaction of Tollens' reagent with propanal using the symbol [O] to represent the reagent.

b) Use the oxidation numbers of the metal ions and atoms to show that propanal reduces Tollens' reagent.

22 Write half-equations for the reduction of:

a) copper(II) ions in alkaline conditions to copper(I) oxide as in Fehling's test

b) $[Ag(NH_3)_2]^+$ ions in Tollens' reagent to silver.

23 Hydrolysis of A, C_4H_9Cl, with hot, aqueous sodium hydroxide produces B, $C_4H_{10}O$.

Heating B with an acidic solution of potassium dichromate(VI) and distilling off the product as it forms gives C, C_4H_8O.

C gives a yellow precipitate with 2,4-dinitrophenylhydrazine and forms a silver mirror when warmed with Tollens' reagent.

Identify compounds A, B and C.

Identifying an unknown carbonyl compound

Figure 17.2.20 shows stages in making, purifying and identifying a carbonyl compound.

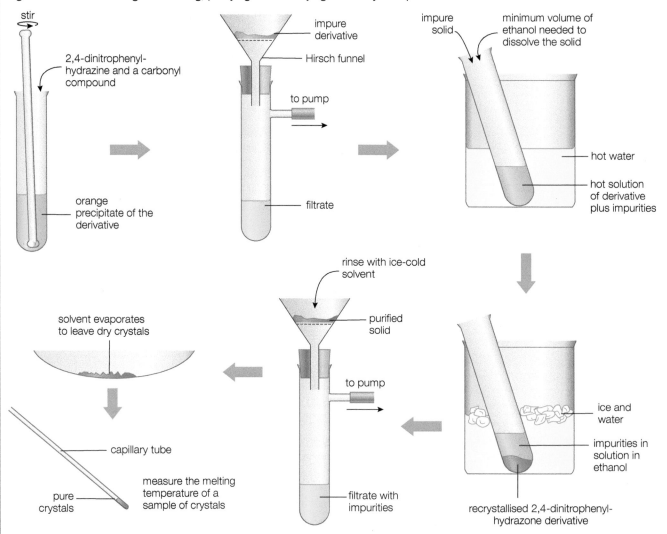

Figure 17.2.20 Making a pure crystalline derivative of a carbonyl compound.

1 Why is it necessary to purify the derivative before measuring its melting temperature?

2 Explain how the procedure illustrated in Figure 17.2.20 removes soluble impurities from the derivative.

3 In this instance, ethanol is the solvent used for recrystallising the derivative. What determines the choice of solvent?

4 When measuring the melting temperature, what are the signs that the derivative is pure?

5 Identify the carbonyl compound that forms a 2,4-dinitrophenylhydrazone that melts at 115 °C. The carbonyl compound boils at 80 °C and does not give an orange precipitate with Fehling's solution.

6 Suggest how the procedure outlined in Figure 17.2.20 could be modified so that insoluble impurities were also removed. Discuss how any changes you suggest might affect the yield of crystals obtained.

The triiodomethane reaction

A compound containing the CH_3CO- group produces a yellow precipitate of triiodomethane (iodoform) when warmed with a mixture of iodine and sodium hydroxide. The test shows the presence of a methyl group next to a carbonyl group in an organic molecule.

Iodine in sodium hydroxide reacts to form iodate(I) ions (see Section 4.11 in Student Book 1). These ions oxidise secondary alcohols to ketones. So alcohols containing the CH_3CHOH- group will also produce a yellow precipitate of triiodomethane. These include all methyl secondary alcohols and one primary alcohol, ethanol.

The reaction takes place in two main steps, substitution of iodine for hydrogen then hydrolysis (see Figure 17.2.21). An overall equation for the reaction is given in Figure 17.2.22.

Tip

An alternative reagent for the triiodomethane reaction is a mixture of potassium iodide and sodium chlorate(I).

Tip

The triiodomethane reaction involves breaking a carbon-carbon bond. It can be used to shorten a carbon chain at room temperature by removing a methyl group.

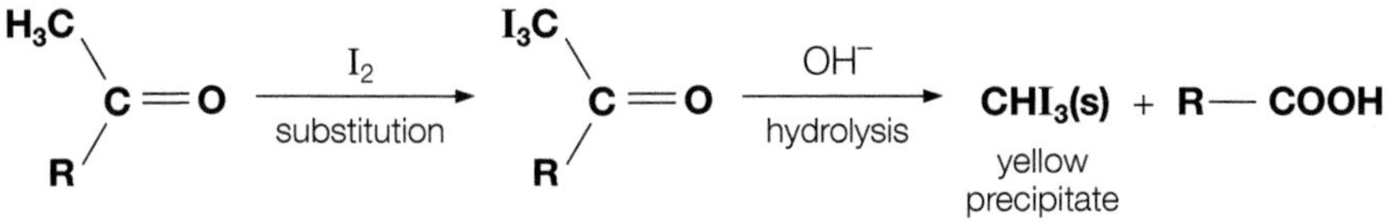

Figure 17.2.21 The two steps in the triiodomethane reaction.

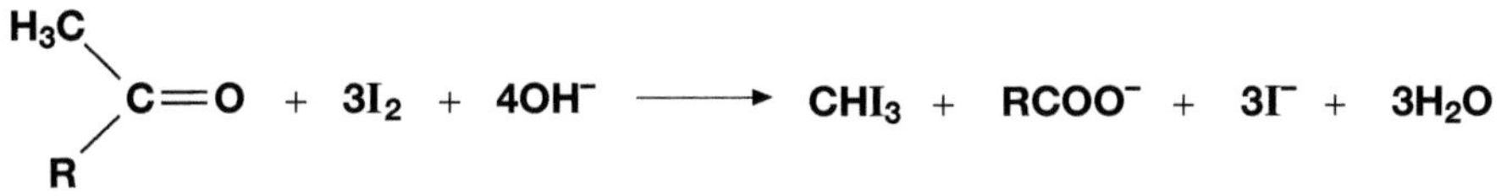

Figure 17.2.22 An overall equation for the triiodomethane reaction.

Test yourself

24 a) Explain why alcohols with the group CH_3CHOH- also give a positive result with the triiodomethane reaction.

   b) Explain why a mixture of iodine with sodium hydroxide solution is chemically equivalent to a mixture of potassium iodide and sodium chlorate(I) solutions.

25 Name and write the displayed formulae of the two isomers of $C_5H_{10}O$ that form a yellow precipitate when they react with iodine in the presence of alkali.

26 Which is the only aldehyde to undergo the triiodomethane reaction?

27 a) Write equations for:

   i) the reaction of iodine with hydroxide ions to form I^- ions, IO^- ions and water

   ii) the reaction of IO^- ions with the ketone $RCOCH_3$ to form $RCOCI_3$ and hydroxide ions

   iii) the reaction of $RCOCI_3$ with hydroxide ions to form CHI_3 and $RCOO^-$ ions.

   b) Show that combining these three equations produces the overall equation given in Figure 17.2.22.

Exam practice questions

1 Copy and complete the table to show three different reactions of propanal. *(6)*

Reactant	Reagent	Organic product	
		Name	**Displayed formula**
CH₃CH₂CHO	Tollens' reagent		
CH₃CH₂CHO		Propanoic acid	
CH₃CH₂CHO	LiAlH₄		

2 Citral and β-ionone are perfume chemicals. β-ionone is one of the chemicals in the oil extracted from violets.

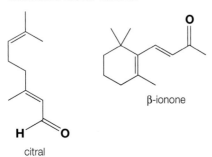

β-ionone

citral

a) Work out the molecular formula of citral. *(1)*

b) i) Name the two functional groups in citral. *(2)*

ii) Name the functional group which is present in β-ionone but not in citral. *(1)*

c) Describe the observations you would expect with each of the two compounds on:

i) warming them with Fehling's solution *(2)*

ii) mixing them with a solution of 2,4-dinitrophenylhydrazine *(2)*

iii) warming them with a mixture of iodine and sodium hydroxide. *(2)*

d) i) Draw the skeletal formula of the product of treating β-ionone with LiAlH₄ in dry ether. *(1)*

ii) Draw the skeletal formula of the compound formed when citral is warmed with an acidic solution of potassium dichromate(VI). *(1)*

e) i) Which of the two compounds have *E/Z* isomers? *(1)*

ii) Draw an *E/Z* isomer of one of the two compounds. *(2)*

iii) Which of the two compounds has optical isomers? *(1)*

3 An optically active compound X with the formula $C_3H_6O_3$ reacts with sodium carbonate giving off a colourless gas. Oxidation of X gives Y which reacts with 2,4-dinitrophenylhydrazine but not with Fehling's solution.

Account for these observations and give the structures of X and Y. *(5)*

4 The molar mass of a hydrocarbon W is $56\,\text{g mol}^{-1}$ and it contains 85.7% carbon. W reacts with hydrogen bromide to form X. Heating X under reflux with aqueous sodium hydroxide produces Y. Heating Y with an acidified solution of potassium dichromate(VI) converts it to Z. Z gives a yellow precipitate with 2,4-dinitrophenylhydrazine but it does not give a precipitate with Benedict's reagent.

a) Identify W, X, Y and Z and give your reasoning. *(8)*

b) Write equations for the reactions mentioned of W, X and Y. *(3)*

c) Name the mechanisms for the reactions of W and X. *(2)*

5 Consider the following pairs of compounds:

a) pentanal and pentan-3-one

b) pentan-3-one and pentan-3-ol

c) pentan-3-ol and pentan-2-ol.

i) Describe how you could distinguish between the compounds in each pair, using a test-tube reaction. In each case, state a reagent and describe what you would observe when it is added to each compound. *(9)*

ii) Describe how you could distinguish between the compounds in each pair using a physical or a spectroscopic method other than the using the infrared fingerprint region. In your answer, you should use data from the Edexcel Data booklet. *(6)*

6 a) Write a half-equation for the oxidation of propanal to propanoic acid. *(1)*

b) Write a half-equation for the oxidation of propanal to the propanoate ion by Tollens' reagent in alkaline conditions. *(1)*

c) Write a half-equation for the oxidation of cyclohexanone to hexanedioic acid. *(1)*

7 a) Use curly arrows to show the mechanism of the reaction of propanal with HCN. *(4)*

b) The reaction is usually performed in the presence of KCN at a pH of about 5. Explain why the rate of this reaction is slow if the pH is either too high or too low. *(4)*

c) Explain why the product obtained is optically inactive. *(2)*

8 Glucose is a carbohydrate. Glucose molecules usually exist in a ring form. In solution, about 1% of the molecules exist in an open chain form.

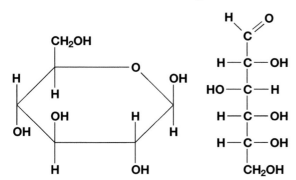

a) i) What is the molecular formula of glucose? *(1)*

ii) By inspection of the formula suggest a reason why glucose and related compounds are called carbohydrates. *(1)*

b) Explain why glucose is a solid at room temperature and is also very soluble in water. *(4)*

c) Sugars such as glucose are classified as aldoses or ketoses based on the type of carbonyl group present in the open chain forms of the molecules. Is glucose an aldose or a ketose? *(1)*

d) i) What would you expect to observe on mixing a solution of glucose with a solution of 2,4-dinitrophenylhydrazine? *(1)*

ii) This reaction is slower with glucose than with compounds such as propanal. Suggest a reason for this. *(2)*

e) i) What would you expect to observe on warming a mixture of a solution of glucose and Tollens' reagent? *(1)*

ii) Why is glucose classified as a reducing sugar? *(1)*

f) Suggest how the open chain molecule reacts to form the ring structure. *(5)*

9 Describe the mechanisms for the reactions of propene with bromine, and propanone with HCN. Identify similarities and differences between the two mechanisms with reference to the nature of the bonds and bond breaking, the reagents involved, the formation of intermediates and the overall effects of the change. *(10)*

10 a) Write an equation for the reaction of butanone with $LiAlH_4$ in dry ether, using [H] to represent the reducing agent. *(1)*

b) Explain why:

i) $LiAlH_4$ reacts with carbonyl compounds but not with alkenes although both contain a double bond *(3)*

ii) dry ether rather than water is used as a solvent with $LiAlH_4$. *(2)*

c) An alternative to $LiAlH_4$ for the reduction of aldehydes and ketones is $NaBH_4$, sodium tetrahydridoborate(III). State which of $LiAlH_4$ and $NaBH_4$ is the stronger reducing agent and justify your answer *(3)*

11 a) Compound P has the formula $C_4H_6O_2$ and an unbranched carbon chain. P reacts with HCN to form a compound Q, $C_6H_8O_2N_2$. P is easily oxidised by acidified potassium dichromate(VI) to an acidic compound R, $C_4H_6O_4$. When 1.0 g R is dissolved in water and titrated with $1.00 \, mol \, dm^{-3}$ sodium hydroxide, the mean titre is $16.90 \, cm^3$. Suggest structures for P, Q and R and explain the reactions. *(9)*

b) A compound Z contains 64.3% carbon, 7.1% hydrogen and 28.6% oxygen by mass. The mass–to–charge ratio of the molecular ion in its mass spectrum is 56. Z reduces Fehling's solution to give an orange precipitate. When Z reacts with hydrogen in the presence of a nickel catalyst, 0.1 g Z reacts with $85.4 \, cm^3$ hydrogen under conditions in which 1 mol gas occupies $24 \, dm^3$. Suggest a structure for Z and account for the observations. *(8)*

12 Propanoic acid can be prepared in a four-step synthesis starting from 2-bromobutane.

Identify the three intermediate compounds and give the reagents needed for each step. *(7)*

17.3.1 Carboxylic acids

Occurrence

Carboxylic acids are compounds with the formula R–COOH where R represents an alkyl group, aryl group or a hydrogen atom. The carboxylic acid group –COOH is the functional group which gives the acids their characteristic properties. Some carboxylic acids are found naturally in insects (Figure 17.3.1) and in plants (Figure 17.3.2). Many organic acids are instantly recognisable by their odours. Ethanoic acid, for example, gives vinegar its taste and smell. Butanoic acid is responsible for the foul smell of rancid butter, while the body odour of goats is a blend of the three unbranched organic acids with 6, 8 and 10 carbon atoms.

> **Tip**
>
> The first two sections of this chapter revisit work on the oxidation of primary alcohols covered in Section 6.3.8 in Student Book 1. Some of the 'Test yourself' questions are also designed to help revise ideas from the first year of the A Level course.

Figure 17.3.1 The traditional names for organic acids were based on their natural origins. The original name for methanoic acid was formic acid because it was first obtained from red ants and the Latin name for 'ant' is formica. This red wood ant can spray attackers with methanoic acid (magnification ×5).

Figure 17.3.2 Many vegetables contain ethanedioic acid, which is commonly called oxalic acid. The level of the acid in rhubarb leaves is high enough for it to be dangerous to eat the leaves. The acid kills by lowering the concentration of calcium ions in blood to a dangerously low level.

Names and structures

The carboxylic acid group can be regarded as a carbonyl group, C=O, attached to an –OH group (see Figures 17.3.3 and 17.3.4), but is better seen as a single functional group with distinctive properties.

Chemists name carboxylic acids by changing the ending of the corresponding alk**ane** to **-oic** acid. So ethane becomes ethanoic acid.

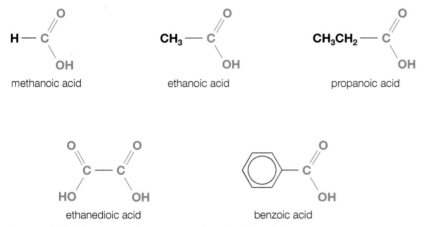

methanoic acid   ethanoic acid   propanoic acid

ethanedioic acid   benzoic acid

Figure 17.3.3 Names and structures of carboxylic acids.

Carboxylic acids form a wide range of derivatives, each with their own characteristics. This is illustrated by the derivatives of ethanoic acid in Figure 17.3.5. All of these compounds contain the **acyl group**, CH_3CO-.

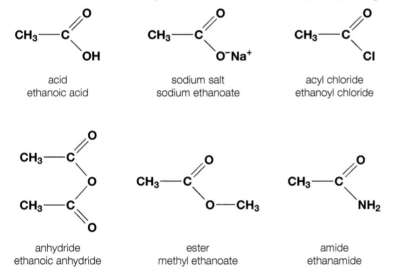

acid
ethanoic acid

sodium salt
sodium ethanoate

acyl chloride
ethanoyl chloride

anhydride
ethanoic anhydride

ester
methyl ethanoate

amide
ethanamide

Figure 17.3.5 Compounds related to carboxylic acids.

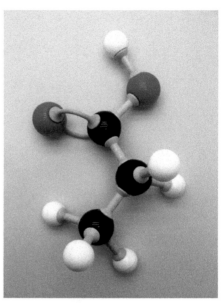

Figure 17.3.4 A ball-and-stick model of a carboxylic acid.

Test yourself

1 Write out the structural formulae and give the IUPAC names of the three carboxylic acids which were traditionally derived from the Latin word 'caper', meaning goat: caproic acid (6C), caprylic acid (8C) and capric acid (10C).

2 Give the molecular formula, skeletal formula and name of the acid shown in Figure 17.3.4.

3 Carboxylic acids and esters are structural isomers.

   a) Name the type of structural isomerism they show

   b) Draw the structure of

      i) the ester which is an isomer of ethanoic acid

      ii) the carboxylic acid which is an isomer of methyl ethanoate.

Physical properties

Even the simplest acids such as methanoic acid and ethanoic acid are liquids at room temperature because of hydrogen bonding between the carboxylic acid groups. Carboxylic acids with more than eight carbon atoms in the chain are solids. Benzoic acid is also a solid at room temperature.

Carbon–oxygen bonds are polar. There is also the possibility of hydrogen bonding between water molecules and the –OH groups and oxygen atoms in carboxylic acid molecules. This means that the simplest acids are soluble in water. However, the solubility falls as the non-polar carbon chain length increases and both hexanoic and benzoic acids are only very slightly soluble at room temperature.

Test yourself

4 Draw a diagram to show hydrogen bonding between ethanoic acid molecules and water molecules.

5 In a non-polar solvent, ethanoic acid molecules dimerise through hydrogen bonding.

   a) Suggest a reason why the acid dimerises in a non-polar solvent but not in water.

   b) Draw a diagram to show an ethanoic acid dimer with two hydrogen bonds between the molecules.

6 a) Explain why sodium ethanoate is a solid at room temperature while ethanoic acid is a liquid.

   b) Explain why benzoic acid is insoluble in cold water whereas sodium benzoate is soluble.

17.3.2 Preparation of carboxylic acids

In the laboratory, carboxylic acids are normally made by oxidising primary alcohols or aldehydes (Section 17.2.5). The usual oxidising agent is an acidic solution of potassium dichromate(VI).

Carboxylic acids can also be made by hydrolysing nitriles. The reagent for speeding up the hydrolysis can either be a solution of a strong acid or a solution of a strong base.

$$RCN + 2H_2O + HCl \rightarrow RCOOH + NH_4Cl$$

$$RCN + H_2O + NaOH \rightarrow RCOONa + NH_3$$

Figure 17.3.6 Hydrolysis of nitriles produces carboxylic acids or their salts.

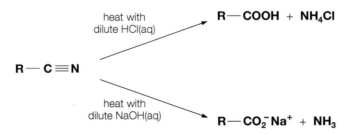

17.3.3 Reactions of carboxylic acids

Reactions as acids

Carboxylic acids are weak acids (see Section 12.4). They are only slightly ionised when they dissolve in water.

$$CH_3COOH(aq) \rightleftharpoons CH_3COO^-(aq) + H^+(aq)$$

The aqueous hydrogen ions in the solutions of these compounds mean that they show the characteristic reactions of acids with metals, bases and carbonates.

Carboxylic acids are sufficiently acidic to produce carbon dioxide when added to a solution of sodium carbonate or sodium hydrogencarbonate (Figure 17.3.7). This reaction distinguishes carboxylic acids from weaker acids such as phenols (see Section 18.1.8).

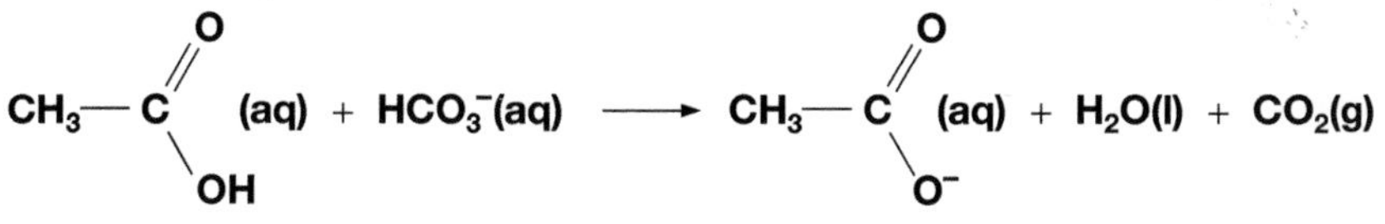

$$CH_3-\overset{\displaystyle O}{\overset{\|}{C}}\diagdown_{OH}(aq) + HCO_3^-(aq) \longrightarrow CH_3-\overset{\displaystyle O}{\overset{\|}{C}}\diagdown_{O^-}(aq) + H_2O(l) + CO_2(g)$$

Figure 17.3.7 The reaction of ethanoic acid with hydrogencarbonate ions.

Citric acid is the weak acid found in the juice of all citrus fruits (Figure 17.3.8). Citric acid contains three carboxylic acid functional groups and has the molecular formula of $C_6H_8O_7$.

Figure 17.3.8 Citrus fruits including lemons, grapefruits, limes, clementines and oranges.

Tip

Carboxylic acids are not readily oxidised (except by combustion) as they are the end products of the oxidation of primary alcohols and aldehydes. However, methanoic acid can be oxidised by acidified potassium manganate(VII) to carbonic acid, H_2CO_3, which decomposes to give carbon dioxide and water. Investigation of the displayed formula of methanoic acid should indicate why it can be oxidised to carbonic acid.

Reduction

Carboxylic acids are much harder to reduce than carbonyl compounds. However, they can be reduced to primary alcohols by the powerful reducing agent lithium tetrahydridoaluminate(III), $LiAlH_4$ (Figure 17.3.9). The reagent is suspended in dry ether (ethoxyethane). Adding dilute acid after the reaction is complete destroys any excess reducing agent.

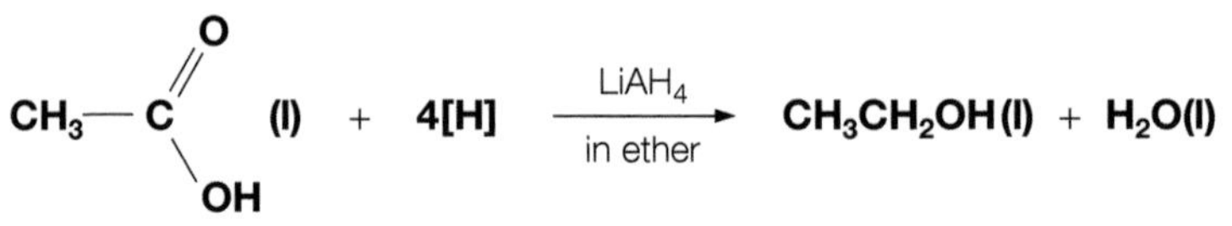

Figure 17.3.9 The reduction of ethanoic acid with $LiAlH_4$.

$$CH_3-C(O)(OH)(l) + 4[H] \xrightarrow[\text{in ether}]{LiAlH_4} CH_3CH_2OH(l) + H_2O(l)$$

Reaction with phosphorus(v) chloride

Phosphorus(v) chloride or phosphorus pentachloride, PCl_5, reacts vigorously with carboxylic acids at room temperature. The reaction replaces the −OH group with a chlorine atom, forming an acyl chloride (Figure 17.3.10). The other product is hydrogen chloride gas, which fumes in moist air.

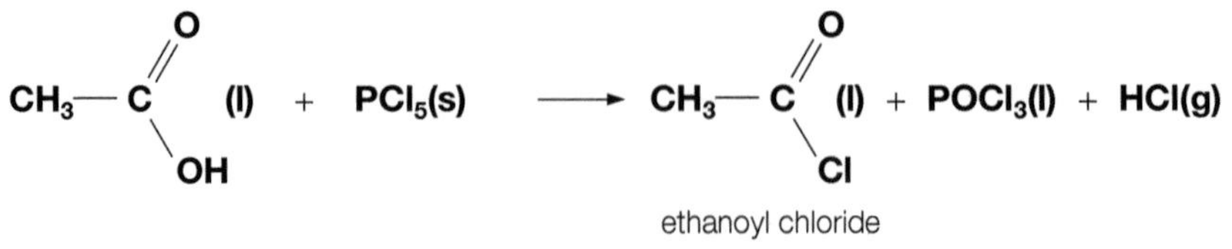

Figure 17.3.10 The reaction of PCl_5 with ethanoic acid.

$$CH_3-C(O)(OH)(l) + PCl_5(s) \longrightarrow CH_3-C(O)(Cl)(l) + POCl_3(l) + HCl(g)$$

ethanoyl chloride

Tip

When treated with PCl_5, the O−H group in an alcohol and the O−H group in a carboxylic acid behave identically. In most other reactions of carboxylic acids, the C=O group modifies the properties of the O−H group.

Test yourself

12 $LiAlH_4$ contains the $[AlH_4]^-$ ion.
  a) Draw a dot-and-cross diagram to show the bonding in the ion.
  b) Deduce the shape of the ion and justify your answer.

13 Write the structural formula and give the name of the product of the reaction of:
  a) butanoic acid with $LiAlH_4$
  b) propanoic acid with PCl_5.

14 a) How can the reaction of carboxylic acids with PCl_5 be used as a test-tube test and what does the test show?
  b) Alcohols and carboxylic acids both give a positive result when tested with PCl_5. How can they be distinguished?

Esterification

Carboxylic acids react with alcohols to form esters (see Section 17.3.5). The two organic compounds are mixed and heated under reflux in the presence of a small amount of a strong acid catalyst such as concentrated sulfuric acid (Figure 17.3.11).

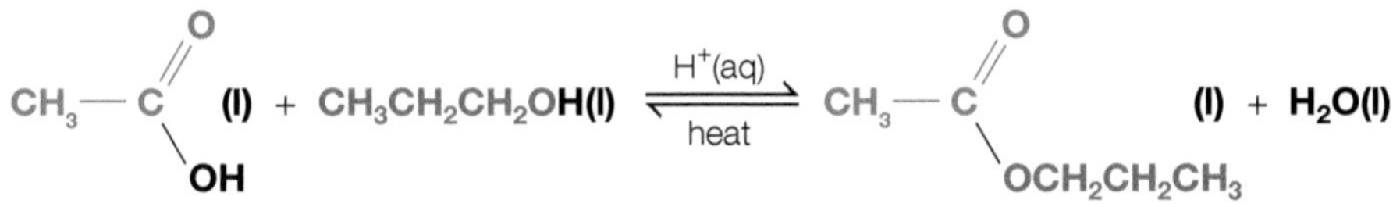

Figure 17.3.11 The formation of the ester propyl ethanoate from ethanoic acid and propan-1-ol.

This reaction is reversible. The conditions for reaction have to be arranged to increase the yield of the ester. One possibility is to use an excess of either the acid or the alcohol, depending on which is the more available or cheaper. Using more concentrated sulfuric acid than needed for its catalytic effect can also help because the acid reacts with the water formed. However, the amount of catalyst added is small so this effect is limited. In some esterification reactions it is possible to distil off either the ester or the water as they form, which encourages the reaction to go to completion.

Test yourself

15 Use Le Chatelier's principle to discuss methods which could be used to increase the yield of an ester formed from an acid and an alcohol.

16 Figure 17.3.11 shows that the $\mathbf{H_2O}$ molecule formed in the esterification reaction is made up of the **–OH** group from the acid and the **H** atom from the alcohol (all in black type). How did chemists confirm that this happened rather than the alternative use of H from the acid and OH from the alcohol? (Hint: see Section 6.1.8 in Student Book 1.)

Preparation of an ester

The sequence of diagrams in Figure 17.3.12 shows the procedure for preparing a small sample of an ester.

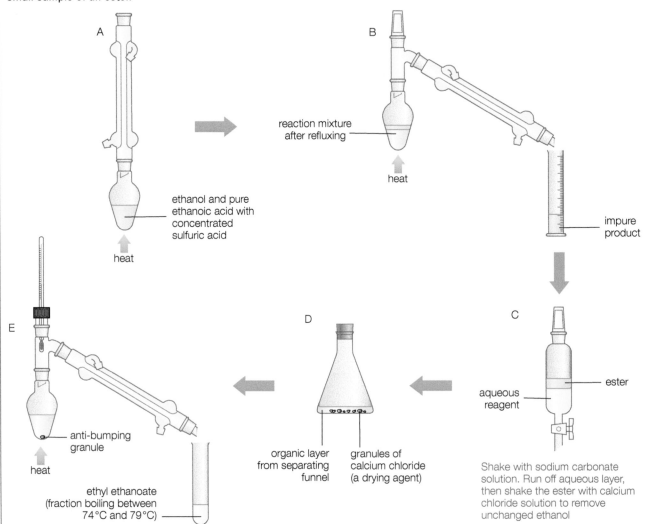

Figure 17.3.12 Stages in the preparation of an ester.

1 Identify what is happening at each of the stages A, B, C, D and E.

2 Write an equation for the reaction which forms the ester, and name the product.

3 What is the purpose of the concentrated sulfuric acid?

4 What are the visible signs of reaction during stage C and what practical precautions are necessary during this stage?

5 A volatile by-product distils off in the boiling range 35–40 °C before the ester in stage E. Suggest a structure for this by-product, which has the molecular formula $C_4H_{10}O$.

6 Calculate the percentage yield if the actual yield is 50 g from 40 g ethanol and 52 g ethanoic acid.

> **Tip**
>
> Refer to Practical skills sheet 3, 'Assessing hazards and risks', which you can access via the QR code for Chapter 17.3 on page 321.

17.3.4 Acyl chlorides

Chemists value acyl chlorides as reactive compounds for synthesis, both on a small laboratory scale and on a large scale in industry. These compounds cause acylation and often provide the easiest way to make important products such as esters.

> **Key term**
>
> **Acylation** is a reaction which substitutes an acyl group for a hydrogen atom. The H atom may be part of an –OH group, an –NH₂ group or a benzene ring.

Reaction with water

Ethanoyl chloride is a colourless liquid that fumes as it reacts with moisture in the air. The reaction between ethanoyl chloride and water is violent at room temperature and forms both ethanoic acid and hydrogen chloride.

$$CH_3COCl(l) + H_2O(l) \rightarrow CH_3COOH(l) + HCl(g)$$

Although hydrogen chloride is a colourless gas, it reacts with moisture in the air to form a mist of droplets of hydrochloric acid. However, the formation of hydrogen chloride can be shown more clearly by its reaction with ammonia vapour to form a white smoke of ammonium chloride (Figure 17.3.13).

$$NH_3(g) + HCl(g) \rightarrow NH_4Cl(s)$$

Reaction with alcohols

Ethanoyl chloride and other acyl chlorides also react rapidly with alcohols at room temperature to form esters (Figure 17.3.14). The reaction of acyl chlorides with alcohols is fast and not reversible and is much preferred as a way to prepare esters to the slow and reversible reaction of acids with alcohols (Section 17.3.3).

Figure 17.3.13 Ethanoyl chloride reacts vigorously with water, releasing hydrogen chloride. A smoke of ammonium chloride forms when an ammonia-soaked glass rod is held in these fumes.

> **Tip**
>
> The rapid hydrolysis of acyl chlorides in moist air makes them difficult to store and use on a large scale. Because of this, acid anhydrides are sometimes used in industry instead of acyl chlorides. Anhydrides are less reactive so are easier to store, and they also produce carboxylic acids rather than the more corrosive fumes of hydrochloric acid when they react. Anhydrides are also cheaper than acyl chlorides. (See Core practical 16: The preparation of aspirin, in Section 17.3.5.)

CH₃—C(=O)—Cl + CH₃CH₂CH₂OH ⟶ CH₃—C(=O)—OCH₂CH₂CH₃ + HCl

Figure 17.3.14 The formation of an ester from ethanoyl chloride and propan-1-ol.

Reaction with ammonia

Acyl chlorides react rapidly with concentrated ammonia to form amides (see Section 18.2.5). For example, when ethanoyl chloride is carefully added to a concentrated aqueous solution of ammonia, a vigorous reaction takes place producing fumes of hydrogen chloride and ammonium chloride plus a residue of ethanamide.

$$CH_3COCl(l) + NH_3(aq) \rightarrow CH_3CONH_2(s) + HCl(g)$$
$$\text{ethanamide}$$

$$HCl(g) + NH_3(g) \rightarrow NH_4Cl(s)$$
$$\text{ammonium chloride}$$

Reaction with amines

Amines react with acyl chlorides to form N-substituted amides (see also Section 18.2.3). This, and other reactions of ethanoyl chloride, are summarised in Figure 17.3.15.

$$CH_3COCl(l) + CH_3CH_2NH_2(aq) \rightarrow CH_3CONHCH_2CH_3(s) + HCl(g)$$
$$\text{N-ethyl ethanamide}$$

The pain reliever paracetamol is an N-substituted amide (see Activity: Paracetamol – an alternative to aspirin, in Section 18.2.3).

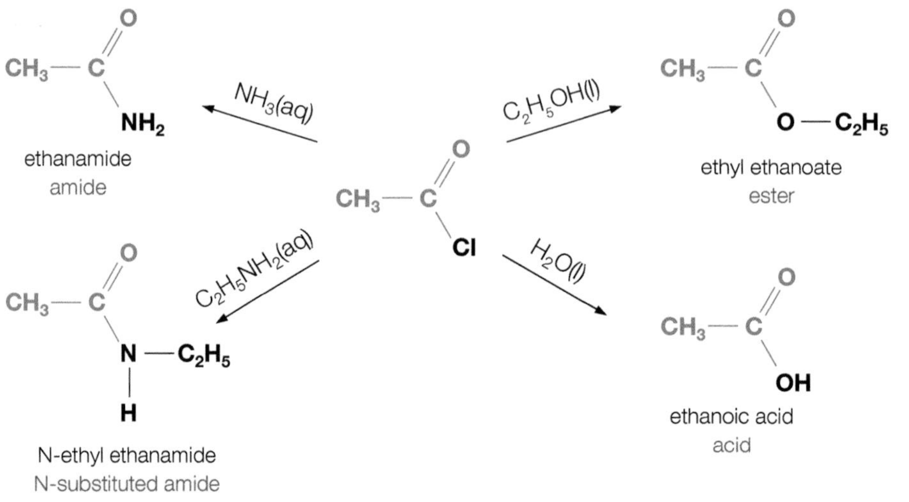

Figure 17.3.15 A summary of the reactions of ethanoyl chloride. All the reactions happen quickly at room temperature. The ethanoyl group in the products is shown in red.

ethanamide
amide

N-ethyl ethanamide
N-substituted amide

ethyl ethanoate
ester

ethanoic acid
acid

Test yourself

17 Write an equation for the reaction between ethanoyl chloride and water. Show that this is an example of hydrolysis.

18 Write an equation for the formation of ethanamide from ethanoyl chloride to show why two moles of ammonia are required for the reaction with one mole of the acyl chloride.

19 Draw the structure and name the product of the reaction of propanoyl chloride and butylamine.

20 The ester, propyl propanoate, can be prepared by reacting propan-1-ol with either propanoic acid or propanoyl chloride. Write an equation for each method and discuss any advantages and disadvantages of using propanoyl chloride.

17.3.5 Esters

Occurrence and uses

Many of the sweet-smelling compounds found in perfumes and fruit flavours are esters. Some drugs used in medicine are esters, including aspirin, paracetamol and the local anaesthetics novocaine and benzocaine. The insecticides malathion and pyrethrin are also esters. Compounds with more than one ester link include fats and oils, as well as polyester fibres. Other esters are important as solvents and plasticisers.

Some esters have odours that resemble particular fruit flavours. Examples are propyl ethanoate (pear), ethyl butanoate (pineapple), octyl ethanoate (orange), 2-methylpropyl ethanoate (apple). Natural fruit flavours are complex mixtures of esters and other compounds, including carboxylic acids. Artificial flavourings using only a few esters are therefore unlikely to replicate natural flavours exactly.

Names and structures

The general formula for an ester is RCOOR′, where R and R′ are alkyl or aryl groups.

The name of an ester is in two parts derived from the acid and the alcohol used to prepare the ester.

The parent carboxylic acid gives the ending of the name. For instance, esters of ethanoic acid all contain the CH_3COO group, so all have names which end in ethanoate.

The rest of the ester molecule, the R′, comes from the alcohol and is an alkyl or aryl group with a name such as methyl or ethyl (Figure 17.3.16).

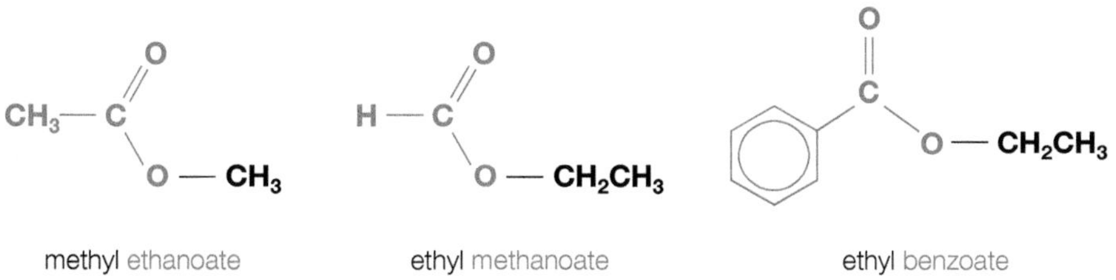

methyl ethanoate   ethyl methanoate   ethyl benzoate

Figure 17.3.16 The names and structures of some esters.

Physical properties

Esters such as ethyl ethanoate are volatile liquids and only slightly soluble in water. All esters contain polar C=O and C–O bonds, but they do not contain O–H bonds and therefore are unable to form hydrogen bonds to each other. This makes esters much more volatile than acids or alcohols with similar M_r values.

Esters with short carbon chains are slightly soluble in water. However, as the non-polar carbon chain length increases, the attractions between the polar bonds in esters and water molecules become insufficient to cause overall solubility.

Test yourself

21 Give the name and displayed formulae of the esters formed when:

   a) butanoic acid reacts with propan-1-ol
   b) ethanoic acid reacts with methanol
   c) ethanoic acid reacts with butan-1-ol.

22 Draw each ester:

   a) propyl ethanoate

   b) ethyl propanoate
   c) 2-methylpropyl ethanoate.

23 Explain, in terms of intermolecular forces, why:

   a) the boiling temperature of ethyl ethanoate is similar to that of ethanol but lower than that of ethanoic acid
   b) ethyl ethanoate is less soluble in water than either ethanol or ethanoic acid.

Hydrolysis reactions

Hydrolysis splits an ester into an alcohol and an acid (or the salt of an acid). Acids or bases can catalyse the hydrolysis.

Hydrolysis catalysed by an acid is a reversible reaction. (Figure 17.3.17). It is the reverse of the reaction used to synthesise esters from carboxylic acids (see Section 17.3.3).

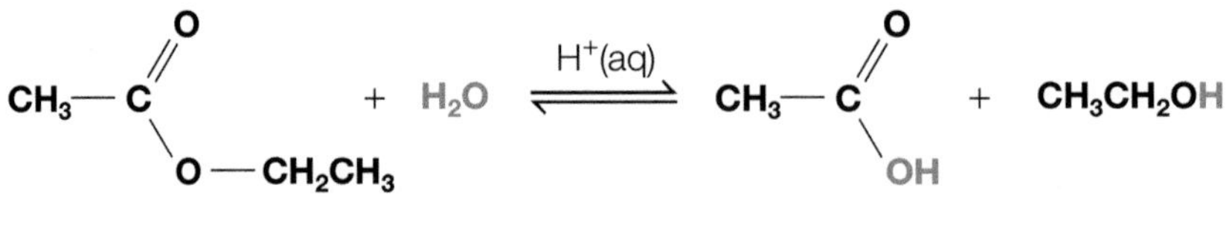

Figure 17.3.17 Hydrolysis of an ester. These are the products when an ester is heated with an excess of dilute acid, such as hydrochloric acid. This reaction is reversible.

$$CH_3-C\overset{O}{\underset{O-CH_2CH_3}{}} + H_2O \underset{}{\overset{H^+(aq)}{\rightleftharpoons}} CH_3-C\overset{O}{\underset{OH}{}} + CH_3CH_2OH$$

ester   acid   alcohol

Base catalysis is generally more efficient because it is not reversible (Figure 17.3.18). This is because the acid formed loses its proton by reacting with excess alkali. This turns it into a negative ion which does not react with the alcohol.

Figure 17.3.18 The result of hydrolysing ethyl ethanoate by heating it with an aqueous alkali such as sodium hydroxide. The salt and alcohol produced do not react with each other, so this reaction is not reversible.

$$CH_3-C\overset{O}{\underset{O-CH_2CH_3}{}} + OH^- \longrightarrow CH_3-C\overset{O}{\underset{O^-}{}} + CH_3CH_2OH$$

ester   salt   alcohol

Test yourself

24 Identify the products of heating:
   a) propyl butanoate with dilute hydrochloric acid
   b) ethyl methanoate with aqueous sodium hydroxide.
25 Under acid conditions the reaction of ethyl ethanoate with water is reversible.
   a) What conditions favour the hydrolysis of the ester?
   b) How do these conditions compare with those for the synthesis of the ester?

Tip

In Core practical 16 the purity of the aspirin formed is checked using melting temperature data. The purity of the aspirin can also be checked by chromatography (see Section 19.5).

For practical guidance, refer to Practical skills sheet 11, 'Synthesis of an organic solid', which you can access via the QR code for Chapter 17.3 on page 321.

The preparation of aspirin

From ancient times, the bark of willow trees and meadowsweet flowers had been used to relieve pain and reduce fevers. In the nineteenth century, chemists identified that material in the willow bark was converted in the body to salicylic acid (2-hydroxybenzoic acid), which they named after *Salix*, the willow tree genus.

For several years, salicylic acid was given to patients to relieve pain. Although it was effective, the compound tasted bitter and caused irritation of the mouth and stomach, so alternatives were investigated. The sodium salt of salicylic acid was found to relieve pain but tasted unpleasant and frequently made patients vomit.

In 1897, Felix Hoffman, working for the Bayer Company, synthesised the ester acetylsalicylic acid, which was effective, tasted less unpleasant and was less irritating. He called his product aspirin from *A* (for acetyl) and *spirin* (from the name *Spirea ulmaria* for the meadowsweet flower).

Aspirin sold extremely well and its production is thought to mark the start of the modern pharmaceutical industry.

Preparation of aspirin

To prepare aspirin, salicylic acid (2-hydroxybenzoic acid) is reacted with ethanoic anhydride. A small amount of phosphoric acid is used as a catalyst to speed up the reaction (Figure 17.3.19). The following is a summary of the procedure to prepare and purify aspirin.

Preparation

A Add 2.0 g of 2-hydroxybenzoic acid to a flask and, in a fume cupboard, add 4.0 cm³ of ethanoic anhydride (density = 1.08 g cm⁻³) and 5 drops of 85% phosphoric acid.

B Attach a reflux condenser to the flask and, using a water bath, heat the mixture under reflux for 10 minutes.

C Remove the flask for the water bath, allow to cool for a few minutes then carefully add 10 cm³ water down the condenser.

D Cool the flask in ice-water and leave until crystallisation is complete. It may be necessary to stir with a glass rod to start crystallisation.

E Filter off the impure solid using a Buchner funnel.

Recrystallisation

F Transfer the impure solid to a flask and insert a reflux condenser.

G Add about 5 cm³ of ethanol down the condenser and heat the flask in a water bath until the crystals dissolve. If the solid does not all dissolve, add a further 5 cm³ of ethanol and continue to warm the solution.

H When the solid has all dissolved, add about 20 cm³ of warm water. Allow the solution to cool slowly and crystals will form. Then cool the flask in ice-water.

I Filter off the crystals using a Buchner funnel. Wash the crystals with about 10 cm³ of ice-cold ethanol then suck air through the solid for about 15 minutes to dry it as much as possible.

J Weigh the pure, dry aspirin.

K Measure the melting temperature of the aspirin produced.

Questions

1 Identify the particularly hazardous chemicals used in the preparation and purification and state the precautions needed when using them.

2 Confirm that the ethanoic anhydride is in excess and calculate the theoretical yield of aspirin.

3 If a student produced 1.8 g aspirin, calculate the percentage yield.

4 Suggest why a reflux condenser is used in steps B and G.

5 Why can crystallisation be sudden during stirring in step D?

6 State why the volume of ethanol used in the recrystallisation is as low as possible.

7 Why is the flask cooled in ice in step H?

8 Why was the suction turned off when the ice-cold ethanol was added in step I?

9 Give practical details in step J to describe how the mass of the aspirin is found.

10 The melting point of pure aspirin is 138–140 °C. Give a reason in each case why a measured melting point might be lower than 138 °C or higher than 140 °C.

11 Suggest why old aspirin tablets that have been exposed to moisture often smell of vinegar.

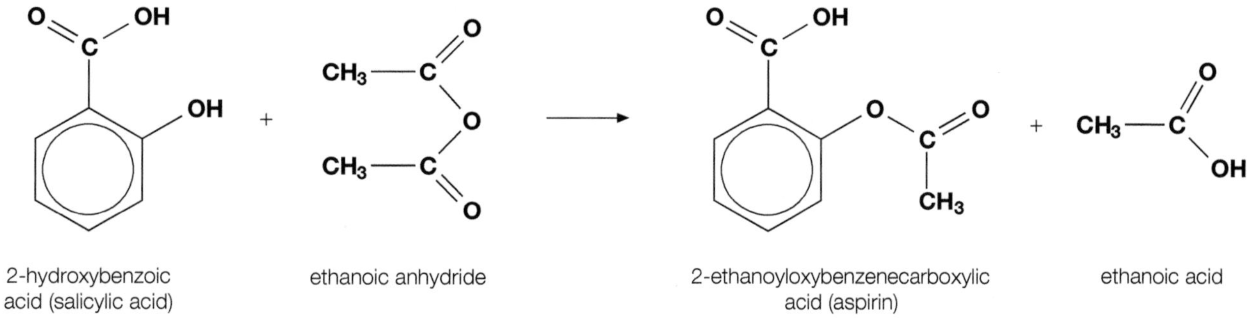

2-hydroxybenzoic acid (salicylic acid) + ethanoic anhydride → 2-ethanoyloxybenzenecarboxylic acid (aspirin) + ethanoic acid

Figure 17.3.19 Formation of aspirin from salicylic acid.

Triglycerides – fats, oils and fatty acids

Fats and vegetable oils are esters of naturally occurring, long-chain carboxylic acids, often called fatty acids, with the alcohol propane-1,2,3-triol, better known as glycerol. The three —OH groups in a glycerol molecule can form three ester links with fatty acids, giving rise to triglycerides (Figure 17.3.20). The fatty acids may be saturated or unsaturated, see Table 17.3.1.

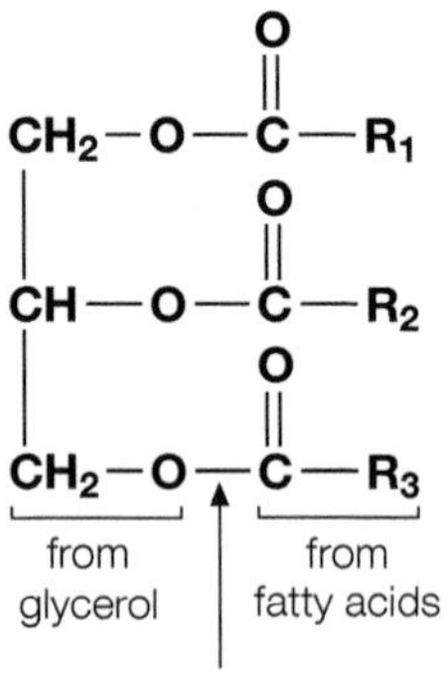

Figure 17.3.20 The general structure of a triglyceride. In natural fats and vegetable oils the hydrocarbon chains, R_1, R_2 and R_3, may all be the same or they may be different.

Table 17.3.1 Examples of fatty acids. The *cis–trans* system for naming geometric isomers is still generally used for fatty acids.

Fatty acid	Chemical name	Formula
Palmitic	Hexadecanoic	$CH_3(CH_2)_{14}COOH$
Stearic	Octadecanoic	$CH_3(CH_2)_{16}COOH$
Oleic	*cis*-Octadec-9-enoic	$CH_3(CH_2)_7CH=CH(CH_2)_7COOH$
Linoleic	*cis, cis*-Octadec-9,12-dienoic	$CH_3(CH_2)_4CH=CHCH_2CH=CH$ $(CH_2)_7COOH$

Fats are solid at around room temperature (below 20 °C) and contain triglycerides with a high proportion of saturated fatty acids. Solid triglycerides are generally found in animals. In lard, for example, the main fatty acids are palmitic acid (28%), stearic acid (8%) and about 56% oleic acid.

Triglycerides with unsaturated fatty acids have lower melting temperatures and have to be cooler before they solidify. Triglycerides of this kind occur in plants and are liquids at around room temperature. In a vegetable oil such as olive oil, the main fatty acids are oleic acid (80%) and linoleic acid (10%).

It is accepted that saturated fats contribute to heart disease, so there is emphasis on eating unsaturated and polyunsaturated fats rather than saturated animal fats.

Alkaline hydrolysis of triglycerides produces soaps and glycerol. Soaps are the sodium or potassium salts of fatty acids and help to remove greasy dirt because they have an ionic (water-loving) head and a long hydrocarbon (water-hating) tail. Most toilet soaps are made from a mixture of animal fat and coconut palm oil. Soaps from animal fat are less soluble and longer lasting. Soaps from palm oils are more soluble so that they lather quickly but also wash away more quickly. Bars of soap also contain a dye and perfume, together with an antioxidant to stop the soap and air combining to make irritant chemicals.

In a similar reaction to hydrolysis, if a triglyceride is heated with methanol in the presence of a base catalyst, a trans-esterification reaction takes place. Methyl esters of the fatty acids are produced together with glycerol. These esters are used as the renewable fuel, biodiesel.

1 Write definitions for the terms 'triglyceride' and 'fatty acid'.
2 Classify the acids in Table 17.3.1 as saturated or unsaturated.
3 Give the name for the relevant acids in Table 17.3.1 using *E*-, *Z*- terminology.
4 Draw the skeletal formulae of palmitic acid and of linoleic acid.
5 Use your answers to Question 4 to explain why triglycerides of palmitic acid are solids at room temperature whereas triglycerides of linoleic acid are liquid oils.
6 Write an equation for the alkaline hydrolysis of the triglyceride in Figure 17.3.20 to form glycerol and soap.
7 Write an equation for the reaction of the triglyceride in Figure 17.3.20 with methanol to form glycerol and three methyl esters. Use your equation to explain why the process is called 'trans-esterification'.
8 Suggest a reason why the product of reacting a vegetable oil with methanol produces a much better fuel than the original oil itself.
9 Suggest a reason why the alcohol and the triglyceride must be very dry to ensure a good yield of biodiesel.
10 Why is it increasingly important to develop new raw materials for making biofuels in order to avoid using vegetable oils or corn starch?

17.3.6 Polyesters

Polyesters are polymers in which the monomers are linked together by ester groups. The formation of polyesters involves a series of condensation reactions where water molecules are lost between the monomer molecules as they react to create ester links (see Section 17.3.5). So polyesters are a type of condensation polymer (see also Section 18.2.9).

Polyesters are formed by condensation reactions:

- between acids with two carboxylic acid groups and alcohols with at least two −OH groups
- or between monomers that have both a carboxylic acid group and an −OH group.

The most common polyester is formed from ethane-1,2-diol (ethylene glycol) and benzene-1,4-dicarboxylic acid (terephthalic acid).

This is used to make the plastic containers for fizzy drinks (Figure 17.3.21), where the polymer is called PET. The initials are short for the traditional name for the polymer, which is polyethylene terephthalate.

PET is also used as fibres to make clothing (Figure 17.3.22), where it is often called simply 'polyester' or by the commercial name Terylene® (from terephthalic acid). Fabrics made from polyester are hard wearing, washable and relatively cheap. In a third form, called Mylar®, the polyester is used as plastic sheets. Because of the polymer's strength, but low density, these sheets can be used to make aviation balloons or sails for hang-gliders.

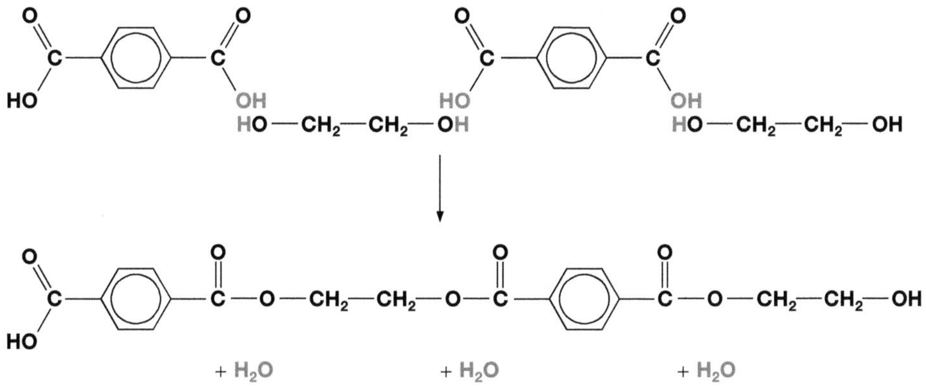

Figure 17.3.23 Condensation polymerisation to produce the polyester Terylene®.

The condensation reactions shown in Figure 17.3.23 can be repeated again and again to produce a polymer with the repeat unit shown in Figure 17.3.24.

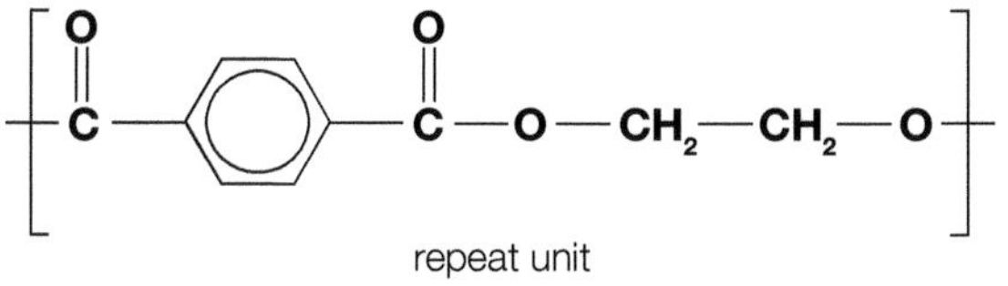

Figure 17.3.24 The repeat unit of Terylene®.

Key terms

A **condensation reaction** is a reaction in which molecules join together by splitting off a small molecule such as water.

A **condensation polymer** is a polymer formed by a series of condensation reactions.

Tip

Diacyl chlorides can also be used instead of diacids.

Figure 17.3.21 Bottles for sparkling drinks are made of the polyester, PET. After heat treatment, this polymer is impermeable to gases.

Figure 17.3.22 The blazer, tie, shirt and trousers that this schoolboy is wearing may all contain polyester (Terylene®).

Perhaps the most important development in polyester chemistry in recent years concerns poly(2–hydroxypropanoic acid), commonly called poly(lactic acid) or PLA. Poly(lactic acid) is possibly the most useful and most versatile of the new biodegradable plastics. It is already used in such diverse goods as plant pots, disposable nappies and absorbable surgical sutures (stitches).

Poly(lactic acid) is manufactured by the condensation polymerisation of lactic acid, a single monomer that contains both a carboxylic acid group and an alcohol group (Figure 17.3.25).

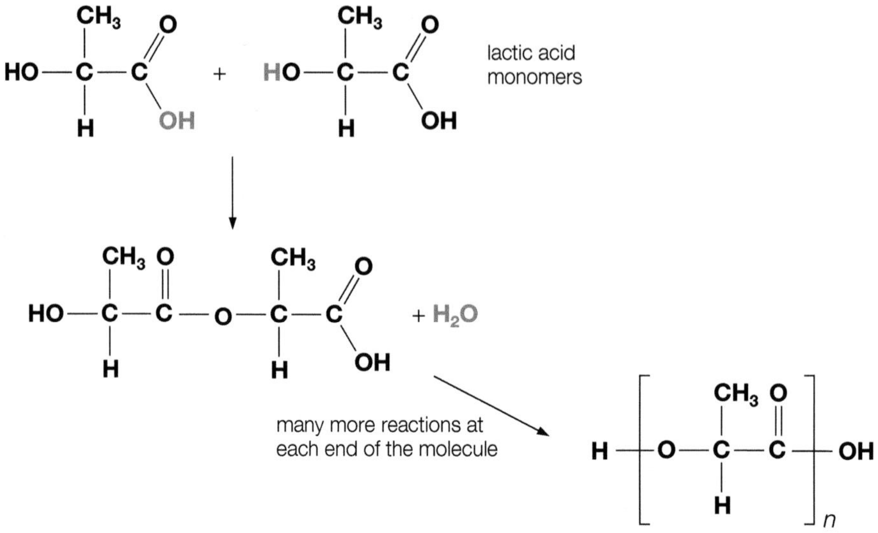

Figure 17.3.25 The synthesis of poly(lactic acid) by condensation polymerisation.

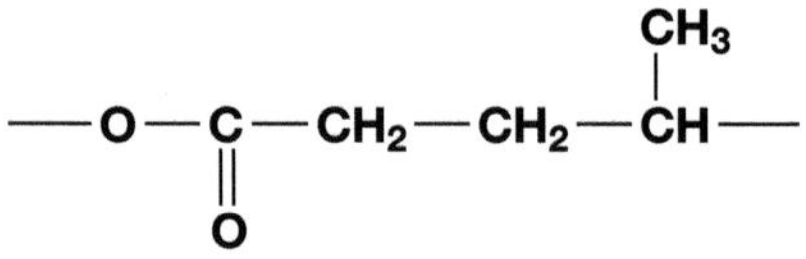

Figure 17.3.26 The repeat unit of a polyester.

Test yourself

26 Draw the repeat unit of the polyester made from propane-1,3-diol and pentanedioic acid.

27 Name the monomer used to make the polymer represented by the repeat unit in Figure 17.3.26.

28 a) Identify the types of intermolecular force that act:

   i) between the chains in polyesters

   ii) between the chains in polyalkenes.

   b) Explain why polyesters are generally biodegradeable whereas polyalkenes are not.

Exam practice questions

1 Account for the fact that butane, propan-1-ol, ethanoic acid and methyl methanoate all have similar relative molecular masses, but their boiling temperatures are 273 K, 371 K, 391 K and 305 K, respectively. *(4)*

2 Copy and complete the diagram below, using propanoic acid as the example, to summarise reactions that form or use carboxylic acids and their derivatives.
  a) Each box should contain the type of compound with the name and formula of the example. *(6)*
  b) Each arrow should be labelled with the type of reaction together with the reagents and conditions. *(8)*

3 The structure of an ester is shown:
$$CH_3CH=CHCH_2CH_2COOCH_3$$

Give the structures of the organic products formed when this ester reacts with:
  a) a solution of bromine in hexane at room temperature *(1)*
  b) hydrogen in the presence of a nickel catalyst at 140 °C *(1)*
  c) aqueous sodium hydroxide when heated under reflux. *(1)*

4 a) Describe two examples of test tube reactions you could use to show the similarities and differences between ethanoic acid and hydrochloric acid. *(3)*
  b) Explain, with the help of equations, the observations you have described in (a). *(6)*

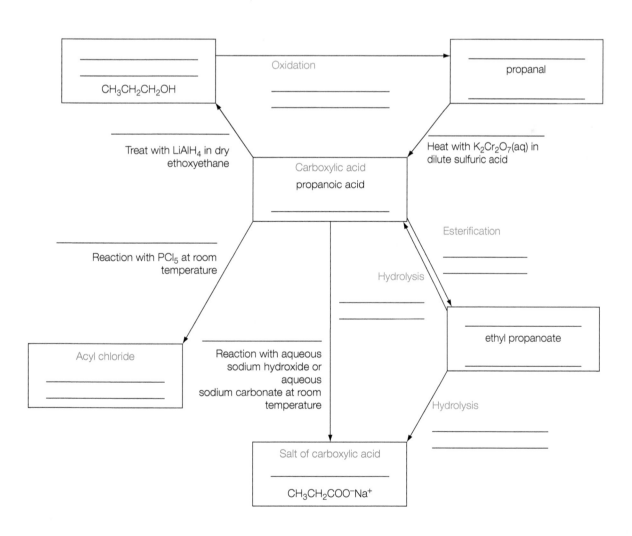

5 Predict the names and structures of the main organic products of each of these reactions:

  a) propanoic acid and phosphorus(v) chloride *(1)*

  b) butanoic acid and $LiAlH_4$ in ethoxyethane *(1)*

  c) pentyl ethanoate and hot, aqueous sodium hydroxide *(2)*

  d) ethanoic acid and aqueous calcium hydroxide *(1)*

  e) ethanoyl chloride and propan-2-ol. *(1)*

6 Ibuprofen is a painkiller with this skeletal formula:

  a) What is the molecular formula of ibuprofen? *(1)*

  b) Suggest whether or not ibuprofen is soluble in water. Explain your answer in terms of intermolecular forces. *(3)*

  c) Draw the displayed formula of the organic products formed when ibuprofen reacts with:

  i) dilute sodium hydroxide solution *(1)*

  ii) ethanol and a little sulfuric acid on warming. *(1)*

7 $1 cm^3$ of ethanol and $1 cm^3$ of ethanoic acid were added to a test tube followed by 3 drops of concentrated sulfuric acid. The test tube was then warmed in a hot water bath for few minutes, after which the contents of the tube were poured into a beaker containing about $50 cm^3$ of cold water. An immiscible liquid with a glue-like smell floated on top of the water in the beaker.

  a) Write an equation for the reaction, showing the structures of the organic compounds involved. *(2)*

  b) State the role of the sulfuric acid. *(1)*

  c) Why was the test tube not heated directly using a Bunsen burner? *(2)*

  d) Name the substance with the glue-like smell. *(1)*

  e) Why was the mixture poured into cold water before testing the smell? *(2)*

8 a) Explain the meaning of the term 'condensation polymerisation'. *(2)*

  b) A section of a polyester is shown.

  Give the structures of a pair of compounds that can react to form this polyester and name the functional groups in the compounds. *(4)*

  c) State one important use of polyester polymers and state the properties of the polymer on which the use depends. *(2)*

9 Three isomeric acids W, X and Y have the molecular formula $C_8H_6O_4$. They all contain a benzene ring. One mole of each acid reacts with two moles of sodium hydroxide.

When each of the acids is heated separately, W and X melt without decomposing. Acid Y decomposes at about $250°C$ to form Z, $C_8H_4O_3$.

Compound X is used to make a polyester with ethane-1,2-diol in which the polymer molecules are linear. Suggest structures for W, X, Y and Z and justify your answers. *(6)*

10 A, B and C are three isomers of $C_2H_4O_2$.

A reacts with sodium carbonate to produce an effervescence of a colourless gas.

B reacts with Fehling's solution to give a red precipitate.

C is almost insoluble in water, but reacts with dilute sodium hydroxide to form soluble products.

Deduce the structure of each isomer and give its name. Explain your deductions. *(6)*

11 Consider this reaction sequence:

$$C_3H_7CN \xrightarrow{step1} C_3H_7COOH \xrightarrow{step2} X$$
$$\xrightarrow[NH_3]{step3} CH_3CH_2CH_2CONH_2$$

  a) Give the reagents and conditions for steps 1 and 2. *(2)*

  b) Give the name and structure of X. *(2)*

  c) Give the displayed formula and name of the product of step 3. *(2)*

12 Ester **B** is used as a solvent for paint strippers. It is formed from acid **A** as shown in the following equation.

$$\begin{array}{c} \text{COOH} \\ | \\ \text{COOH} \end{array} + 2C_2H_5OH \rightleftharpoons \begin{array}{c} \text{COOC}_2\text{H}_5 \\ | \\ \text{COOC}_2\text{H}_5 \end{array} + 2H_2O$$

acid **A**   ester **B**

$$\Delta H = -20 \, \text{kJ mol}^{-1}$$

In an experiment, 0.50 mol of acid **A** was mixed with 0.80 mol of ethanol and a small amount of concentrated sulfuric acid, and the mixture left to reach equilibrium at a given temperature. The equilibrium mixture formed contained 0.27 mol of ester **B**. The total volume of the mixture was $V\,\text{dm}^3$.

a) Calculate the amounts of each of the other three substances present in the equilibrium mixture with ester **B**. *(3)*

b) Hence calculate a value for K_c for the equilibrium at this temperature and give its units. State why the volume V need not be known. *(4)*

c) State the effect, if any, on the value of K_c of increasing the temperature of the equilibrium mixture. Justify your answer. *(3)*

d) Name ester **B** and suggest a precaution necessary when using it as a paint stripper. *(2)*

13 The four isomeric esters W, X, Y and Z with molecular formula $C_4H_8O_2$ were separately heated under reflux with sodium hydroxide. The mixtures formed were then distilled.

The distillates from W and from X gave a yellow precipitate when treated with iodine and sodium hydroxide, but those from Y and Z did not.

After cooling, the solutions remaining in the distillation flasks were acidified with dilute sulfuric acid.

Solutions in the flasks from W and Z decolourised potassium manganate(VII), but those from X and Y did not.

a) Draw the structures of the four isomeric esters. *(4)*

b) Identify W, X, Y and Z and explain your deductions. *(6)*

Arenes – benzene compounds

18.1.1 Arenes

Arenes are hydrocarbons – such as benzene, methylbenzene and naphthalene. They are ring compounds in which there are delocalised electrons. The simplest arene is benzene. Traditionally chemists have called the arenes 'aromatic' ever since the German chemist Friedrich Kekulé was struck by the fragrant smell of oils such as benzene. In their modern name 'arene', the '**ar-**' comes from **ar**omatic and the ending '**-ene**' points to the fact that they are unsaturated hydrocarbons, like the alkenes. However, the chemistry of arenes is different from that of alkenes in many ways.

Benzene is an important and useful chemical. It was first isolated in 1825 by the fractional distillation of whale oil, which was commonly used for lighting homes. Later, it was obtained by the fractional distillation of coal tar. Today, it is obtained by the catalytic reforming of fractions from crude oil.

Many important compounds, including painkillers such as aspirin, paracetamol and ibuprofen, antiseptics such as Dettol® and TCP® (Figures 18.1.1 and 18.1.2) and polymers such as Terylene® and polystyrene, contain the remarkably stable ring of six carbon atoms, the benzene ring, in their structures.

4-chloro-3,5-dimethylphenol
(Dettol)

2,4,6-trichlorophenol
(TCP)

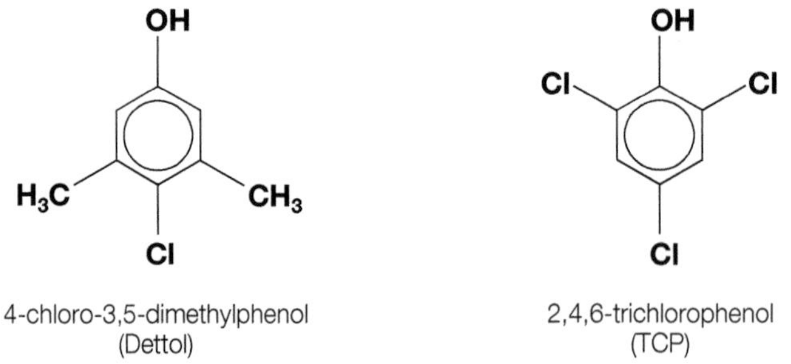

Figure 18.1.2 The antiseptics Dettol and TCP both contain a benzene ring in their structure.

18.1.2 The structure of benzene

Friedrich Kekulé played a crucial part in our understanding of the structure of benzene as the result of a dream. The dream helped Kekulé to propose a possible structure for benzene which had an empirical formula of CH and a molecular formula of C_6H_6. Kekulé had been working on the problem of the structure of benzene for some time. Then, one day in 1865, while dozing in front of the fire, he dreamed of a snake biting its own tail. This inspired him to think of a ring structure for benzene (Figure 18.1.3).

Key term

Arenes are hydrocarbons with a ring or rings of carbon atoms in which there are delocalised electrons.

Delocalised electrons are bonding electrons that are not fixed between two atoms in a bond but shared between three or more atoms.

Figure 18.1.1 The antiseptics used in some throat sprays are similar in structure to Dettol and TCP.

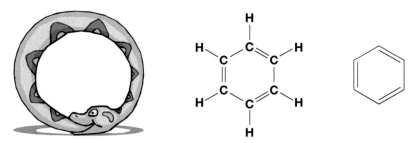

Figure 18.1.3 Kekulé's snake and his structural and skeletal formulae for the structure of benzene. Kekulé's formula would have the systematic name cyclohexa-1,3,5-triene.

Kekulé's structure explained many of the properties of benzene. It was accepted for many years, but still left some problems.

The absence of isomers of 1,2-dichlorobenzene

Kekulé's structure suggests that there should be two isomers of 1,2-dichlorobenzene, one in which the chlorine atoms are attached to carbon atoms linked by a single carbon–carbon bond, the other in which the chlorine atoms are attached to carbon atoms linked by a double carbon–carbon bond (Figure 18.1.4).

In practice, it has never been possible to separate two isomers of 1,2-dichlorobenzene or any other 1,2-disubstituted compound of benzene. To get round this problem, Kekulé suggested that benzene molecules might somehow alternate rapidly between the two possible structures, but this failed to satisfy his critics.

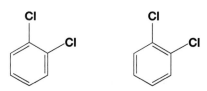

Figure 18.1.4 Possible isomers of 1,2-dichlorobenzene.

The bond lengths in benzene: X-ray diffraction data

The Kekulé structure shows a molecule with alternating single and double bonds. This would imply that three of the bonds are similar in length to the carbon–carbon single bond in alkanes while the other three are similar in length to the carbon–carbon double bond in alkenes. X-ray diffraction studies show that the carbon atoms in a benzene molecule are at the corners of a regular hexagon (Figure 18.1.5). All the bonds are the same length, shorter than single bonds but longer than double bonds (Figure 18.1.6).

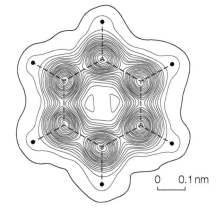

Figure 18.1.5 Electron density map of benzene.

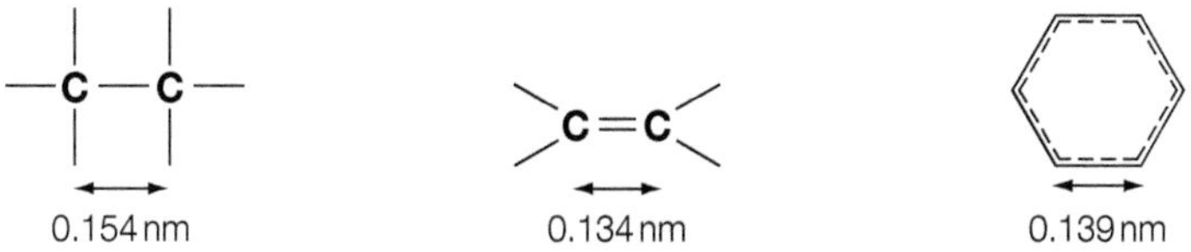

Figure 18.1.6 Carbon-carbon bond lengths in ethane, ethene and benzene.

The resistance to reaction of benzene

An inexperienced chemist looking at the Kekulé structure might expect benzene to behave chemically like a very reactive alkene and to take part in addition reactions with bromine, hydrogen bromide and similar reagents. Benzene does not do this. The compound is much less reactive than alkenes and its characteristic reactions are substitutions, not additions.

Key term

Hydrogenation is the reaction
of hydrogen gas with a molecule
containing a double bond. Hydrogen
adds across the double bond which
becomes saturated.

The stability of benzene: thermochemical data

A study of enthalpy (energy content) changes show that benzene is more stable than expected for a compound with the Kekulé formula. This conclusion is based on a comparison of the enthalpy changes of hydrogenation of benzene and cyclohexene.

Cyclohexene is a cyclic hydrocarbon with one double bond. Like other alkenes, it adds hydrogen in the presence of a nickel catalyst at 140°C to form cyclohexane.

The enthalpy change of the reaction, $\Delta H^{\ominus}$, is $-120\,kJ\,mol^{-1}$ (Figure 18.1.7).

Figure 18.1.7 Hydrogenation of cyclohexene.

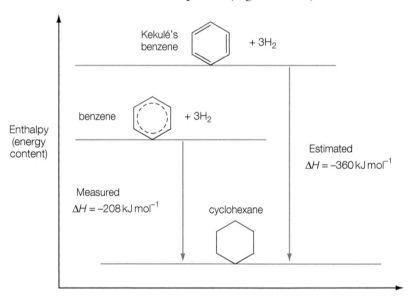

So, if benzene has three carbon–carbon double bonds, as in Kekulé's structure, we might reasonably predict that $\Delta H^{\ominus}$ for the hydrogenation of benzene should be $-360\,kJ\,mol^{-1}$. But, when the hydrogenation is carried out, the measured enthalpy change is only $-208\,kJ\,mol^{-1}$.

The measured enthalpy change is much less exothermic than the estimated value. This suggests that the addition of hydrogen to benzene does not involve normal double bonds and that benzene is actually lower in energy and much more stable than expected (Figure 18.1.8).

Figure 18.1.8 Comparing the measured enthalpy change of hydrogenation of benzene with the estimated enthalpy change of hydrogenation for Kekulé's structure.

Bonding in benzene: infrared data

The fact that benzene does not have carbon–carbon bonds like ethane or ethene is also reflected in its infrared absorption spectrum. IR spectra provide unique fingerprints of compounds in which each bond shows characteristic absorptions at specific frequencies in the infrared region of the electromagnetic spectrum. In IR spectra, the frequencies are normally shown as wavenumbers in units of cm^{-1}, i.e. the number of waves in a centimetre rather than the number of waves per second.

Look closely at the infrared spectrum of benzene in Figure 18.1.9 and that of oct-1-ene, $CH_3(CH_2)_5CH{=}CH_2$ in Figure 18.1.10.

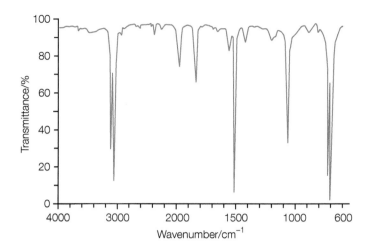

Figure 18.1.9 The infrared absorption spectrum of benzene.

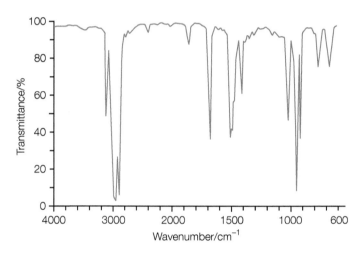

Figure 18.1.10 The infrared absorption spectrum of oct-1-ene.

Notice that benzene does not have the typical strong absorptions of C–H bonds in $-CH_2$ and $-CH_3$ groups in the wavenumber range 2962–2853 cm^{-1}, nor the C=C absorption of an alkene, like oct-1-ene, just below 1700 cm^{-1}.

Instead, and unlike alkanes and alkenes, benzene has strong absorptions at about 3050 cm^{-1} and 750 cm^{-1}. All this provides further evidence that benzene does not have normal C–C or C=C bonds in its structure.

Test yourself

1 Assume that the empirical formula of benzene is CH. What further information is needed to show that its molecular formula is C_6H_6? What methods do chemists use to obtain this information?

2 Draw one possible structure for C_6H_6 that is not a ring. Why does this structure not fit with Kekulé's structure for benzene?

3 An arene consists of 91.3% carbon.

a) What is the empirical formula of the arene?

b) What is the molecular formula of the arene if its molar mass is 92 g mol^{-1}?

c) Draw the structure of the arene.

18.1.3 Delocalisation in benzene

The accumulation of the evidence discussed in Section 18.1.2 led to increased activity in the search for a more accurate model for the structure of benzene. One model was to treat the carbon–carbon bonds in benzene as halfway between single and double bonds and draw them with a full line and a dashed line side-by-side, as in Figure 18.1.6. This model explains the absence of isomers of 1,2-dichlorobenzene, the equal carbon–carbon bond lengths in benzene and also its resistance to reaction. In recent years, the bonding between carbon atoms in benzene has been simplified to a circle inside a hexagon:

Despite this evidence, benzene may be represented in equations or mechanisms either by the Kekulé structure or by the simplified structure above.

Although the simplified structure allows an improved understanding of the properties of benzene, a better insight comes from considering its electronic and orbital structure.

Figure 18.1.11 shows benzene with normal covalent sigma bonds (σ bonds) between its carbon and hydrogen atoms. Each carbon atom uses three of its electrons to form three σ bonds with its three neighbours. This leaves each carbon atom with one electron in an atomic p orbital.

Figure 18.1.11 Sigma bonds in benzene, with one electron per carbon atom remaining in a p orbital.

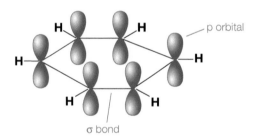

These six p electrons do not pair up to form three carbon–carbon double bonds (consisting of a σ bond plus a π bond) as in the Kekulé structure. Instead, they are shared evenly between all six carbon atoms, giving rise to circular clouds of negative charge above and below the ring of carbon atoms (Figure 18.1.12). This is an example of a delocalised π electron system, which occurs in any molecule where the conventional structure shows alternating double and single bonds. Within the π electron system, the electrons are free to move anywhere.

Molecules and ions with delocalised electrons, in which the charge is spread over a larger region than usual, are more stable than might otherwise be expected. In benzene, this accounts for the compound being $152 \, kJ \, mol^{-1}$ more stable than expected for the Kekulé structure.

The development of ideas concerning the structure of benzene illustrates the way in which theories develop and get modified as new knowledge becomes available.

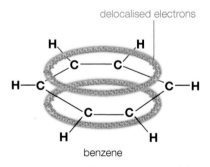

Figure 18.1.12 Representation of the delocalised π bonding in benzene. The circle in a benzene ring represents six delocalised electrons. This way of showing the structure explains the shape and stability of benzene.

4 a) Look carefully at Figure 18.1.8. How much more stable is real benzene than Kekulé's structure for benzene?

   b) Predict the enthalpy changes for the complete hydrogenation of cyclohexa-1,4-diene and of cyclohexa-1,3-diene. Justify your predictions.

5 How does the model of benzene molecules with delocalised π electrons account for the following?

   a) The benzene ring is a regular hexagon.

   b) There are no isomers of 1,2-dichlorobenzene.

   c) Benzene is less reactive than cycloalkenes.

6 When chemists thought that benzene had alternate single and double bonds, it was sometimes named cyclohexa-1,3,5-triene. Why is this now an unsatisfactory systematic name for benzene?

18.1.4 Naming arenes

The name 'benzene' comes from *gum benzoin*, a natural product containing benzene derivatives. These derivatives of benzene are named either as substituted products of benzene or as compounds containing the phenyl group, C_6H_5-. The names and structures of some derivatives of benzene are shown in Table 18.1.1.

Table 18.1.1 The names and structures of some derivatives of benzene.

Systematic name	Substituent group	Structure
Chlorobenzene	Chloro, $-Cl$	C_6H_5-Cl
Nitrobenzene	Nitro, $-NO_2$	$C_6H_5-NO_2$
Methylbenzene	Methyl, $-CH_3$	$C_6H_5-CH_3$
Phenol	Hydroxy, $-OH$	C_6H_5-OH
Phenylamine	Amine, $-NH_2$	$C_6H_5-NH_2$

The names used for compounds with a benzene ring can be confusing. The phenyl group C_6H_5- is used to name many compounds in which one of the hydrogen atoms in benzene has been replaced by another atom or group. The use of phenyl in this way dates back to the first studies of benzene. At this time 'phene' was suggested as an alternative name for benzene, based on a Greek word for 'giving light'. The name 'phene' was suggested because benzene had been discovered in the tar formed on heating coal to produce gas for lighting.

When more than one hydrogen atom is substituted, numbers are used to indicate the positions of substituents on the benzene ring (Figure 18.1.13). The ring is numbered to get the lowest possible numbers. In phenyl compounds, such as phenol and phenylamine, the $-OH$ and $-NH_2$ groups are assumed to occupy the 1 position.

> **Tip**
>
> The name 'benzyl' has been used to represent the group $C_6H_5CH_2-$ as in benzyl chloride, $C_6H_5CH_2Cl$. Good practice now avoids this by using the systematic name for this compound, (chloromethyl)benzene, to remove any confusion with phenyl.

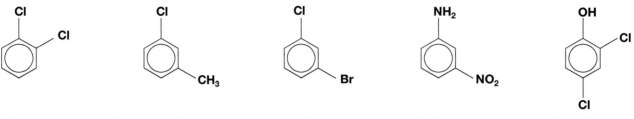

1,2-dichlorobenzene   1-chloro-3-methylbenzene   1-bromo-3-chlorobenzene   3-nitrophenylamine   2,4-dichlorophenol

Figure 18.1.13 Naming disubstituted products of benzene, phenylamine and phenol.

Test yourself

7 Why is the middle compound in Figure 18.1.13 named 1-bromo-3-chlorobenzene and not 1-chloro-3-bromobenzene?

8 An old bottle of chemical was found with the label m-dinitrobenzene. Draw the structure and give the systematic name of this compound.

9 Name each of these disubstituted arenes.

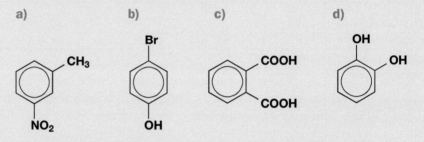

a)   b)   c)   d)

10 Draw and name the isomers of $C_6H_4Cl_2$, $C_6H_3Cl_3$, $C_6H_2Cl_4$ and C_6HCl_5. (Beware of duplicates!)

18.1.5 The properties and reactions of benzene and arenes

Arenes are non-polar compounds with weak London forces between their molecules. The boiling temperatures of arenes depend on the size of the molecules. The bigger the molecules, the higher the boiling temperatures. Benzene and methylbenzene are liquids at room temperature, while naphthalene is a solid (Figure 18.1.14).

Figure 18.1.14 Three arenes: benzene, methylbenzene and naphthalene.

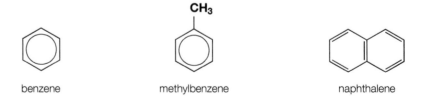

benzene   methylbenzene   naphthalene

Arenes, like other hydrocarbons, do not mix with water, but they do mix freely with non-polar solvents such as cyclohexane.

The reactions of arenes

Arenes burn in air. Unlike straight-chain alkanes and alkenes of similar molar mass, they burn with a very smoky flame because of the high ratio of carbon to hydrogen in their molecules (Figure 18.1.15).

Benzene and other arenes are similar to alkenes in having a prominent electron-dense region. In arenes this is a delocalised ring of π electrons, while in alkenes it is a single localised π bond. Because of this similarity, both arenes and alkenes react with electrophiles in many of their reactions. However, the similarity ends there, because the overall reactions of arenes involve substitution, unlike those of alkenes, which involve addition.

It is quite easy to see why addition reactions are difficult for arenes. If, for example, benzene reacted with bromine in an addition reaction, the ring of delocalised π electrons would be broken (Figure 18.1.16). Given the stability associated with the delocalised π system, this would require much more energy than that needed to break the one double bond in ethene. So, arenes largely undergo substitution rather than addition reactions in order to retain their delocalised π electrons.

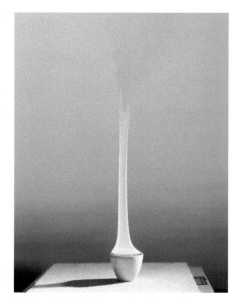

Figure 18.1.15 A sample of methylbenzene burning in air showing the yellow flame and very smoky fumes.

Figure 18.1.16 Benzene does not combine with bromine in an addition reaction because its ring of delocalised π electrons would be broken.

Test yourself

11 The circle in a benzene ring represents six delocalised electrons. Some structures for naphthalene show circles in both rings. Suggest why many chemists prefer to see naphthalene drawn as in Figure 18.1.14 and not with two circles.

12 Explain the existence of weak attractive forces between benzene molecules, which are uncharged and non-polar.

13 a) Explain why benzene does not mix with water.

   b) Name a solvent, other than cyclohexane, with which you would expect benzene to mix freely.

14 Write equations for:

   a) the complete combustion of benzene

   b) the incomplete combustion of methylbenzene to form carbon.

Tip

Some addition reactions to arenes do occur, but if sufficient energy is provided to start these reactions, further addition then occurs until saturated compounds are formed. See hydrogenation reactions in Section 18.1.7.

18.1.6 Electrophilic substitution reactions of benzene

Halogenation – bromination and chlorination

Addition of bromine to an alkene can occur when bromine water alone is added with no need for a catalyst. The π electrons induce a dipole in the bromine molecule and the $\delta+$ bromine atom is a sufficiently strong electrophile for reaction to take place. However, by contrast, the greater stability of the benzene ring means that the induced dipoles and the $\delta+$ bromine atoms in Br_2 are not sufficiently electrophilic to react with benzene.

Key terms

Electrophilic substitution reactions in arenes involve the replacement of a hydrogen atom following attack by an electrophile.

Chemists sometimes use the term **halogen carrier** to describe substances such as iron(III) bromide and aluminium chloride which catalyse the reaction of benzene with chlorine or bromine.

Electrophilic substitution by bromine can occur (Figure 18.1.17) but, as in all electrophilic substitution reactions of benzene, the first step of the reaction involves use of a catalyst to produce a stronger electrophile.

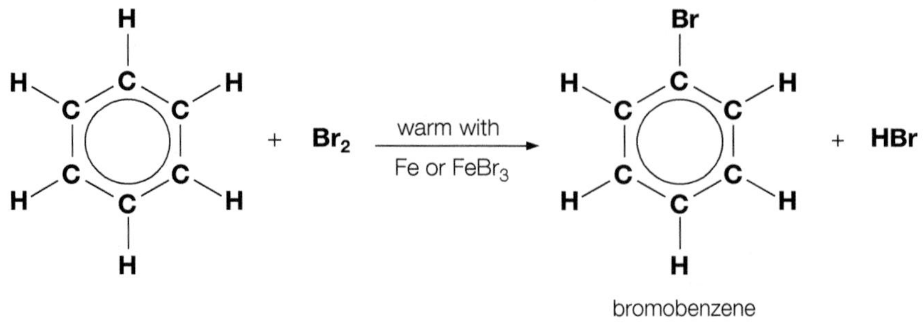

Figure 18.1.17 The reaction of benzene with bromine.

When benzene is warmed with bromine in the presence of iron filings, the bromine first reacts with the iron to form iron(III) bromide.

$$2Fe(s) + 3Br_2(l) \rightarrow 2FeBr_3(s)$$

The iron(III) bromide then acts as a catalyst for the reaction of bromine with benzene by polarising further bromine molecules until a positive Br^+ ion is formed. This ion acts as the electrophile.

$$\overset{\delta+}{Br}-\overset{\delta-}{Br} + FeBr_3 \rightarrow Br-Br\cdots\cdots FeBr_3 \rightarrow Br^+ + FeBr_4^-$$

The Br^+ ion is a reactive electrophile, which is strongly attracted to the delocalised electrons in benzene. As it approaches the benzene ring, the Br^+ ion forms a covalent bond to one of the carbon atoms using two electrons from the π system (Figure 18.1.18). This step produces an intermediate cation.

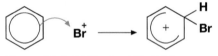

Figure 18.1.18 Electrophilic Br^+ ions use two of the delocalised electrons in benzene to form an intermediate cation.

Tip

In these intermediate cations, the positive charge is delocalised over five carbons. The other carbon, the one at which substitution occurs, is attached to four atoms, so it is saturated and therefore not part of the electron delocalisation.

This intermediate then breaks down to form bromobenzene as two electrons are returned from the C–H bond to the π system and the stable, delocalised ring is restored (Figure 18.1.19). At the same time, an H^+ ion is released from the intermediate cation. This H^+ ion immediately combines with the Br^- ion released in stage 1 to form hydrogen bromide.

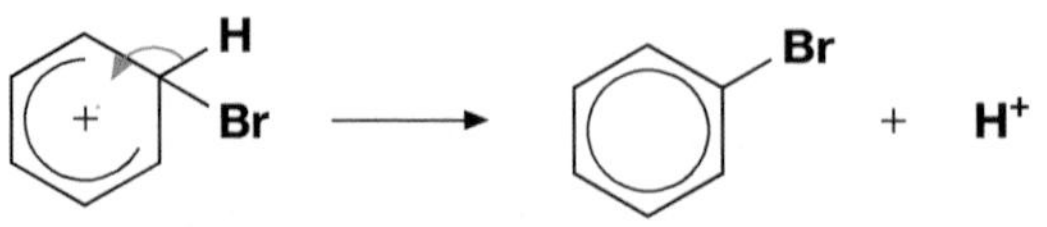

Figure 18.1.19 The intermediate cation breaks down to form bromobenzene.

A similar reaction occurs when benzene is warmed with chlorine in the presence of iron, iron(III) chloride or aluminium chloride. The catalysts are often referred to as **halogen carriers**.

Nitration

Nitration of benzene, and other arenes, is important because it produces a range of important products including dyes and powerful explosives such as TNT (trinitrotoluene, now called 1-methyl-2,4,6-trinitrobenzene) (Figures 18.1.20 and 18.1.21).

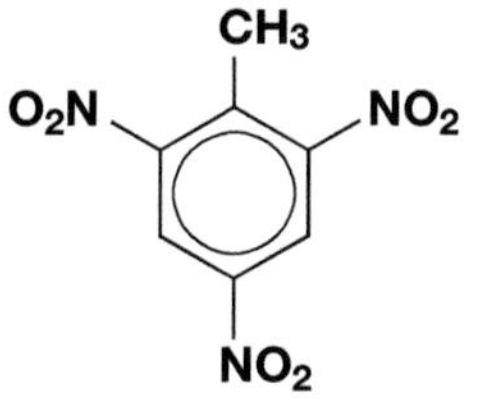

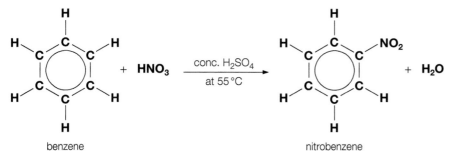

Figure 18.1.21 The structure of TNT.

Figure 18.1.20 Nitrated organic compounds, like TNT (trinitrotoluene) and nitroglycerine, are useful explosives in demolition, mining, tunnelling and road building.

When benzene is warmed to about 55 °C with concentrated nitric acid in the presence of concentrated sulfuric acid, the major product is yellow, oily nitrobenzene (Figure 18.1.22).

benzene + HNO$_3$ $\xrightarrow[\text{at 55 °C}]{\text{conc. H}_2\text{SO}_4}$ nitrobenzene + H$_2$O

Figure 18.1.22 The nitration of reaction of benzene.

An electrophilic substitution reaction occurs in which hydrogen is replaced by a nitro group, $-NO_2$. If the reaction mixture is heated above 55 °C, further nitration occurs forming dinitrobenzene.

At 55 °C, concentrated nitric acid on its own reacts very slowly with benzene, and concentrated sulfuric acid by itself has practically no effect. However, in a mixture of the two, sulfuric acid reacts with nitric acid to produce the nitronium ion, NO_2^+, which is a very reactive electrophile.

Key term

A **nitration** reaction of an arene is an electrophilic substitution where a hydrogen atom is replaced by a nitro group, $-NO_2$.

Sulfuric acid is a stronger acid than nitric so initial protonation of nitric acid occurs.

$$HNO_3 + H_2SO_4 \rightarrow H_2NO_3^+ + HSO_4^-$$

The protonated acid then loses water to form the nitronium ion.

$$H_2NO_3^+ \rightarrow NO_2^+ + H_2O$$

An overall equation is:

$$HNO_3 + H_2SO_4 \rightarrow NO_2^+ + HSO_4^- + H_2O$$

The NO_2^+ ion is a reactive electrophile. It replaces a hydrogen in the benzene ring in a two-step electrophilic substitution mechanism (Figure 18.1.23) similar to that which occurs in the bromination of benzene.

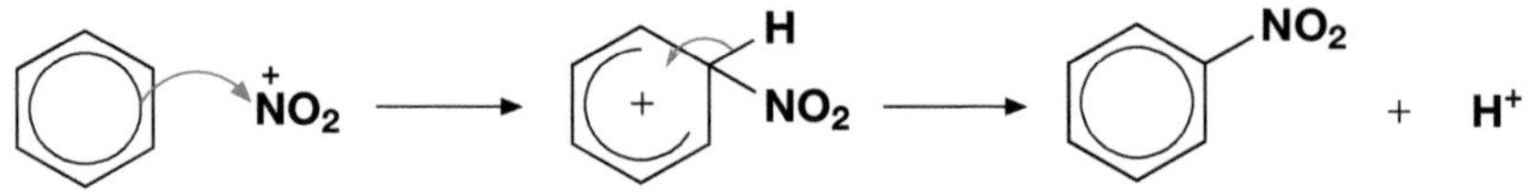

Figure 18.1.23 Electrophilic substitution mechanism for the nitration of benzene.

Tip

In this reaction, concentrated HNO_3 acts as a base and accepts a proton from the stronger acid, H_2SO_4.

Tip

Make sure that the correct bonding is shown. It is the nitrogen atom that is bonded to a carbon atom in the ring, not an oxygen atom, so the bond to the ring from the nitro group must always start at N, not O.

Test yourself

17 Explain why dilute nitric acid does not react with benzene.

18 Write an overall equation for the formation of the nitronium ion in which the water produced is also protonated.

19 Three possible isomers of dinitrobenzene can be produced. One of these isomers is called 1,2-dinitrobenzene.
- **a)** Draw and name the structures of the other two dinitrobenzenes.
- **b)** Why is there no isomer called 1,6-dinitrobenzene?

20 Why is TNT mixed with a compound containing a high proportion of oxygen, such as potassium nitrate, when it is used as an explosive?

21 Write **a)** an overall equation and **b)** a mechanism for the formation of 1-methyl-4-nitrobenzene from methylbenzene.

Alkylation and acylation (Friedel–Crafts reactions)

The **Friedel–Crafts reaction** is an important method for substituting an alkyl group or an acyl group for a hydrogen atom in an arene. This reaction was discovered and developed jointly by the French organic chemist Charles Friedel (1832–1899) and the American, James Crafts (1839–1917). The reaction is used on both a laboratory and an industrial scale.

Key term

The **Friedel-Crafts reaction** is an electrophilic substitution where a hydrogen atom in an arene is replaced by an alkyl or an acyl group following attack by a carbocation or an acylium ion electrophile.

In a Friedel–Crafts reaction, a halogenoalkane or an acyl chloride is refluxed with an arene in the presence of aluminium chloride as catalyst. For example, if benzene is refluxed with chloromethane and aluminium chloride, a substitution reaction occurs forming methylbenzene (Figure 18.1.24).

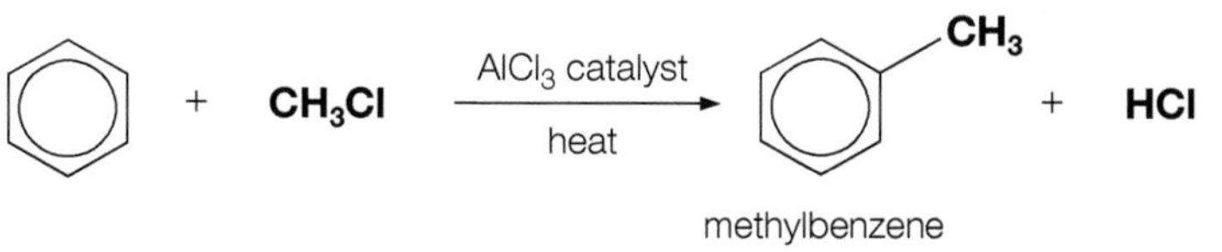

Figure 18.1.24 Friedel-Crafts alkylation of benzene with chloromethane forming methylbenzene.

A similar reaction occurs when benzene is refluxed with the acyl chloride, ethanoyl chloride, plus aluminium chloride as a catalyst. This time, the product is phenylethanone, also known as methylphenylketone (Figure 18.1.25).

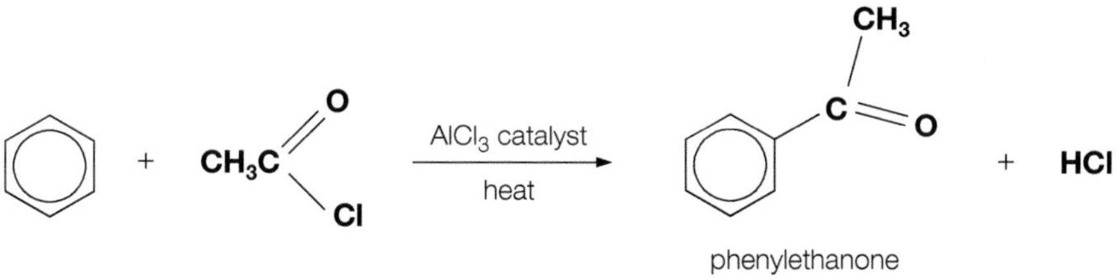

Figure 18.1.25 Friedel-Crafts acylation of benzene with ethanoyl chloride forming phenylethanone.

In Friedel–Crafts reactions, the aluminium chloride plays the important catalytic role in creating the electrophiles which attack benzene. So, when chloromethane is mixed with aluminium chloride, $AlCl_3$ molecules remove Cl^- ions from polar $^{\delta+}CH_3-Cl^{\delta-}$ molecules, allowing reactive carbocations, CH_3^+ ions, to act as electrophiles.

$$^{\delta+}CH_3-Cl^{\delta-} + AlCl_3 \rightarrow CH_3^+ + AlCl_4^-$$
$$\text{electrophile}$$

These reactive CH_3^+ electrophiles then attack the delocalised π system of benzene molecules to form an intermediate cation, which breaks down producing methylbenzene and H^+ ions (Figure 18.1.26).

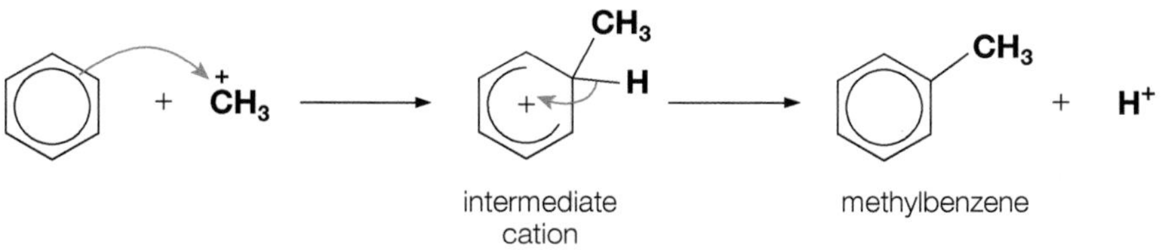

Figure 18.1.26 The reaction of CH_3^+ electrophiles with benzene in the Friedel-Crafts alkylation reaction to produce methylbenzene.

Finally, the aluminium chloride catalyst is regenerated as H^+ ions released in the electrophilic substitution react with $AlCl_4^-$ ions.

$$H^+ + AlCl_4^- \rightarrow HCl + AlCl_3$$

Tip

Acylium ions are formed in mass spectrometers as fragment ions of compounds such as ketones or acid derivatives (Chapter 19).

In a Friedel–Crafts acylation reaction, an acyl chloride, RCOCl, reacts with aluminium chloride to from an acylium ion (Figure 18.1.27). This ion acts as the electrophile in the two-step electrophilic substitution reaction (Figure 18.1.28).

$$R{-}\underset{\underset{O}{\|}}{C}{-}Cl + AlCl_3 \longrightarrow R{-}\underset{\underset{O}{\|}}{C^+} + AlCl_4^-$$

an acylium ion

Figure 18.1.27 Formation of an acylium ion.

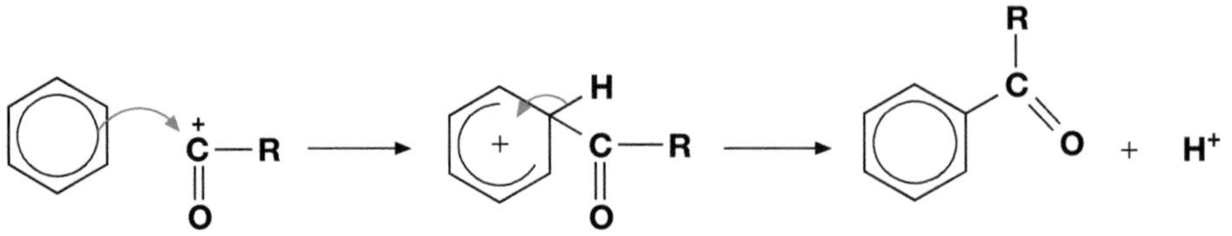

Figure 18.1.28 Electrophilic substitution mechanism for Friedel-Crafts acylation of benzene.

Tip

In Friedel-Crafts alkylation, the initial product contains an alkyl group attached to a benzene ring. Alkyl groups are electron releasing, so the electron density on the ring is greater than on benzene, which makes further substitution likely.

In Friedel-Crafts acylation, the initial product contains a carbonyl group attached to a benzene ring. The carbonyl group withdraws electron density from the benzene ring, so further substitution is unlikely.

Test yourself

22 This question is about the Friedel–Crafts reaction between benzene and ethanoyl chloride, CH_3COCl, in the presence of aluminium chloride, $AlCl_3$, as catalyst.

a) Write an equation to show how $AlCl_3$ molecules react with polarised CH_3COCl molecules to produce reactive acylium ions.

b) Write an equation for the electrophilic substitution of benzene by acylium ions to produce phenylethanone.

c) Write an equation to show how molecules of the aluminium chloride catalyst are regenerated.

23 **a)** Why must the reaction mixture be completely dry during a Friedel–Crafts reaction?

b) Draw the structures of the products of a Friedel–Crafts reaction of benzene with:

i) 2-iodo-2-methylpropane

ii) propanoyl chloride.

c) Suggest a reason for using an iodoalkane instead of a chloroalkane in a Friedel–Crafts reaction.

18.1.7 Addition reaction of benzene with hydrogen

The characteristic reactions of benzene involve substitution because this type of reaction retains the delocalised π electron system with its associated stability. However, addition reactions involving disruption of the π electron system do occur. Benzene, like alkenes, will undergo addition with hydrogen in the presence of a nickel catalyst, but at considerably higher temperatures (Figure 18.1.29).

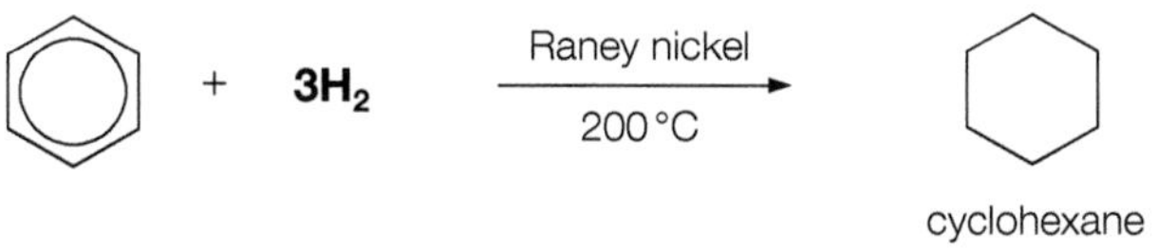

$$\text{benzene} + 3H_2 \xrightarrow[200\,°C]{\text{Raney nickel}} \text{cyclohexane}$$

Figure 18.1.29 The addition of hydrogen to benzene forming cyclohexane.

A higher temperature is needed with benzene in order to break up the stable π electron system and allow addition to occur. A special finely divided form of nickel, called Raney nickel, is also used because this has an extremely high surface area. The addition of hydrogen to benzene is thought to occur in three stages. First, the formation of cyclohexa-1,3-diene, then cyclohexene and finally cyclohexane.

The catalytic hydrogenation of benzene is important industrially in the manufacture of cyclohexane, which is used to make nylon.

Tip

In the presence of ultraviolet light, chlorine will add to benzene in a free-radical reaction to form a mixture of chlorinated cyclohexanes including 1,2,3,4,5,6-hexachlorocyclohexane, shown below. There are several isomers of $C_6H_6Cl_6$, one of which has been used as commercial insecticide, but such use is now restricted because of concerns about its toxicity to humans.

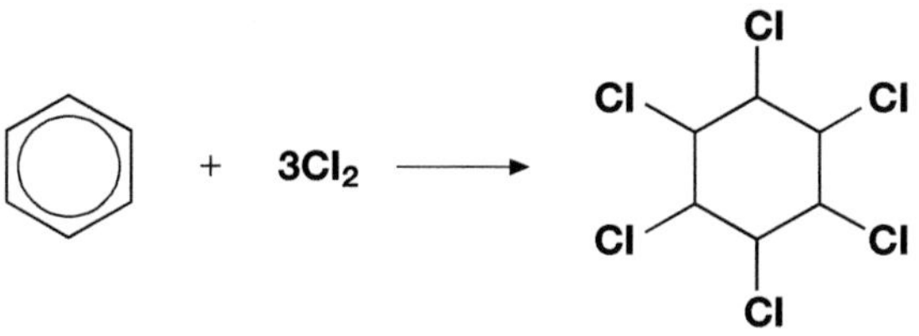

$$\text{benzene} + 3Cl_2 \longrightarrow \text{1,2,3,4,5,6-hexachlorocyclohexane}$$

Test yourself

24 a) Why is Raney nickel used in the manufacture of cyclohexane from benzene?

   b) Write equations to show the three stages in the hydrogenation of benzene via cyclohexa-1,3-diene and cyclohexene to form cyclohexane.

   c) Why do cyclohexa-1,3-diene and cyclohexene react more readily with hydrogen than benzene?

25 Explain why 1,2,3,4,5,6-hexachlorocyclohexane shows geometric isomerism.

Studying the reaction of benzene with chlorine

Figure 18.1.30 shows the apparatus that might once have been used to prepare chlorobenzene by heating benzene with chlorine gas in the presence of iron filings. This preparation is now banned in teaching laboratories in schools and colleges.

1 Why is this reaction banned in the teaching laboratories of schools and universities?

2 a) Why should the reaction be carried out in a fume cupboard?

 b) Why is a hotplate used?

 c) Why is the oil bath at 70 °C?

3 During the reaction, iron reacts with chlorine to form iron(III) chloride, which then acts as an electron-pair acceptor, first polarising the Cl_2 molecules as $Cl^{\delta+}$–$Cl^{\delta-}$ and then forming the electrophile Cl^+.

 Write equations to show:

 a) the formation of iron(III) chloride

 b) the polarisation of Cl_2 as $Cl^{\delta+}$–$Cl^{\delta-}$ by iron(III) chloride and the formation of Cl^+.

4 Write a mechanism for the reaction of the electrophile Cl^+ with benzene to form an intermediate cation in a first step, and then the formation of chlorobenzene and H^+ in a second step.

5 The reaction shown in Figure 18.1.30 is just as effective if aluminium chloride or iron(III) chloride are used in place of iron. The substances $FeCl_3$ and $AlCl_3$ are often described as catalysts and halogen carriers.

 a) Why are these substances described as halogen carriers for the reaction?

 b) Why is it correct to describe aluminium chloride and iron(III) chloride as catalysts?

 c) Why is it incorrect to describe iron as a catalyst for the reaction?

6 Aluminium chloride acts as a catalyst for the chlorination of benzene by polarising Cl_2 molecules in the same way as iron(III) chloride.

 a) Do you think aluminium chloride is likely to be more or less effective than iron(III) chloride?

 b) Explain your answer to part (a).

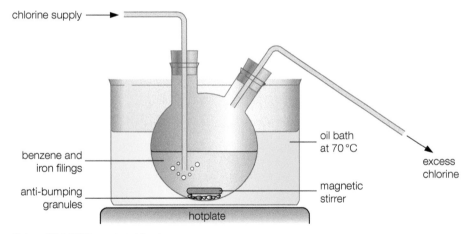

chlorine supply

benzene and iron filings

anti-bumping granules

hotplate

oil bath at 70 °C

excess chlorine

magnetic stirrer

Figure 18.1.30 Preparing chlorobenzene.

Figure 18.1.31 Crystals of phenol.

18.1.8 Phenol

Phenol is an example of a compound with a functional group directly attached to a benzene ring. In phenol, the functional group is −OH. Experiments show that the −OH group affects the behaviour of the benzene ring while the benzene ring modifies the properties of the −OH group. As a result of this, phenol has some distinctive and useful properties.

As expected, the −OH group gives rise to hydrogen bonding in phenol and therefore much stronger intermolecular forces than in benzene. This results in phenol being a solid at room temperature (Figure 18.1.31).

The –OH group in phenol also allows it to hydrogen bond with water. As a result of this, phenol dissolves slightly in water and much more readily in alkalis, such as sodium hydroxide solution (Figure 18.1.32), with which it forms a soluble ionic compound. In this reaction, phenol is behaving as an acid. However, phenol does not ionise significantly in water and it does not react with carbonates to produce carbon dioxide.

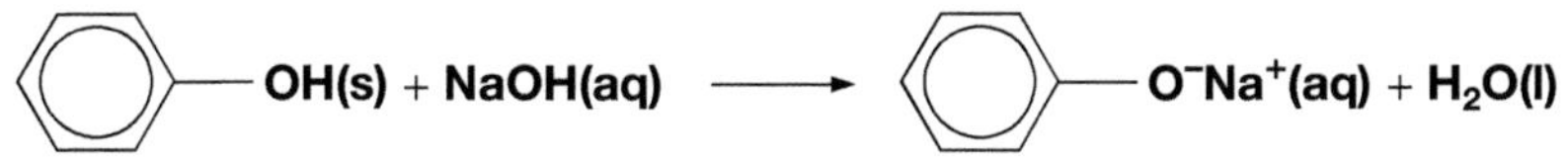

Figure 18.1.32 Phenol reacts with aqueous sodium hydroxide to form a colourless solution of sodium phenoxide.

The derivatives of phenol are named in a similar fashion to those of benzene, by numbering the carbon atoms in the benzene ring starting from the –OH group (Figure 18.1.33).

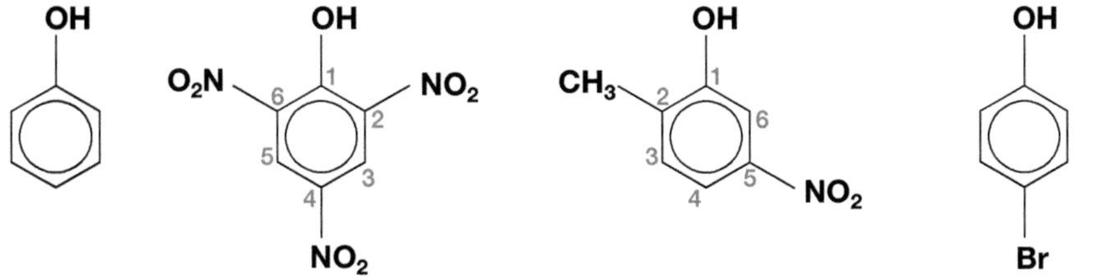

phenol | 2,4,6-trinitrophenol | 2-methyl-5-nitrophenol (not 6-methyl-3-nitrophenol) | 4-bromophenol

Figure 18.1.33 The structure of phenol and other substituted phenols.

Test yourself

26 Explain, in terms of intermolecular forces, why:

  a) phenol is a solid while benzene is a liquid at room temperature

  b) phenol, unlike benzene, is slightly soluble in water

  c) phenol does not mix with water as freely as ethanol.

27 What would you expect to observe on heating phenol until it burns?

28 Identify one way in which the –OH group behaves similarly in phenol and ethanol, and one way in which it behaves differently.

29 **a)** What would you expect to observe if you added enough dilute hydrochloric acid to a solution of phenol in sodium hydroxide to make the mixture acidic?

  b) Explain the reaction that occurs.

18.1.9 Reactions of the benzene ring in phenol

Comparing the reactivity of benzene and phenol

The –OH group in phenol activates its benzene ring and makes it more reactive than benzene itself. A lone pair of electrons on the –OH group interacts with the delocalised electrons in the benzene ring, releasing electrons into the ring and increasing the electron density in the ring. This makes electrophilic attack easier so, as a result, electrophilic substitution takes place under much milder conditions with phenol than with benzene.

Reaction with bromine

An aqueous solution of phenol reacts readily with bromine water to produce an immediate white precipitate of 2,4,6-tribromophenol as the orange/yellow bromine colour fades. The reaction is rapid at room temperature with bromine water. There is no need to heat the mixture or use a catalyst (Figure 18.1.34).

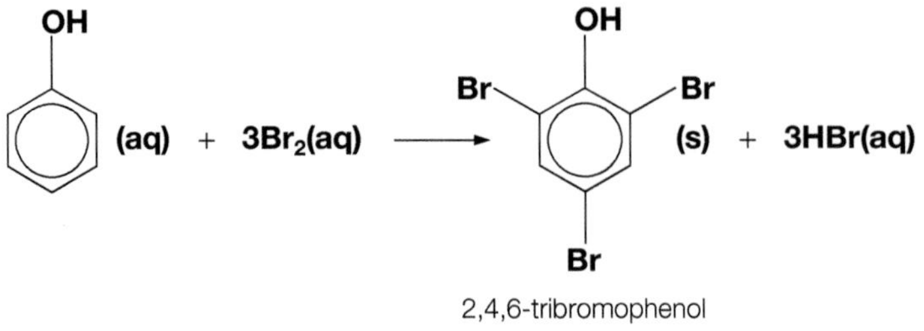

Figure 18.1.34 The reaction of phenol with bromine.

Reaction with nitric acid

Dilute nitric acid reacts rapidly with phenol at room temperature to form a brown mixture. The main products of the reaction are 2-nitrophenol and 4-nitrophenol (Figure 18.1.35). Notice how the conditions for nitrating phenol are so much milder than those needed to nitrate benzene.

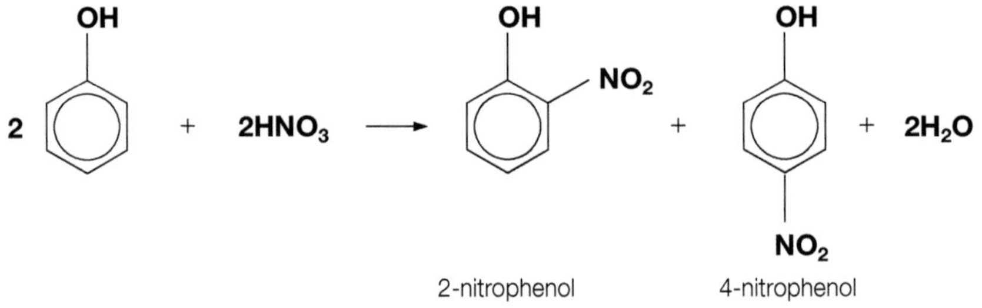

Figure 18.1.35 Nitrating phenol with dilute nitric acid.

> ### Tip
>
> The compound 4-nitrophenol is an important intermediate in the production of paracetamol (see the Activity: Paracetamol – an alternative to aspirin, in Section 18.2.3).

Test yourself

30 a) Write an equation for the reaction of chlorine with phenol and name the organic product.

   b) Explain why the reaction of phenol with chlorine does not require a catalyst whereas the chlorination of benzene does.

Manufacturing phenol

Phenol is manufactured from benzene, propene and oxygen in two stages. The process is known as the cumene process (Figure 18.1.36).

The first stage of the process involves the acid-catalysed electrophilic substitution of benzene with propene to form cumene.

The second stage involves the air oxidation of cumene. This produces equimolar amounts of phenol and propanone, a valuable co-product. About 100 000 tonnes of phenol are manufactured each year in the UK using the cumene process.

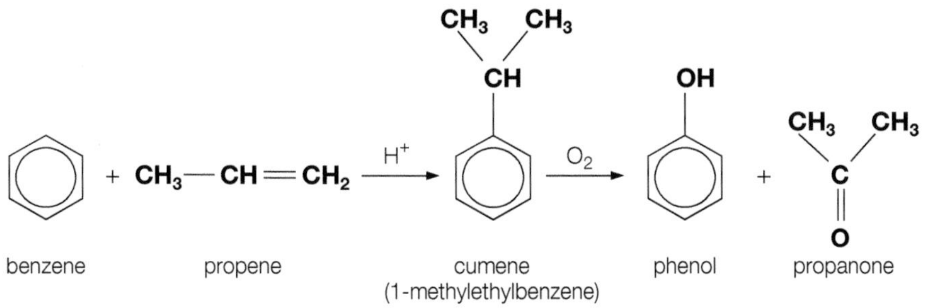

Figure 18.1.36 The manufacture of phenol by the cumene process.

1 Benzene and propene are obtained for the cumene process from crude oil. What processes, starting with crude oil, are used to produce:
   a) benzene
   b) propene?
2 In the first stage of the cumene process, H^+ ions react with propene to produce electrophiles.
   a) Write the formulae of two possible electrophiles produced when H^+ ions react with propene.
   b) Explain why one of these electrophiles is more stable than the other.
   c) Name and draw the structure of a second possible product of this first stage besides cumene.
3 a) Write a mechanism for the reaction of the more stable electrophile, identified in Question 2(b), with benzene to produce cumene.
   b) Why is this reaction described as acid-catalysed?
4 Write an equation for the second stage of the process in which cumene is oxidised to phenol and propanone.
5 The actual yield in the cumene process is 85%. Calculate the mass of benzene required to manufacture 1 tonne of phenol and the mass of propanone formed at the same time.

Exam practice questions

1 a) Describe the structure and bonding in benzene. *(6)*

b) Explain, in terms of structure and bonding, why benzene and ethene react differently with electrophiles. *(4)*

2 Benzene reacts with bromomethane, CH_3Br, to form $C_6H_5CH_3$. The reaction involves electrophilic substitution.

a) Explain the term 'electrophilic substitution'. *(2)*

b) Name the product and identify a catalyst that could be used for the reaction. *(2)*

c) Outline a mechanism for the reaction, including curly arrows. Include also an equation for the step that forms the attacking electrophile. *(3)*

d) Give the names of the two chemists who are associated with this type of reaction of benzene. *(1)*

3 Benzene is one of the most important aromatic compounds in industry. The flow chart below shows the formation of three useful products from benzene.

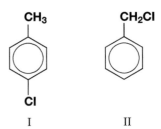

a) Name substances A to F. *(6)*

b) Reactions to form D and nitrobenzene involve electrophilic substitution.

   i) Write the formula of the electrophile in each case. *(2)*

   ii) Give one condition needed for the nitration of benzene to form nitrobenzene, and outline the mechanism of the reaction using curly arrows where appropriate. *(4)*

   iii) Suggest one use for the compounds formed by the nitration of arenes. *(1)*

c) i) Suggest the conditions used to produce A from benzene. *(1)*

   ii) Suggest a possible use for A. *(1)*

d) Substance D can be dehydrogenated to form phenylethene (styrene) which is used to manufacture poly(phenylethene).

   i) Write an equation for the formation of poly(phenylethene) from phenylethene. *(1)*

   ii) State one use of poly(phenylethene). *(1)*

4 a) Chlorobenzene can be produced from benzene and chlorine with a suitable catalyst.

   i) Name the catalyst. *(1)*

   ii) Describe briefly how chlorobenzene could be prepared. *(3)*

b) Under suitable conditions methyl benzene can be used to make the halogen-containing compounds I and II shown below.

CH_3 CH_2Cl

      Cl

    I   II

   i) Name compounds I and II. *(2)*

   ii) State the type of reaction that has occurred in each case. *(2)*

c) Only one of the compounds reacts on warming with aqueous sodium hydroxide.

   i) Identify which compound reacts and justify your answer. *(3)*

   ii) Draw the structure of the organic product formed with excess sodium hydroxide solution and state the type of reaction that has occurred. *(2)*

5 A student prepared a sample of methyl 3-nitrobenzoate by nitration of methyl benzoate according to the following instructions.

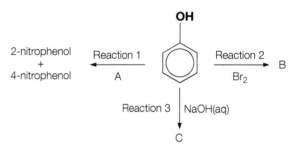

A Put $5\,cm^3$ of methyl benzoate into a $100\,cm^3$ conical flask and carefully add $1\,cm^3$ of concentrated sulfuric acid. When the methyl benzoate has dissolved, cool the mixture in ice.

B Prepare the nitrating mixture in a separate boiling tube by carefully adding $4\,cm^3$ of concentrated nitric acid to $4\,cm^3$ of concentrated sulfuric acid and cool this mixture in the ice as well.

C Add the nitrating mixture dropwise to the methyl benzoate solution in sulfuric acid. Stir the mixture using a thermometer and use the ice bath to keep the temperature between 5 and $15\,°C$. When the addition is complete, remove the flask from the ice and let it stand at room temperature for 15 minutes.

D After 15 minutes, pour the reaction mixture over about $50\,g$ of crushed ice and stir until all the ice has melted and crystals of methyl 3-nitrobenzoate have formed.

E Filter the solid by suction using a Buchner funnel, wash the solid with cold water and then transfer it to a small conical flask.

F Add $25\,cm^3$ of ethanol to the solid and warm the contents of the conical flask to about $50\,°C$ by immersing it in a beaker of hot water.

G Once the impure solid has dissolved, pure methyl 3-nitrobenzoate can be recovered by cooling the solution in an ice bath and then collecting the crystals that form by suction filtration.

H Dry the crystals using absorbent paper and weigh them.

a) Methyl 3-nitrobenzoate can be nitrated further to form dinitro- and trinitro-derivatives. Give two ways in which the procedure tried to prevent this. *(2)*

b) Name two nitro-compounds that may contaminate the impure methyl 3-nitrobenzoate. *(2)*

c) State why the impure methyl 3-nitrobenzoate was washed with water before recrystallisation. *(1)*

d) Explain why impurities that are only present in small amounts are removed during recrystallisation. *(2)*

e) i) State why some methyl 3-nitrobenzoate is lost during recrystallisation. *(1)*
   ii) State how the procedure tried to minimise this loss. *(2)*

f) What is the effect in step H of cooling the solution in ice rather than allowing it to cool slowly? *(1)*

g) Assuming that excess nitric acid is used, calculate the theoretical yield of methyl 3-nitrobenzoate. (Density of methyl benzoate $= 1.1\,g\,cm^{-3}$) *(4)*

h) If $3.5\,g$ methyl 3-nitrobenzoate are formed, calculate the percentage yield. *(1)*

6 Three reactions of phenol are summarised in the flow diagram below.

OH

2-nitrophenol
+
4-nitrophenol

Reaction 1
A

Reaction 2
Br₂
B

Reaction 3 | NaOH(aq)

C

a) Name reagent A. *(1)*

b) Draw the structural formulae of B and of C. *(2)*

c) Under suitable conditions, nitrophenols can be converted to dinitrophenols.
   i) Suggest the reaction conditions for converting nitrophenols to dinitrophenols. *(2)*
   ii) One possible dinitrophenol is 2,3-dinitrophenol. Write the names of all the other possible dinitrophenols. *(3)*

7 A synthesis of 1-phenylpropene from benzene is shown below.

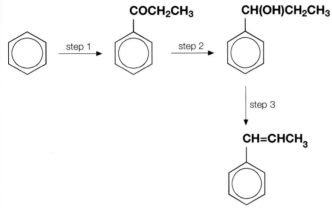

a) i) Identify a reagent and a catalyst for step 1. *(2)*

ii) Write an equation for the reaction between them to form an electrophile and draw a mechanism for the reaction of this electrophile with benzene. *(3)*

b) Identify a reagent for step 2. Name the type of mechanism involved. *(2)*

c) Identify a reagent for step 3. Name the type of reaction involved. *(2)*

d) Discuss the stereochemistry of the compounds formed in steps 2 and 3. *(6)*

8 The drug L-dopa is used in the treatment of Parkinson's disease. Its structural formula is:

HO, CH$_2$ COOH
CH
HO NH$_2$

a) Write the structural formulae of the sodium salts formed when L-dopa reacts with:

i) excess sodium hydroxide solution *(1)*

ii) excess sodium carbonate solution. *(1)*

b) Predict the structural formula of the organic ion produced when L-dopa reacts with excess dilute hydrochloric acid. *(1)*

c) The benzene ring in L-dopa is more reactive with electrophiles than that in benzene.
Explain why the benzene ring in L-dopa is more susceptible to attack by electrophiles than benzene itself. *(2)*

d) In the human body, an enzyme called 'L-dopa decarboxylase' eliminates the carboxylic acid group from L-dopa to produce the primary amine, dopamine.

i) What is a primary amine? *(1)*

ii) Draw the structural formula of dopamine. *(1)*

9 The following data are provided to answer this question.

Mean bond energy of C=C = 612 kJ mol^{-1}
Mean bond energy of C–C = 347 kJ mol^{-1}
Mean bond energy of C–H = 413 kJ mol^{-1}

Enthalpy change of atomisation of H$_2$(g) = 218 kJ mol^{-1} (Hint: per mol of atoms formed.)

Enthalpy change of atomisation of C(s) = 715 kJ mol^{-1}

a) Use the data provided to calculate the enthalpy change of formation of gaseous Kekulé benzene. *(7)*

b) The experimental value for the enthalpy change of formation of gaseous benzene is +82 kJ mol^{-1}.
Compare this value with the value you calculated in part (a) and state a reason for the difference. *(3)*

10 a) The pK_a values for ethanol and phenol are 16.0 and 10.0, respectively.
Compare the acid strengths of the two compounds and explain why there is such a large difference. *(6)*

b) Both ethanol and phenol react with ethanoyl chloride to form esters.
Draw and name the two esters. *(2)*

18.2.1 The structures and names of amines

Amines are nitrogen compounds in which one or more of the hydrogen atoms in ammonia, NH_3, has been replaced by an alkyl or an aryl group. The number of these groups determines whether the compound is a primary amine, a secondary amine or a tertiary amine. If one H atom in ammonia is replaced by an alkyl or aryl group, the compound is a primary amine. If two H atoms in ammonia are replaced, the compound is a secondary amine, and if all three H atoms in ammonia are replaced, the compound is a tertiary amine.

The amine group is present in amino acids. As a result of this, the amine group plays an important part in metabolism and is part of the structure of many medicinal drugs (Figure 18.2.1).

Chemists have two systems for naming amines.

Figure 18.2.1 The active constituent of asthma inhalers is salbutamol, which contains the amine functional group.

> **Tip**
>
> Note that the terms 'primary', 'secondary' and 'tertiary' do not have the same meaning with amines as they do with halogenoalkanes or alcohols. For halogenoalkanes or alcohols, the term depends on the number of alkyl or aryl groups attached to the carbon bearing the halogen or OH group.

Simple amines

Simple amines are treated as a combination of the alkyl or aryl group followed by the ending **-amine**. So, $CH_3CH_2CH_2CH_2NH_2$ is butylamine, $C_6H_5NH_2$ is phenylamine and $CH_3CH_2NHCH_3$ is ethylmethylamine. The prefixes di- and tri- are used when there are two or three of the same alkyl or aryl group (Figure 18.2.2).

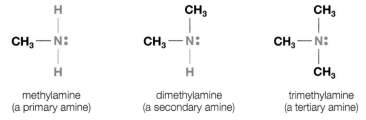

methylamine
(a primary amine)

dimethylamine
(a secondary amine)

trimethylamine
(a tertiary amine)

Figure 18.2.2 The structures and names of primary, secondary and tertiary amines containing the methyl group.

More complex amines

The prefix 'amino' is used in compounds that have a second functional group, such as amino acids. So the systematic name for H_2NCH_2COOH is aminoethanoic acid. The prefix 'diamino' is used for compounds containing two amino groups, such as 1,2–diaminoethane, $H_2NCH_2CH_2NH_2$.

Test yourself

1 Draw the structures of:

  a) diethylamine

  b) ethylmethylpropylamine

  c) 1,6-diaminohexane

  d) 1,2-diaminopentane, which contributes to the smell of rotting flesh and has the common name cadaverine

  e) 1-phenyl-2-aminopropane, an amphetamine that is an addictive stimulant.

2 Salbutamol is the active ingredient in asthma inhalers. Its structure is shown below.

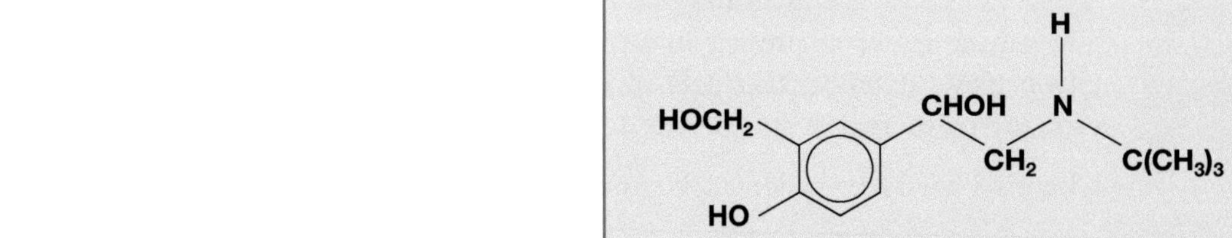

  a) Is its amine group primary, secondary or tertiary?

  b) What other functional groups does salbutamol contain?

3 a) Draw the structures and name all the primary amine isomers of $C_4H_9NH_2$.

  b) Draw the structures of all the secondary and tertiary amines that are isomers of $C_4H_9NH_2$.

4 Classify the halogenoalkanes, alcohols and amines below as primary, secondary or tertiary.

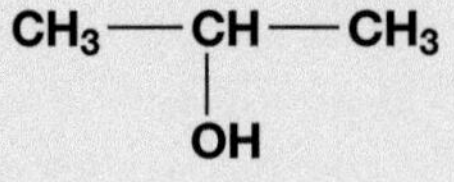

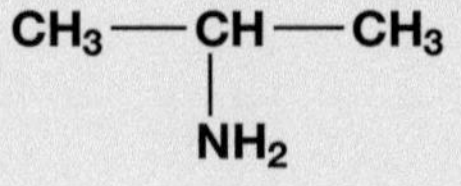

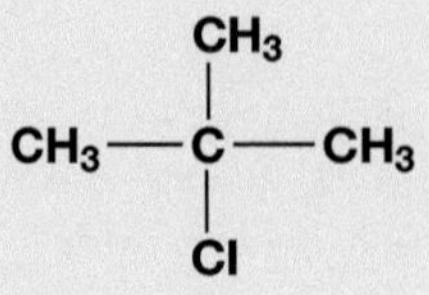

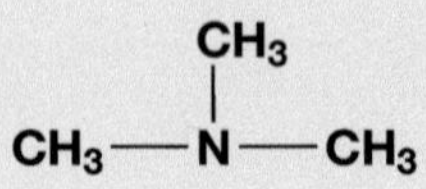

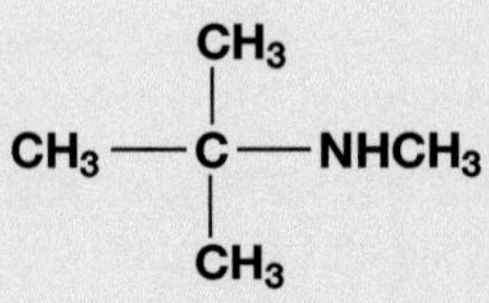

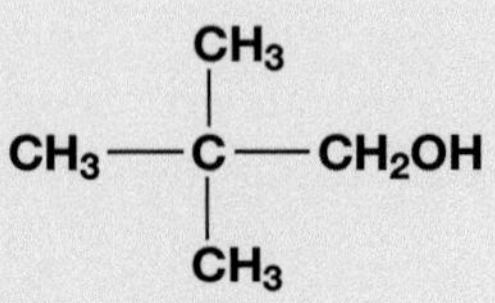

18.2.2 The structures and names of amides

Amides are nitrogen compounds derived from carboxylic acids in which an $-NH_2$ group replaces the $-OH$ group. The chemistry of amides is important because amide groups link up the monomers in proteins and in synthetic polymers such as nylon and Kevlar®. The link is also found in many medicinal drugs (Figure 18.2.3).

Amides have the general structure $R-\underset{\underset{O}{\|}}{C}-NH_2$

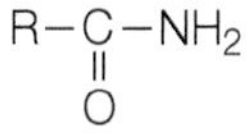

and the amide functional group is $-\underset{\underset{O}{\|}}{C}-\underset{\underset{H}{|}}{N}-$

Tip

Amides were once called acid amides to reflect their relationship to acids. Take care not to confuse the amine functional group, which has no C=O bond, with the amide functional group, which does contain C=O.

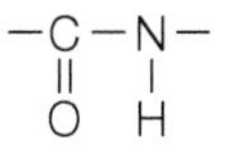

Amides are named using the suffix **-amide** after a stem that indicates the number of carbon atoms in the molecule including that in the C=O group (Figure 18.2.4).

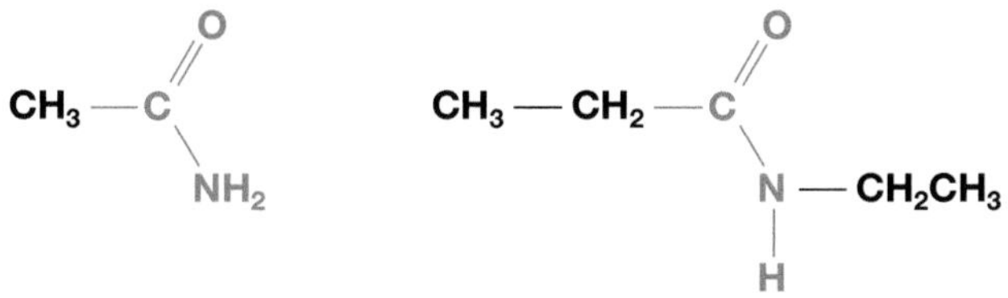

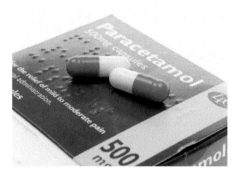

Figure 18.2.3 Paracetamol molecules contain the amide group.

ethanamide   N-ethylpropanamide

Figure 18.2.4 The structures and names of amides. Note that N-ethylpropanamide has an ethyl group substituted for one of the hydrogen atoms of the $-NH_2$ group. The prefix N indicates this and should be included in the name.

Test yourself

5 Draw the structures of:

a) butanamide   b) N-methylpentanamide

c) hexanediamide   d) N-phenylethanamide.

18.2.3 The properties and reactions of amines

The physical and chemical properties of the simplest amines are similar to those of ammonia. So, methylamine and ethylamine are gases at room temperature and they smell like ammonia, though with a fishy character (Figure 18.2.5). Propylamine and butylamine are liquids at room temperature.

Like ammonia, alkyl amines with short hydrocarbon chains dissolve readily in water because they can hydrogen bond with it. Phenylamine, with its large non-polar benzene ring, is only slightly soluble in water.

Figure 18.2.5 The smell of fish is partly due to ethylamine.

Amines as bases

Primary amines, like ammonia, can act as Brønsted–Lowry bases. The lone pair of electrons on the nitrogen atom of ammonia and amines is a proton (H^+ ion) acceptor.

Reaction with water

Ammonia acts as a Brønsted–Lowry base and a small proportion of the dissolved molecules react with water to form a weakly alkaline solution containing hydroxide ions.

$$NH_3(aq) + H_2O(l) \rightleftharpoons NH_4^+(aq) + OH^-(aq)$$

Butylamine and other amines act as Brønsted–Lowry bases in a similar manner by removing H^+ ions from water molecules to form an alkaline solution containing hydroxide ions (Figure 18.2.6).

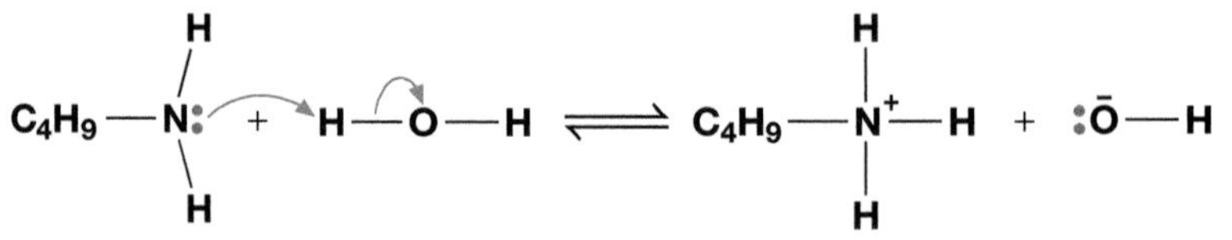

Figure 18.2.6 The reaction of butylamine with water.

$$C_4H_9NH_2(aq) + H_2O(l) \rightleftharpoons C_4H_9NH_3^+(aq) + OH^-(aq)$$

butylamine   butylammonium ion

As with ammonia, the reaction of amines with water is reversible so alkyl amines are also weak bases, although stronger than ammonia. This is because the alkyl group is electron releasing and increases the electron density on the lone pair on the nitrogen. This effect makes the lone pair more attractive to protons than the lone pair on the nitrogen in ammonia. The equilibrium in Figure 18.2.6 lies further to the right than the equilibrium involving ammonia.

By contrast, phenylamine is a much weaker base than ammonia because the lone pair in phenylamine is delocalised into the π cloud of the benzene ring (Figure 18.2.7) and is less attractive to protons than the lone pair in ammonia. Therefore, the equilibrium for the reaction of phenylamine with water lies further to the left than that for ammonia.

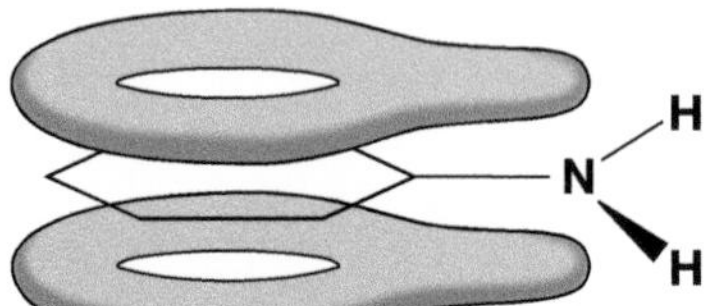

$$C_6H_5NH_2(l) + H_2O(l) \rightleftharpoons C_6H_5NH_3^+(aq) + OH^-(aq)$$

Reaction with acids – formation of salts

Amines react even more readily with acids than they do with water. The lone pair on the nitrogen atom rapidly accepts an H^+ ion from the acid to form a substituted ammonium salt.

$$C_4H_9NH_2(g) + HCl(g) \rightarrow C_4H_9NH_3^+Cl^-(s)$$

butylamine   butylammonium chloride

When the vapour of gaseous amines such as ethylamine reacts with hydrogen chloride gas, the product, ethylammonium chloride, forms as a white smoke. The smoke settles as a white solid (Figure 18.2.8).

$$CH_3CH_2NH_2(g) + HCl(g) \rightarrow CH_3CH_2NH_3^+Cl^-(s)$$

ethylamine   ethylammonium chloride

Figure 18.2.7 Delocalisation of the lone pair into the π cloud in phenylamine.

Tip

Note that the shorthand C_4H_9 used here to represent the carbon chain in butylamine can also represent several other arrangements of the carbon chain.

Tip

The pK_a values of their conjugate acids (see Section 12.2) give a measure of the strength of ammonia (pK_a = 9.25), butylamine (pK_a = 10.61) and phenylamine (pK_a = 4.62) as bases. A higher pK_a value corresponds to a stronger base.

This reaction is very similar to that of ammonia with hydrogen chloride to form ammonium chloride.

$$NH_3(g) + HCl(g) \rightarrow NH_4^+Cl^-(s)$$

Phenylamine, $C_6H_5NH_2$, is only slightly soluble in water, but it dissolves in concentrated hydrochloric acid very easily. This is because it reacts with H^+ ions in the acid to form phenylammonium ions which are soluble in the aqueous mixture.

$$C_6H_5NH_2(l) + H^+(aq) \rightarrow C_6H_5NH_3^+(aq)$$

If a strong base, such as sodium hydroxide, is added to the aqueous phenylammonium ions, H^+ ions are removed from the phenylammonium ions and yellow, oily phenylamine reforms.

$$C_6H_5NH_3^+(aq) + OH^-(aq) \rightarrow C_6H_5NH_2(l) + H_2O(l)$$
$$\text{phenylamine}$$

Figure 18.2.8 The vapours from ethylamine solution and concentrated hydrochloric acid react to form a white smoke of ethylammonium chloride.

Test yourself

6 Methylamine, like ammonia, mixes with and dissolves in water whatever proportions of the two are mixed together. Why is this?

7 a) Ethane (boiling temperature −89 °C) and methylamine (boiling temperature −6 °C) have very similar molar masses, but very different boiling temperatures. Why is this?

   b) Consider the boiling temperatures of methylamine (−6 °C), dimethylamine (7 °C) and trimethylamine (4 °C). Why do you think the boiling temperature of trimethylamine, $(CH_3)_3N$, is lower than that of dimethylamine?

8 a) Write equations for the reactions of cyclohexylamine, $C_6H_{11}NH_3$, and phenylamine with water.

   b) Explain why cyclohexylamine is a stronger base than phenylamine.

9 a) Write an equation to show the formation of a salt when propylamine vapour reacts with hydrogen bromide gas.

   b) Explain why the reactants are both gases, but the product is a solid.

10 Write equations for the following reactions and name the products:

   a) methylamine with concentrated sulfuric acid

   b) dimethylamine with concentrated sulfuric acid.

Amines as ligands

When ammonia and amine act as bases and accept a proton, they do so by donating a lone pair of electrons to the proton. Ammonia and amines can also donate a lone pair of electrons to transition metal ions and act as ligands (Section 15.6).

When butylamine is added to aqueous copper(II) sulfate solution, a deep blue solution is formed (Figure 18.2.9). Four butylamine molecules replace four

Figure 18.2.9 The result of adding butylamine to a solution of copper ions. The hydrated copper(II) ions give the light blue colour, while the dark blue colour is due to the formation of a complex between the copper and the butylamine.

water ligands and a deep blue complex is formed, similar to the complex formed by ammonia.

$$4C_4H_9NH_2(aq) + [Cu(H_2O)_6]^{2+}(aq) \rightarrow [Cu(C_4H_9NH_2)_4(H_2O)_2]^{2+}(aq) + 4H_2O(l)$$

Diamines such as 1,2-diaminoethane, $H_2NCH_2CH_2NH_2$, donate two lone pairs and are called bidentate ligands (see Section 15.8).

Key term

An **ammine** is a complex in which one or more ammonia molecules are co-ordinately bonded to a metal ion.

Tip

Note that ammonia, NH_3, in complexes is described as '**ammine**', whereas the $-NH_2$ group in organic compounds such as $C_4H_9NH_2$ is described as 'amine'.

Amines as nucleophiles

Reaction with halogenoalkanes

Amines are nucleophiles as well as bases and ligands, just like ammonia. As nucleophiles, their lone pair of electrons is attracted to any positive ion or positive centre in a molecule.

So, amines react with the δ+ carbon atoms in the C–Hal bond of halogenoalkanes in a nucleophilic substitution reaction. The protonated amine formed in the first step then loses a proton to form the secondary amine, butylmethylamine (Figure 18.2.10).

Tip

In these reactions, the loss of H^+ or the formation of HBr is shown in the equations. In the presence of amines that are bases, these acids form salts, but to avoid complication in the equations the salts are not shown here.

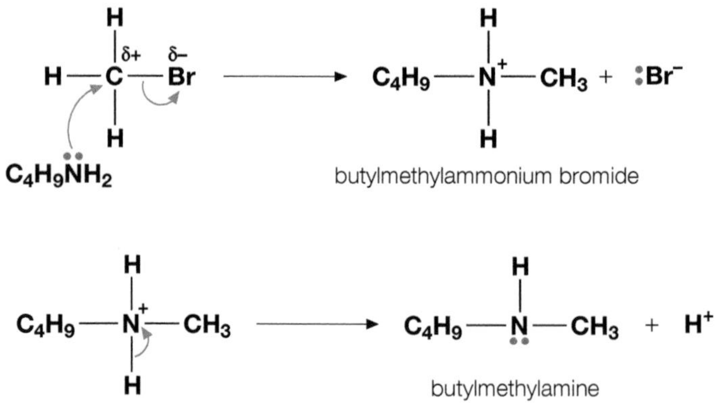

Figure 18.2.10 The reaction of butylamine with bromomethane to form the secondary amine butylmethylamine.

As with the reaction of ammonia with halogenoalkanes (Section 18.2.4), further reaction is possible.

The lone pair on the nitrogen atom of the secondary amine product is more reactive than the lone pair on the primary amine reagent because of the inductive effect of the extra alkyl group. So the secondary amine can also react with the halogenoalkane in a reaction which forms a tertiary amine (Figure 18.2.11).

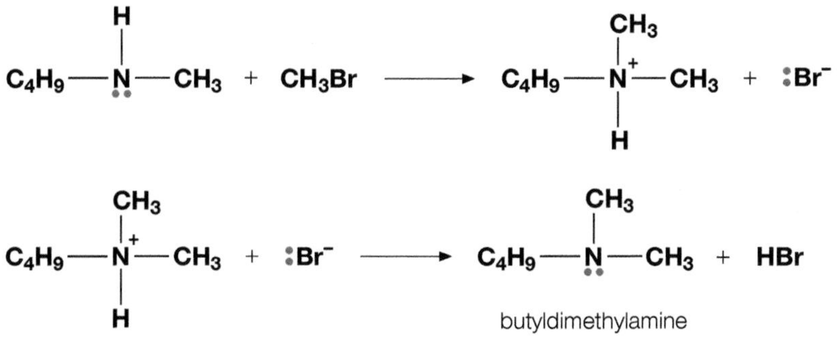

$$\text{C}_4\text{H}_9-\overset{\text{H}}{\underset{..}{\text{N}}}-\text{CH}_3 + \text{CH}_3\text{Br} \longrightarrow \text{C}_4\text{H}_9-\overset{\text{CH}_3}{\underset{\text{H}}{\overset{+}{\text{N}}}}-\text{CH}_3 + :\text{Br}^-$$

$$\text{C}_4\text{H}_9-\overset{\text{CH}_3}{\underset{\text{H}}{\overset{+}{\text{N}}}}-\text{CH}_3 + :\text{Br}^- \longrightarrow \text{C}_4\text{H}_9-\overset{\text{CH}_3}{\underset{..}{\text{N}}}-\text{CH}_3 + \text{HBr}$$

butyldimethylamine

Figure 18.2.11 Formation of the tertiary amine butyldimethylamine.

The tertiary amine similarly can react further to form a quaternary ammonium salt (Figure 18.2.12).

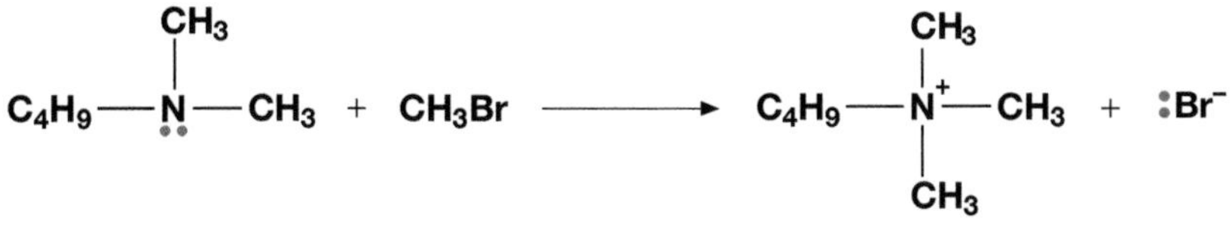

$$\text{C}_4\text{H}_9-\overset{\text{CH}_3}{\underset{..}{\text{N}}}-\text{CH}_3 + \text{CH}_3\text{Br} \longrightarrow \text{C}_4\text{H}_9-\overset{\text{CH}_3}{\underset{\text{CH}_3}{\overset{+}{\text{N}}}}-\text{CH}_3 + :\text{Br}^-$$

butyltrimethylammonium bromide

Figure 18.2.12 Formation of the quaternary ammonium salt butyltrimethylammonium bromide.

> **Key term**
>
> **Quaternary ammonium salt** is an ammonium salt where all four hydrogens are replaced by alkyl or aryl groups.

It is possible to limit further reaction by using an excess of the primary amine so that there is a much greater chance of the primary amine rather than the secondary amine acting as nucleophile with the halogenoalkane molecules.

If an excess of the halogenoalkane is used, the quaternary ammonium salt is the main product.

> **Tip**
>
> Quaternary ammonium salts where two of the alkyl groups are long chains, such as $[(\text{CH}_3(\text{CH}_2)_{17}]_2\text{N}(\text{CH}_3)_2{}^+\text{Cl}^-$, are used in fabric softeners.

Reaction with acyl chlorides

Amines also react as nucleophiles with the $\delta+$ carbon atoms in the $-\overset{\text{O}}{\overset{\|}{\text{C}}}-\text{Cl}$ group of acyl chlorides such as ethanoyl chloride (Figure 18.2.13). The reaction forms an N–substituted amide (see also Section 17.3.4).

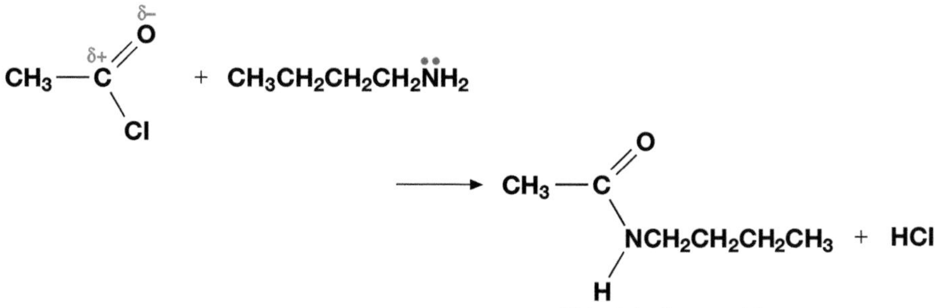

$$\text{CH}_3-\overset{\overset{\delta-}{\text{O}}}{\underset{\text{Cl}}{\overset{\delta+}{\text{C}}}} + \text{CH}_3\text{CH}_2\text{CH}_2\text{CH}_2\overset{..}{\text{N}}\text{H}_2$$

$$\longrightarrow \text{CH}_3-\overset{\text{O}}{\overset{\|}{\text{C}}}\underset{\overset{|}{\text{H}}}{-\text{NCH}_2\text{CH}_2\text{CH}_2\text{CH}_3} + \text{HCl}$$

N-butyl ethanamide

Figure 18.2.13 The reaction of butylamine with ethanoyl chloride.

A reaction of this type is involved in the manufacture of paracetamol; this is discussed in the activity that follows.

Paracetamol – an alternative to aspirin

Aspirin and paracetamol (Figure 18.2.14) are by far the largest selling pain relievers available without a doctor's prescription. However, aspirin is not without hazards. Every year, about 200 people die of aspirin poisoning due to deliberate or accidental overdose. A large proportion of these deaths are of young children who die accidentally after eating the tablets. Aspirin is also recognised as a cause of internal bleeding and gastric ulcers.

Paracetamol has been produced and marketed as a safer alternative to aspirin. It is a good analgesic (pain reliever) without the harmful side-effects of aspirin, as long as the recommended doses are not exceeded.

Compared to other non-prescription pain relievers, paracetamol is much more toxic when an overdose is taken because it can cause potentially fatal liver damage.

The compound from which paracetamol is produced and the physiologically active compound that paracetamol produces in the body is 4-aminophenol. Unfortunately, this is toxic, so its harmful effect is reduced by conversion to its ethanoyl derivative.

Paracetamol can be produced by reacting 4-aminophenol with ethanoyl chloride. In industry, however, ethanoic anhydride, $(CH_3CO)_2O$, is used in preference to ethanoyl chloride because it is cheaper and less vigorous in its reactions.

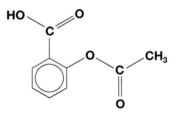

aspirin

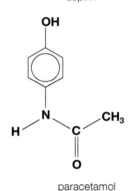

paracetamol

Figure 18.2.14 The structural formulae of aspirin and paracetamol.

1 Draw the structure of 4-aminophenol.
2 **a)** Write an equation for the reaction of 4-aminophenol with ethanoyl chloride to produce paracetamol.
  b) Why does ethanoyl chloride react in this way with 4-aminophenol?
3 When ethanoyl chloride reacts with 4-aminophenol, the –OH group in 4-aminophenol is susceptible to attack as well as the $-NH_2$ group.
  a) Why is the –OH group in 4-aminophenol also susceptible to reaction with ethanoyl chloride?
  b) Write an equation for the reaction of ethanoyl chloride with the –OH group in 4-aminophenol.
4 Fortunately, the $-NH_2$ group in 4-aminophenol is more reactive than the –OH group. So, in industry the reaction conditions can be carefully chosen so that only the $-NH_2$ group is ethanoylated using ethanoic anhydride. Write an equation for the reaction of 4-aminophenol with ethanoic anhydride to produce paracetamol.
5 Aspirin, like paracetamol, is manufactured by ethanoylation using ethanoic anhydride.
  a) What do you understand by the term 'ethanoylation'?
  b) Draw the structure of the compound that is ethanoylated to produce aspirin.
6 **a)** What is the main benefit of aspirin tablets?
  b) Summarise the risks posed by aspirin tablets.
  c) What are the advantages of paracetamol over aspirin?

Test yourself

11 Write the formula of the cobalt-containing ions formed by the reaction of $Co(H_2O)_6^{2+}$ ions with an excess of 1,2-diaminoethane.

12 Write equations for the following reactions:

a) chloromethane with propylamine

b) propanoyl chloride with ethylamine.

13 a) Write an equation and draw a mechanism for the reaction of chloromethane with an excess of ethylamine.

b) Write equations for any further reactions that occur if an excess of chloromethane reacts with ethylamine.

18.2.4 The preparation of amines

Preparing aliphatic amines

From halogenoalkanes

Aliphatic amines can be prepared by heating the corresponding halogenoalkanes in a sealed flask with excess ammonia in ethanol.

During the first step of the reaction ammonia acts as a nucleophile and then in the second step it acts as a base to remove a proton from the salt initially formed (Figure 18.2.15).

Tip

Use of excess ammonia limits the chance of further substitution, so a primary amine is the major product.

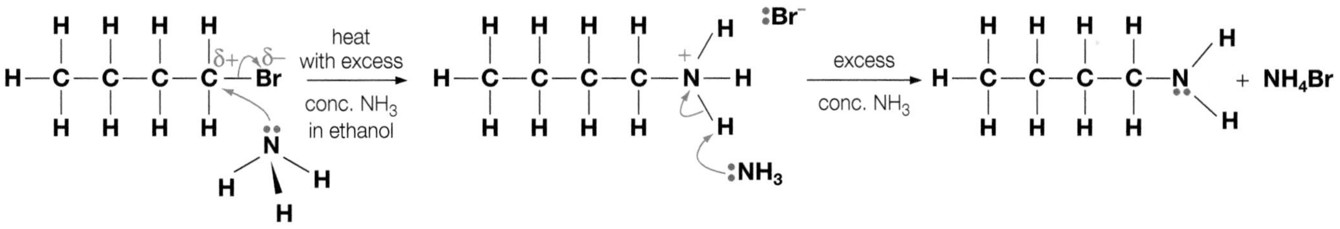

Figure 18.2.15 The preparation of butylamine from 1-bromobutane and excess concentrated ammonia in ethanol.

From nitriles

Reduction of nitriles produces primary amines. Unlike the nucleophilic substitution reaction of ammonia with halogenoalkanes considered above, this reaction produces a pure product as no further reaction can occur. Reduction can be achieved in two ways:

a) Hydrogenation using hydrogen gas in the presence of a nickel catalyst.

$$\underset{\text{butanenitrile}}{CH_3CH_2CH_2C{\equiv}N} + 2H_2 \rightarrow \underset{\text{butylamine}}{CH_3CH_2CH_2CH_2NH_2}$$

b) Reduction using $LiAlH_4$ in ethoxyethane, followed by dilute acid.

$$CH_3CH_2CH_2C{\equiv}N + 4[H] \rightarrow CH_3CH_2CH_2CH_2NH_2$$

Tip

The overall equation is complex, so simplified equations of this sort, using [H] to represent the reducing agent, are accepted in A Level examinations.

Preparing aromatic amines

The usual laboratory method for introducing an amine group into an aromatic compound is a two-step process – first nitration to make a nitro compound and then reduction (Figure 18.2.16). The reduction of the aromatic nitro-compound is achieved by boiling under reflux with tin and concentrated hydrochloric acid (Figure 18.2.17). The aromatic amine dissolves in excess concentrated hydrochloric acid, forming a salt. The free amine can be liberated from the solution by adding sodium hydroxide solution. It is then separated from the mixture by steam distillation.

Figure 18.2.16 The two-step preparation of phenylamine from benzene.

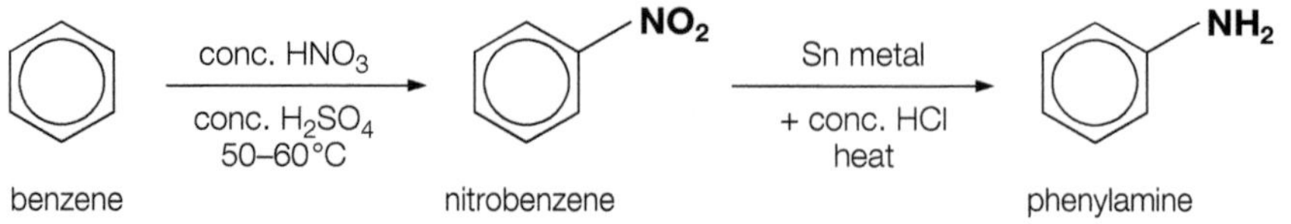

Equations for the reduction of nitrocompounds are usually simplified by using [H] to represent the reducing agent.

$$C_6H_5NO_2 + 6[H] \rightarrow C_6H_5NH_2 + 2H_2O$$

Figure 18.2.17 Reducing nitrobenzene to phenylamine by refluxing with tin and hot concentrated hydrochloric acid.

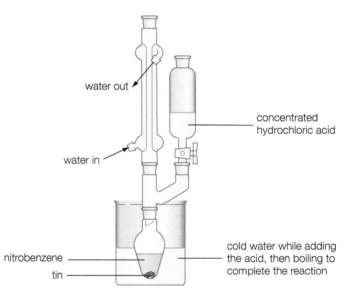

water out

concentrated hydrochloric acid

water in

nitrobenzene

tin

cold water while adding the acid, then boiling to complete the reaction

Tip

Reduction of nitrobenzene to phenylamine is an important reaction in industry notably in the preparation of dyes. Tin is an expensive metal so in industry the cheaper metal iron is used.

Tip

The reaction of a carboxylic acid with ammonia or an amine forms a salt, so in the laboratory, reactions using acyl chlorides are preferred for making amides. However, industrial methods to make polyamides use acids (see Section 18.2.9) rather than acyl chlorides because of the difficulty of storing acyl chlorides and using them on a large scale (see Section 17.3.4).

18.2.5 The preparation of amides

Amides form rapidly at room temperature when acyl chlorides, such as ethanoyl chloride, react with ammonia or with amines. For example, when ethanoyl chloride is carefully added to a concentrated aqueous solution of ammonia, a vigorous reaction takes place producing fumes of hydrogen chloride and ammonium chloride plus a residue of ethanamide.

$$CH_3COCl(l) + NH_3(aq) \rightarrow CH_3CONH_2(s) + HCl(g)$$
$$\text{ethanamide}$$

$$HCl(g) + NH_3(g) \quad \rightarrow \quad NH_4Cl(s)$$
$$\text{ammonium chloride}$$

The preparation of paracetamol discussed in the Activity in Section 18.2.3 involves the synthesis of an N-substituted amide.

18.2.6 Amino acids and proteins

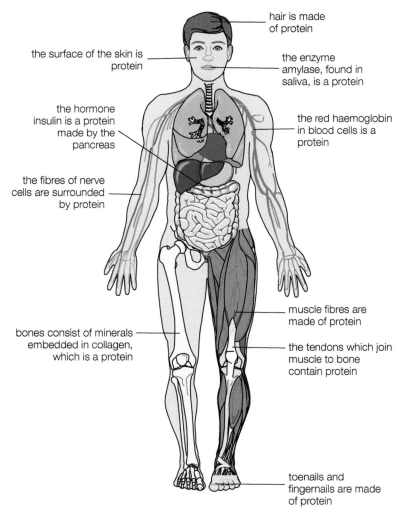

the surface of the skin is protein

hair is made of protein

the enzyme amylase, found in saliva, is a protein

the hormone insulin is a protein made by the pancreas

the red haemoglobin in blood cells is a protein

the fibres of nerve cells are surrounded by protein

bones consist of minerals embedded in collagen, which is a protein

muscle fibres are made of protein

the tendons which join muscle to bone contain protein

toenails and fingernails are made of protein

Figure 18.2.18 Proteins in the human body.

Key terms

Amino acids are compounds that contain both an amino group and a carboxylic acid group.

2-amino acids are amino acids with the formula $RCH(NH_2)COOH$. The amino group is attached to the carbon next to the carboxylic acid group.

Amino acids are compounds that contain two functional groups – the amino group, $-NH_2$, and the carboxylic acid group, –COOH. Amino acids are the monomers that make up proteins, the naturally occurring polymers which comprise 15% of the human body. There are many different protein molecules in our bodies (Figure 18.2.18). Muscles, hair, enzymes and hormones all consist of proteins. Some proteins contain thousands of amino acid units.

Names and formulae

About 20 different amino acids are found widely in naturally-occurring proteins. The names and formulae of six of these are shown in Figure 18.2.19. The simplest amino acid is glycine, $H_2N–CH_2–COOH$.

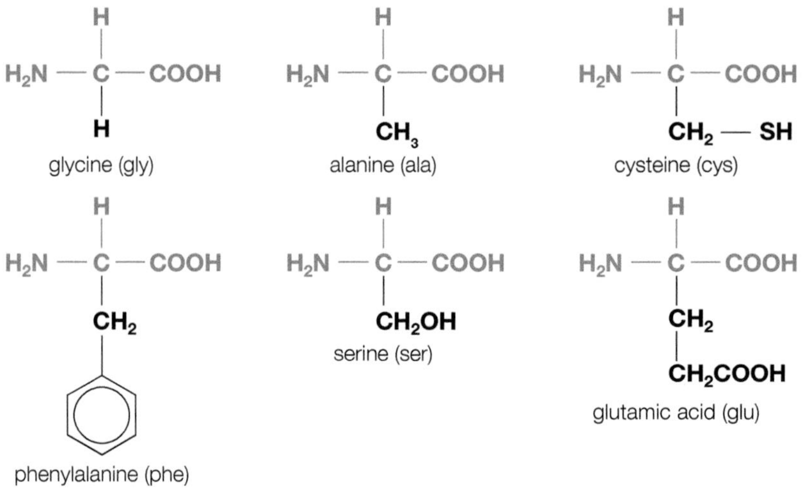

Figure 18.2.19 Six of the amino acids that occur in proteins.

Notice in Figure 18.2.19 that all six formulae have the amino group attached to the carbon atom next to the carboxylic acid group, carbon number 2 in the chain. This is the case with all the amino acids that occur naturally. This carbon number 2 is sometimes described as the alpha (α) carbon atom. So all the amino acids in proteins are **2-amino acids** (or α-amino acids) and their general formula can be written as $RCH(NH_2)COOH$.

R stands for the side groups in different amino acids (Table 18.2.1). The common names and R side groups of several other amino acids are shown on a data sheet headed 'The common names and R side groups of some amino acids', which you can access via the QR code for Chapter 18.2 on page 321.

Table 18.2.1 The R side groups in some amino acids.

Common name	Abbreviated name	R side group
Glycine	gly	H–
Alanine	ala	$CH_3–$
Cysteine	cys	$HS–CH_2–$
Phenylalanine	phe	$C_6H_5–CH_2–$
Aspartic acid	asp	$HOOC–CH_2–$

Many of the natural amino acids have complex structures, so it is simpler and more convenient to use their common names rather than their systematic names. These common names are sometimes abbreviated to a 'three-letter code', which is usually the first three letters in the name. So, H_2NCH_2COOH is normally called 'glycine' rather than 2-aminoethanoic acid and its abbreviated name is 'gly'.

Amino acid structures

All the amino acids that occur in proteins, except glycine, have a central carbon atom attached to four different groups. This is shown clearly in Figure 18.2.19. So, except for glycine, all these amino acids have chiral molecules that can exist as mirror images (Section 17.1.3). The mirror-image forms of the amino acid alanine are shown in Figure 18.2.20. All amino acids found in proteins occur in the L–configuration.

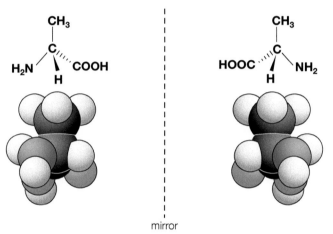

Figure 18.2.20 The mirror-image forms of the amino acid alanine. The mirror images are chiral and cannot be superimposed.

As a result of their chirality, the separate (+) and (−) isomers of all naturally occurring amino acids, except glycine, can rotate the plane of plane-polarised light.

Test yourself

17 State the systematic name for each amino acid:
   a) alanine
   b) phenylalanine
   c) serine.
18 A dipeptide contains two amino acids linked together. How many different dipeptides can be formed from the 20 naturally occurring amino acids?
19 Explain why the amino acid glycine is not chiral.
20 How could you distinguish between samples of the two mirror-image forms of an amino acid by experiment?

18.2.7 The acid–base properties of amino acids

As amino acids carry an amino group, $-NH_2$, and a carboxylic acid group, $-COOH$, they show both the basic properties of primary amines and the acidic properties of carboxylic acids.

The formation of zwitterions

In aqueous solution, carboxylic acid groups ionise producing hydrogen ions, $H^+(aq)$, whereas amino groups are basic and attract hydrogen ions. As a result of this, amino acids form ions in aqueous solution because the amino groups accept hydrogen ions (protons) and the acid groups give them away (Figure 18.2.21). The ions formed are, however, unusual in that they have both positive and negative charges. Chemists call them **zwitterions**, from a German word 'zwitter' meaning 'hybrid or hermaphrodite'.

Figure 18.2.21 Glycine forming a zwitterion.

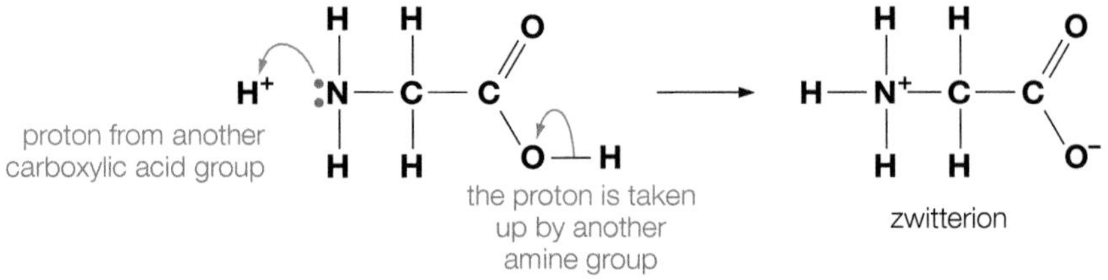

proton from another carboxylic acid group

the proton is taken up by another amine group

zwitterion

An amino acid can only form zwitterions at a particular pH. If the pH is too high, the solution is too alkaline and in these conditions OH^- ions remove H^+ ions from the zwitterions, forming negative ions (Figure 18.2.22 right). On the other hand, if the pH is too low, the solution is too acidic. In this case, H^+ ions react with the zwitterions, producing positive ions (Figure 18.2.22 left). Amino acids can therefore exist in three forms depending on the pH: a cation form, a zwitterion and an anion form. However, at one particular pH, molecules of the amino acid will be in the zwitterion form (Figure 18.2.22 centre) and this pH value is called the **isoelectric point**.

Notice from Figure 18.2.22 that the net charge on an amino acid molecule varies with the pH. The net charge is positive in acid solutions and negative in alkaline solutions. At the isoelectric point, the positive and negative charges balance and the net charge on the zwitterion is zero.

Figure 18.2.22 The ions formed by an amino acid at different pH values.

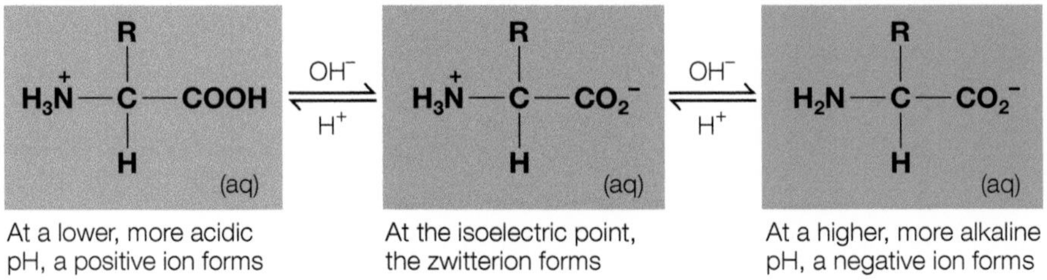

At a lower, more acidic pH, a positive ion forms

At the isoelectric point, the zwitterion forms

At a higher, more alkaline pH, a negative ion forms

All amino acids form zwitterions along the lines described above, but their isoelectric points may differ because of the different character of their R groups. In fact, some amino acids, like glutamic acid and aspartic acid, have two $-COOH$ groups and others have two $-NH_2$ groups, which influences their isoelectric point significantly.

The movement of H^+ ions from the $-COOH$ group of an amino acid to its $-NH_2$ group occurs in solution before the solid amino acid crystallises out. This means that amino acids also exist as zwitterions in the solid state. This ionic character of amino acids accounts for their high solubility in water and their high melting temperatures.

Test yourself

21 The relative molecular masses of butylamine, $CH_3(CH_2)_3NH_2$, propanoic acid, CH_3CH_2COOH, and glycine, H_2NCH_2COOH are very similar. But glycine (melting temperature 262 °C) is a solid at room temperature, whereas butylamine (melting temperature −49 °C) and propanoic acid (melting temperature −21 °C) are liquids. Why is this?

22 a) Write equations to show the reactions of alanine with:

   i) dilute hydrochloric acid

   ii) aqueous sodium hydroxide.

   b) How do the products from alanine of these two reactions differ from the zwitterions of alanine at its isoelectric point?

23 Why do zwitterions of amino acids exist just as readily in the solid state as they do in aqueous solution?

18.2.8 From amino acids to peptides and proteins

Peptides are compounds made by linking amino acids together in chains. The simplest example is a dipeptide with just two amino acids linked together by a peptide bond. Figure 18.2.23 shows the formation of a peptide bond between alanine and glycine to form the dipeptide, 'ala–gly'.

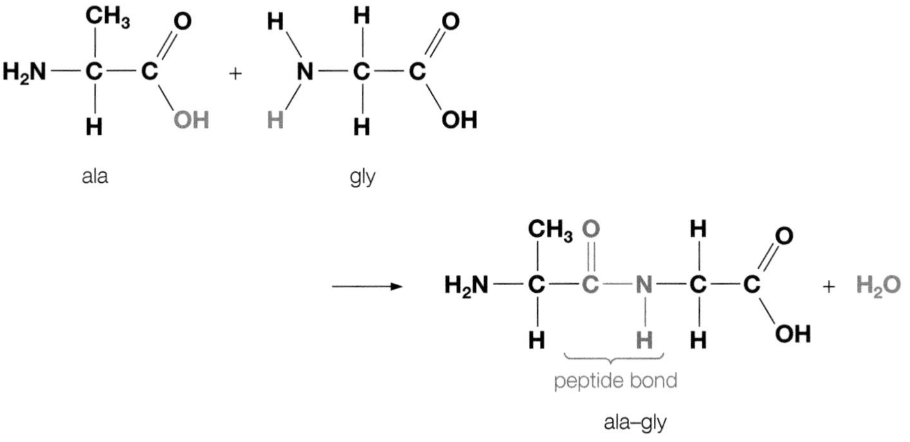

ala   gly

peptide bond

ala–gly

Figure 18.2.23 The formation of a peptide bond between two amino acids.

For chemists, the peptide bond, $-\overset{\displaystyle O}{\underset{}{\overset{\|}{C}}}-\overset{\displaystyle H}{\underset{}{\overset{|}{N}}}-$, is simply an example of the amide bond (Section 18.2.2). However, the tradition in biochemistry is to call it a 'peptide bond' or a peptide link.

Notice in Figure 18.2.23 that when a peptide bond forms between two amino acid molecules, a molecule of water is eliminated at the same time. This is an example of a condensation reaction.

Key terms

Peptides are chains of amino acids linked by peptide bonds.

A peptide bond is an amide link formed when the $-NH_2$ group of one amino acid reacts with the $-COOH$ group of another.

A condensation reaction is a reaction in which molecules join together by splitting off a small molecule such as water.

In **condensation polymerisation**, a polymer is formed by a series of condensation reactions.

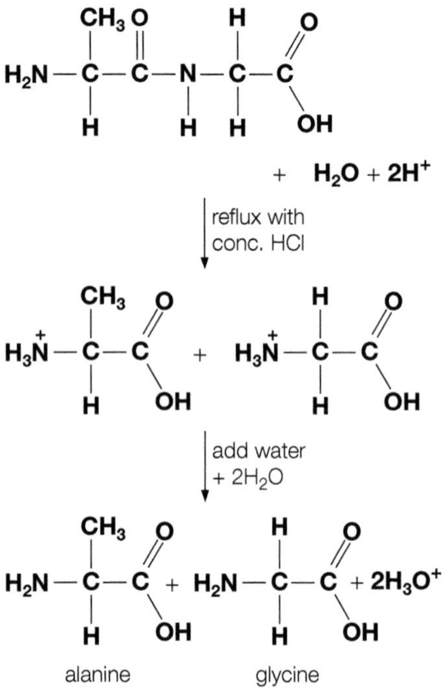

Figure 18.2.24 Hydrolysing a peptide with acid to produce α-amino acids.

Further condensation reactions can occur between the dipeptide and other amino acid molecules to produce polypeptides and eventually proteins. This is what happens when proteins are synthesised from amino acids in our bodies. The overall process is an example of **condensation polymerisation** (Section 17.3.6).

Polypeptides are long-chain peptides. There is no clear dividing line between peptides and polypeptides or between polypeptides and proteins. Some chemists do, however, make a distinction between polypeptides and the longer amino acid chains in proteins. They restrict the definition of polypeptides to chains with 10 to 50 or so amino acids.

The hydrolysis of peptides and proteins

Digestive enzymes in the stomach and small intestine catalyse the hydrolysis of peptide bonds, splitting proteins into polypeptides and then polypeptides into amino acids. Chemists can achieve the same result and hydrolyse the peptide bond by treating proteins and peptides with suitable enzymes, or simply by heating in acidic or alkaline solution.

When proteins and peptides are hydrolysed by refluxing with concentrated hydrochloric acid, the product contains the cation forms of the α-amino acids. These are converted to the α-amino acids on dilution with water (Figure 18.2.24).

After hydrolysis, the mixture of amino acids produced can be separated and identified by chromatography (Chapter 19).

Test yourself

24 Draw the structures of the two dipeptides that can be produced from serine and phenylalanine.

25 Show that splitting a dipeptide into two amino acids is an example of hydrolysis.

26 a) Identify the functional groups in the sweetener, aspartame.

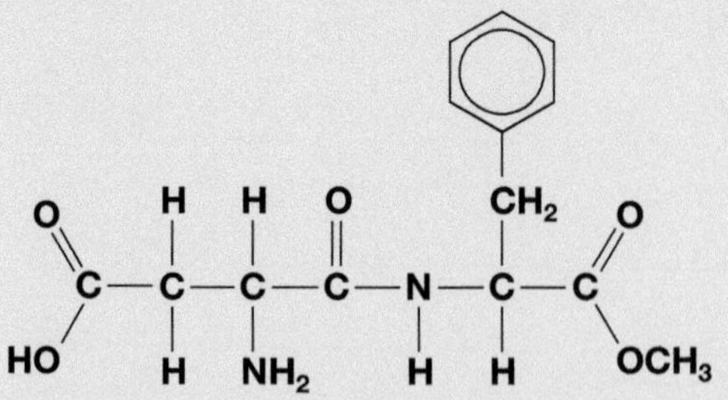

b) How does aspartame differ from a dipeptide?

c) Suggest a reason why aspartame cannot be used to sweeten food that will be cooked.

d) Why do you think that soft drinks sweetened with aspartame carry a warning for people with the genetic disorder that means that they must not eat phenylalanine?

18.2.9 Condensation polymerisation

Peptides are condensation polymers. When amino acids react, water molecules are eliminated and a peptide link is formed. There are two other important classes of condensation polymers: **polyesters** formed from dicarboxylic acids and diols (see Section 17.3.6) and **polyamides** formed from dicarboxylic acids and diamines.

Key terms

Polyesters are polymers with ester links between monomer units.

Polyamides are polymers with amide links between monomer units.

Polyamides

Polyamides are polymers in which the monomers are linked by an amide bond. This is exactly the same as the amide bond in proteins, in which it is usually called the peptide bond (Figure 18.2.23). So, proteins and polypeptides are naturally occurring polyamides.

From your studies earlier in this topic, you will know that polypeptides and proteins are synthesised in living things by condensation reactions between amino acids. In these reactions, the amino group, $-NH_2$, of one amino acid reacts with the carboxylic acid group, $-COOH$, of another amino acid to split out water and form an amide link (Figure 18.2.25). This process is then repeated time after time to produce a polymer (protein) with tens, hundreds or, in some cases, thousands of units.

The first synthetic and commercially important polyamides were various forms of nylon. These were not, however, produced from amino acids. Instead, they were formed by condensation polymerisation between diamines and dicarboxylic acids. One of the commonest forms of nylon is nylon-6,6. This is made by a condensation reaction between 1,6-diaminohexane and hexanedioic acid (Figure 18.2.25). The product is named nylon-6,6 because both monomers contain six carbon atoms.

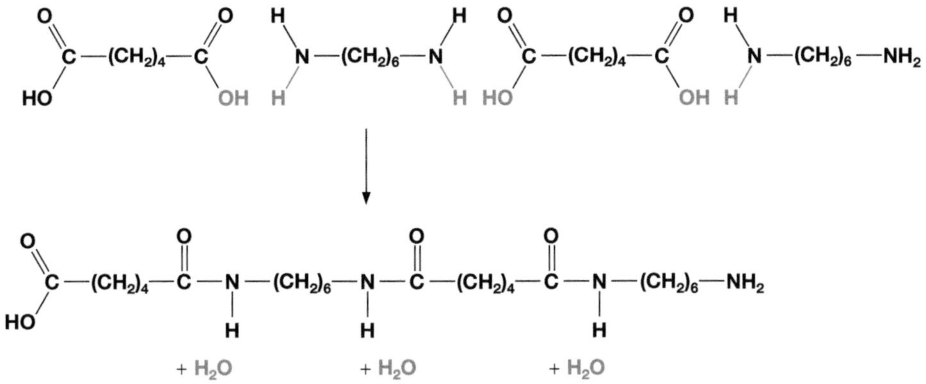

Figure 18.2.25 Condensation polymerisation to make nylon-6,6.

Tip

Early polymer chemists found it easier to synthesise polyamides using separate dicarboxylic acid and diamine molecules rather than have the carboxylic acid functional group and the amine functional group on the same molecule, as nature does in an amino acid.

Nylon-6,6 can be produced more readily in the laboratory using hexanedioyl dichloride in place of the less reactive hexanedioic acid. A solution of hexanedioyl dichloride in cyclohexane is floated on an aqueous solution of 1,6-diaminohexane. Nylon-6,6 forms as a skin at the interface and can be pulled out as fast as it is produced forming a long thread – the 'nylon rope' (Figure 18.2.26).

In this reaction hydrogen chloride molecules are eliminated in the condensation reaction (Figure 18.2.27).

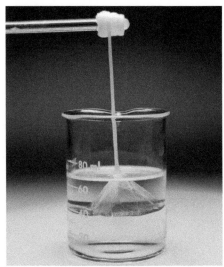

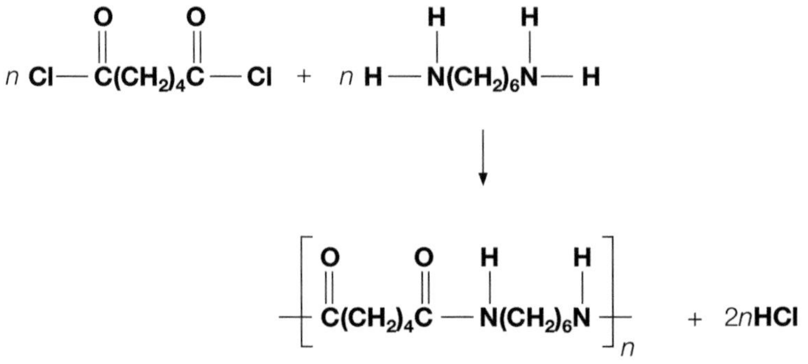

Figure 18.2.26 The nylon rope trick.

$$n \ Cl-\overset{\overset{\displaystyle O}{\|}}{C}(CH_2)_4\overset{\overset{\displaystyle O}{\|}}{C}-Cl \ + \ n \ H-\overset{\overset{\displaystyle H}{|}}{N}(CH_2)_6\overset{\overset{\displaystyle H}{|}}{N}-H$$

$$\left[-\overset{\overset{\displaystyle O}{\|}}{C}(CH_2)_4\overset{\overset{\displaystyle O}{\|}}{C} - \overset{\overset{\displaystyle H}{|}}{N}(CH_2)_6\overset{\overset{\displaystyle H}{|}}{N} - \right]_n \ + \ 2nHCl$$

Figure 18.2.27 The reaction used to make nylon-6,6 in the laboratory.

Although nylon is similar in structure to wool and silk, it does not have the softness of the natural fibres. It is, however, much harder wearing and one of its earliest uses was as a substitute for silk in the manufacture of ladies' stockings (Figure 18.2.28).

Apart from their obvious use in stockings and tights, nylon fibres are used in various forms of clothing. In fact, about 75% of the UK nylon consumption goes on clothing, but its uses are many and varied. Nylon is used to make nylon ropes that don't rot, machine bearings that don't wear out, and it is mixed with wool to make durable carpets.

Nylon is the collective name for polymers with aliphatic hydrocarbon sections linked by amide bonds. They are aliphatic polyamides in which the polar amide bonds are fixed and inflexible, but the non-polar hydrocarbon sections are free to flex, rotate and twist. So, as the hydrocarbon sections become longer, we would expect the nylon polymers to become more flexible with weaker bonding between the molecules and therefore also a lower melting temperature.

This suggests that the properties of polyamides can be modified by changing the length and nature of the hydrocarbon sections. Chemists have followed up these ideas to develop polyamides in which the hydrocarbon sections are aromatic rather than aliphatic. These polymeric **ar**omatic **amid**es are described as aramids. Aramids, such as Kevlar® (Figure 18.2.29), are extremely strong, rigid, fire-resistant and lightweight. Much of the strength is due to the extensive hydrogen bonding between the chains.

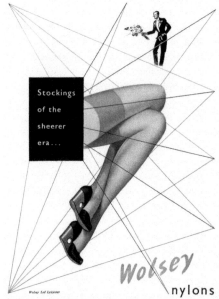

Figure 18.2.28 The American firm Du Pont patented nylon in February 1938. The first nylon stockings went on sale in the USA on 15 May 1940. In New York alone, four million pairs were sold in a few hours.

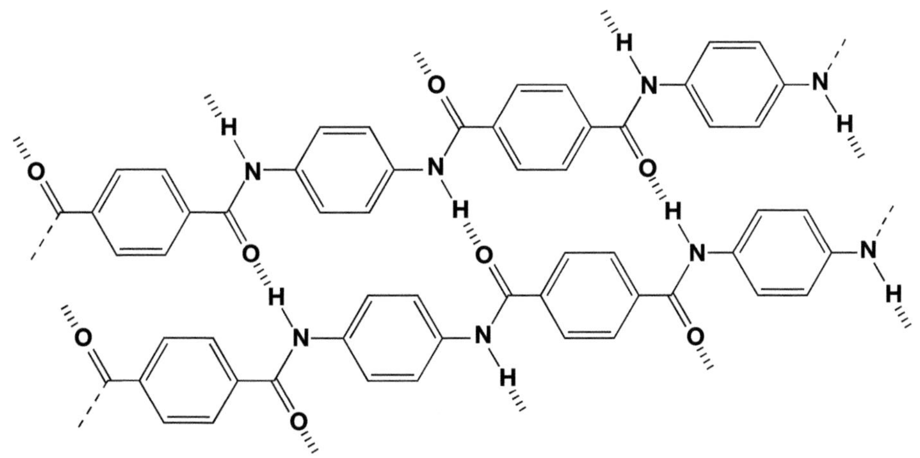

Figure 18.2.29 Polymer chains in Kevlar.

Test yourself

27 a) What type of polymerisation would produce the polymer with a repeat unit like that below?

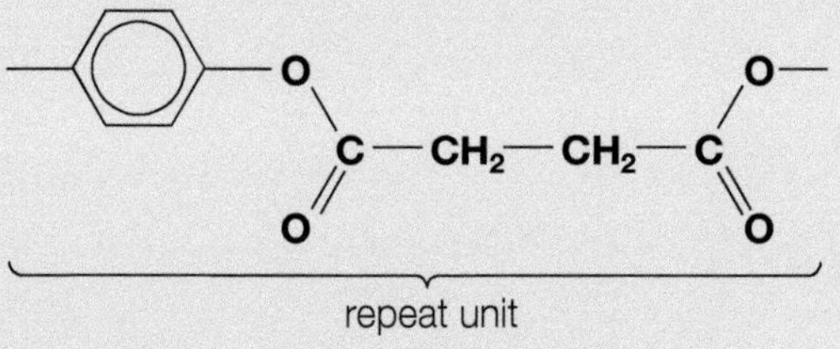

repeat unit

b) Draw the structure of the monomer or monomers that would be used to prepare the polymer.

28 The compound below can form a polymer.

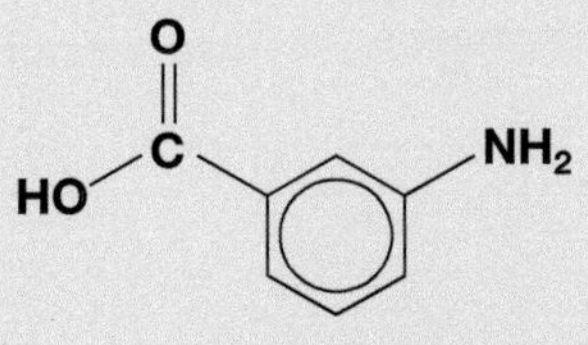

a) Identify the functional groups involved in forming the polymer.

b) What type of polymerisation will the monomer undergo?

c) What other product forms during polymerisation?

d) Draw a short length of the polymer chain showing two repeat units.

29 State two similarities and two differences between the structure of nylon-6,6 and the structure of a protein.

Modelling and synthesising polyamides

Experiments show that nylon polymers with longer hydrocarbon sections to their chains are more flexible than those with shorter sections. Kevlar is similar to nylon-6,6 but with benzene rings rather than aliphatic chains linked by the amide group. The repetition of benzene rings in its structure makes Kevlar exceptionally strong and very inflexible compared with nylon-6,6. Because of this, it is used extensively in tyres, brakes and clutch fittings, in ropes and cables and in protective clothing (Figure 18.2.30).

1 Look closely at the structure of one chain of Kevlar in Figure 18.2.29.
   a) Explain how Kevlar is a condensation polymer of benzene-1,4-dicarboxylic acid and benzene-1,4-diamine.
   b) Weight for weight, Kevlar is five times stronger than steel. This exceptional strength of Kevlar is due to hydrogen bonding between the separate chains. Use Figure 18.2.29 to explain why interchain hydrogen bonding is so strong in Kevlar.
   c) Suggest a reason why Kevlar is made from monomers with functional groups in the 1,4 positions and not from isomers with functional groups in the 1,2 or 1,3 positions.

2 Using a molecular model kit, make one repeat unit for the structure of Kevlar and explore the flexibility of the structure. (Hint: use the Kekulé structure with alternating double and single bonds for the benzene ring.)

Repeat the model-making and flexibility testing with one repeat unit for the structure of nylon-6,6. Why is nylon-6,6 flexible whereas Kevlar is inflexible?

3 A condensation polymer can be prepared by mixing equal amounts of the monomers below at room temperature.

$$H_2N(CH_2)_3CHCH_2NH_2$$
$$\quad\quad\quad |$$
$$\quad\quad CH_2OH$$

$$\underset{Cl}{\overset{O}{\underset{\|}{C}}}(CH_2)_2\underset{COOH}{CHCH_2}\underset{Cl}{\overset{O}{\underset{\|}{C}}}$$

   a) Draw the structure of one repeat unit of the polymer formed from the two monomers.
   b) The polymer forms even more rapidly if the reaction mixture contains sodium carbonate. Why is this?
   c) The polymer molecules obtained at room temperature can be linked to one another (cross-linked) by a second reaction. Explain how this cross-linking can be achieved and state the conditions needed for it to happen.
   d) Explain how the choice of reaction conditions can control the extent of polymerisation and the extent of cross-linking.

Figure 18.2.30 This policeman is wearing a bulletproof jacket made from Kevlar.

18.2.10 Comparing addition and condensation polymers

Formation of the polymers

Although both addition polymerisation (Section 6.2.12 in Student Book 1) and condensation polymerisation result in the formation of long-chain organic molecules, known as polymers, from relatively small organic molecules, known as monomers, there are some clear differences between the two processes.

- The **type of reaction** involved.
As its name suggests, addition polymerisation involves only addition reactions, whereas condensation polymerisation involves addition plus elimination. As monomer units join together, a small molecule, usually water or hydrogen chloride, is eliminated and splits off.

- The **type of links** along the polymer chain.
In addition polymers, the central chain consists of carbon atoms linked by carbon–carbon single bonds. In condensation polymers, the central chain consists of short aliphatic or aryl sections linked by ester groups or amide groups.

- The **type of monomer** involved.
In addition polymerisation, the monomers have molecules with carbon–carbon double bonds. In condensation polymerisation, the monomers have molecules with at least two functional groups which may be the same or different.

- The **conditions for preparation** of the polymers.
In general, addition polymerisations require an initiator together with high temperature and high pressure, unless a catalyst is involved. In contrast, condensation polymerisations do not require initiators and usually occur at a much lower temperature and atmospheric pressure.

Polymer properties

These differences between addition and condensation polymerisation lead to considerable variations in the properties of polymers. Polymeric materials include plastics, fibres and elastomers. As polymer science has grown, chemists and materials scientists have learned how to develop new materials with particular properties.

Some of the ways of modifying the properties of polymers include:

- altering the average length of polymer chains
- changing the structure of the monomer to one with different side groups and different intermolecular forces
- varying the extent of cross-linking between chains
- selecting a monomer that produces a biodegradable polymer
- producing a co-polymer
- adding fillers and pigments
- making composites.

> **Tip**
>
> Most addition polymers are not biodegradeable, but a recent development is the use of poly(ethenol), sometimes called polyvinyl alcohol. Poly(ethenol) is used to make plastic bags that dissolve in water and the soluble capsules containing liquid detergent, which slowly dissolve and release detergent as the washing cycle progresses.
>
> The repeat unit of poly(ethenol) is:
>

Plastics are materials made of long-chain molecules that can be moulded into shapes which are retained.

Elastomers are materials made of long-chain molecules that can be moulded into new shapes but which spring back to their original shape when the pressure is removed.

Co-polymers are polymers made from two or more monomers, each of which could produce a polymer.

Composites are materials made up of two or more recognisable constituents, each of which contributes to the properties of the composite (Figure 18.2.31).

Figure 18.2.31 An electron micrograph of a glass fibre composite showing rods of glass fibre embedded in a polyester matrix. Magnification is × 660.

Exam practice questions

1 Consider the six compounds below.

a) $CH_3CH_2CONH_2$

d) (structure: six-membered ring with NH)

b) $(CH_3CH_2)_3N$

e) (structure: benzene ring with CH_2-NH_2)

c) (structure:
$H_2N-\overset{\overset{\displaystyle H}{|}}{\underset{\underset{\displaystyle CH_3}{|}}{C}}-COOH$)

f) (structure:
$O=\overset{\overset{\displaystyle NH_2}{|}}{\underset{\underset{\displaystyle NH_2}{|}}{C}}$)

Classify the compounds as:
A primary amines
B secondary amines
C tertiary amines
D amides
E amino acids. *(6)*

2 Amines such as butylamine and phenylamine both behave as bases.
a) Give the meaning of the term 'base' and the feature of an amine molecule that causes it to act as a base. *(2)*
b) Give the formula of the salt formed when butylamine reacts with sulfuric acid. *(1)*
c) Explain why butylamine is a stronger base than phenylamine. *(3)*
d) Describe what you would see when an excess of butylamine is added with shaking to an aqueous solution of copper(II) sulfate. State the role of butylamine in the reaction and give the formula of the product. *(3)*

3 An incomplete structure of the dipeptide threonylisoleucine is shown in the diagram.

(structure of dipeptide threonylisoleucine:
$H_2N-\overset{\overset{\displaystyle H}{|}}{C}$ ... $\overset{\overset{\displaystyle H}{|}}{C}-COOH$
with $H-\overset{}{C}-OH$, CH_3 on left; $H-\overset{}{C}-CH_3$, CH_2, CH_3 on right)

a) i) Redraw the structure of the dipeptide, inserting the missing peptide link. *(2)*
   ii) On your structure, circle all the chiral centres. *(3)*
b) What does the presence of a chiral centre tell you about a compound? *(3)*
c) Draw the structures of the products obtained when the dipeptide is refluxed with excess concentrated hydrochloric acid. *(2)*

4 Look closely at the structures of glycine and glutamic acid in Figure 18.2.19.
a) Using glycine as an example, explain the meaning of the term 'the isoelectric point' of an amino acid. *(3)*
b) The isoelectric point of glycine is at pH = 6.0. Why is the isoelectric point of glycine not at pH = 7? *(5)*
c) How do you think the isoelectric point of glutamic acid compares with that of glycine? Explain your answer. *(3)*

5 a) Ethylamine can be prepared by reaction between bromoethane and ammonia.
   i) Write an equation for the reaction involved. *(1)*
   ii) Name the type of reaction taking place. *(1)*
   iii) Draw the structures of two other molecular organic products that may be produced when bromoethane reacts with ammonia. *(2)*
   iv) Explain why these two other organic products may be formed. *(3)*
b) Describe and explain what happens when the apparatus and materials shown in the diagram are set up. *(5)*

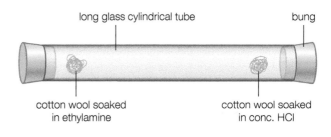

long glass cylindrical tube   bung

cotton wool soaked in ethylamine   cotton wool soaked in conc. HCl

6 The diagram shows a series of reactions beginning with the amine cadaverine. Cadaverine is formed when proteins decompose.

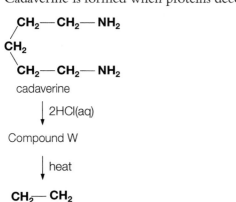

cadaverine

| 2HCl(aq)
↓

Compound W

| heat
↓

N⁺H₂ Cl⁻ + Compound X

| NaOH(aq)
↓

NH + Compound Y + Compound Z

piperidine

a) i) What characteristic physical property of cadaverine would you expect to notice if you were provided with a sample of it? *(1)*

ii) What is the systematic name of cadaverine? *(1)*

iii) Draw the structural formula of compound W. *(1)*

iv) Write the name and formula of compound X. *(2)*

v) Write the formulae of compounds Y and Z. *(2)*

b) Amines are classed as primary, secondary and tertiary.

i) Explain the difference in structure between the three types of amine. *(3)*

ii) Which type(s) do cadaverine and piperidine belong to? *(2)*

c) How will the infrared spectrum of cadaverine compare with that of piperidine? Explain your answer. *(2)*

7 Short sections of the molecular structures of two polymers, A and B, are shown in the diagram at the bottom of the page.

a) Draw the simplest repeat unit for each polymer. *(2)*

b) Draw and name the structural formula of the monomer used to prepare polymer A. *(2)*

c) i) Draw the structural formulae of the two monomers that could be used to prepare polymer B. *(2)*

ii) Name one of the two monomers. *(1)*

d) During the last decade, degradable polymers have been developed to reduce the quantity of plastic waste that is dumped in landfill sites. State and explain why polymer B, which is a polyester, is more likely to be degradable than polymer A. *(4)*

8 a) A compound containing carbon, hydrogen and nitrogen contains 61.0% carbon and 15.3% hydrogen by mass.

i) Calculate its empirical formula *(3)*

ii) What other piece of data is required to deduce its molecular formula? *(1)*

iii) If the molecular formula of the compound is the same as its empirical formula, draw and name all possible structures for the compound. *(4)*

b) Each of the structures you have drawn in part (a)(iii) can act as a base. Predict and explain the relative basic strength of the structures. *(2)*

c) In aqueous solution, tertiary amines are weaker bases than secondary amines. Explain why this is so. *(3)*

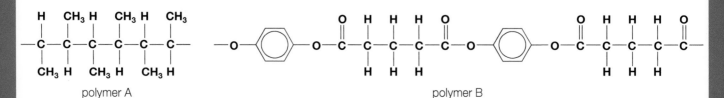

polymer A   polymer B

9 a) Poly(ethenol) is made from ethenyl ethanoate,

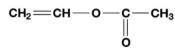

Ethenyl ethanoate is first polymerised to produce poly(ethenyl ethanoate). This is then converted to poly(ethenol) by replacing about 90% of the CH_3COO- groups in poly(ethenyl ethanoate) with $-OH$ groups.

  i) Draw the structure of ethenol and the repeat unit in pure poly(ethenol). *(2)*

  ii) Draw the repeat unit in poly(ethenyl ethanoate). *(1)*

  iii) Calculate the relative molecular mass of a sample of poly(ethenol) containing 2000 monomer units in which 10% of the monomer units contain CH_3COO- groups that have not been replaced by $-OH$ groups. *(4)*

  iv) Why is poly(ethenol) soluble in water? *(2)*

b) The solubility in water of poly(ethenol) depends on the percentage of CH_3COO- groups replaced. Maximum solubility in water occurs when about 88% of the CH_3COO- groups are replaced by $-OH$ groups. If fewer groups are replaced, the poly(ethenol) becomes less soluble. When more CH_3COO- groups are replaced, the solubility decreases and when all the groups are replaced, the poly(ethenol) is insoluble in water.

  i) Why does poly(ethenol) become less soluble if fewer CH_3COO- groups are replaced by $-OH$ groups? *(1)*

  ii) Why do you think the solubility decreases when more than 88% of the CH_3COO- groups are replaced by $-OH$ groups? *(1)*

  iii) When all the CH_3COO- groups are replaced by $-OH$ groups, the poly(ethenol) crystallises and becomes insoluble in water. Suggest what may have happened to make the poly(ethenol) insoluble. *(2)*

10 This question is concerned with the molecular interactions between the polymer chains in various plastics.

a) Tables of data show that the melting temperatures of low density polythene and high density polythene differ by $23\,°C$.

  i) What type of molecular interactions occur between polythene molecules? *(1)*

  ii) Why are the densities of the two polythenes different? *(2)*

  iii) Suggest which type of polythene has the higher melting temperature and explain why. *(3)*

b) The polymer Kevlar is a polyamide with a much higher melting temperature than polythene.

  i) Explain the term 'polyamide'. *(1)*

  ii) Kevlar is manufactured from the monomers benzene 1,4-dicarboxylic acid and benzene 1,4-diamine. Draw the structure of one repeat unit in a molecule of Kevlar. *(1)*

  iii) What is the strongest type of intermolecular force in Kevlar? *(1)*

  iv) Which atoms and groups are involved in the strongest type of intermolecular force in Kevlar? *(2)*

c) Kevlar is a very strong polymer as it contains crystalline regions.

  i) How does the arrangement of Kevlar molecules differ in crystalline regions and non-crystalline regions? *(1)*

  ii) Why do the crystalline regions help to make Kevlar strong? *(1)*

11 The following describes a method for the laboratory preparation of phenylamine.

Read the method and answer the questions which follow.

1 Place 9.0g of tin and $4.2\,cm^3$ nitrobenzene in a round-bottomed flask and attach a condenser.

2 Slowly pour $4\,cm^3$ of concentrated hydrochloric acid down the condenser and shake the flask.

3 Add further $4\,cm^3$ portions of acid up to a total of $20\,cm^3$ with shaking. If the reaction becomes too vigorous, cool the flask in cold water but do not allow the flask to get too cold.

4 When all the acid has been added and the vigour of the reaction has subsided, remove the condenser and heat the flask for 30 minutes in a boiling water bath.

5 If there is a residue of tin, add further hydrochloric acid dropwise until it has dissolved.

6 Cool the flask to room temperature and carefully with shaking add an excess of concentrated sodium hydroxide solution until there is a clear solution.

7 Set up the flask for steam distillation and distil until the distillate is clear.

8 Transfer the distillate to a separating funnel and add $6\,g$ of sodium chloride to saturate the aqueous layer.

9 Run off the phenylamine into a boiling tube, add sodium hydroxide pellets and stopper the tube.

10 Leave to stand until the liquid is clear.

The density of nitrobenzene is $1.20\,g\,cm^{-3}$ and the density of phenylamine is $1.0\,g\,cm^{-3}$.

a) Why is a condenser used in steps 2 and 3? *(1)*

b) In step 3, why should the flask not be too cold? *(1)*

c) Suggest why the flask was heated in step 4 without the condenser. *(1)*

d) In step 6, give two reasons why an excess of sodium hydroxide is needed. *(2)*

e) Suggest why sodium chloride was added to the aqueous layer. *(1)*

f) Was the phenylamine the upper or lower layer in the separating funnel? *(2)*

g) Why were the sodium hydroxide pellets added in step 9? *(1)*

h) Calculate the expected yield of phenylamine starting from $4.2\,cm^3$ of nitrobenzene. *(3)*

i) Write half-equations for the reduction of nitrobenzene to phenylamine in acid conditions and the oxidation of tin to $SnCl_6^{2-}$ ions and combine them to give an overall equation for the reaction. *(3)*

12 Propylamine and ethanamide have similar relative molecular masses, but otherwise they have different properties.

- Ethanamide has an absorption at $1681\,cm^{-1}$ in its infrared absorption spectrum whereas propylamine has no absorption in this region of its spectrum.
- Ethanamide is not basic whereas propylamine is basic.
- Ethanamide is a solid at room temperature whereas propylamine is a liquid.

Explain these differences. *(6)*

Organic synthesis

18.3

18.3.1 Organic synthesis

A lot of the purpose and pleasure of chemistry comes from making new materials such as polymers, perfumes, drugs and dyes. This making of new materials is called synthesis. The synthesis of organic compounds is very important in the research and production of new and useful products. Many features of modern life depend on the skills of chemists and their ability to synthesise new and complex materials. New colours, dyes and fabrics for the fashion industry are synthetic organic molecules. So also are the liquid crystals used in the flat screens of laptops or tablets (Figures 18.3.1 and 18.3.2). These organic compounds in the computer screen have been tailor-made by chemists to respond to an electric field and affect light.

Figure 18.3.1 Liquid crystals photographed through a microscope with polarised light.

Figure 18.3.2 The structure of a molecule which makes up a liquid crystal.

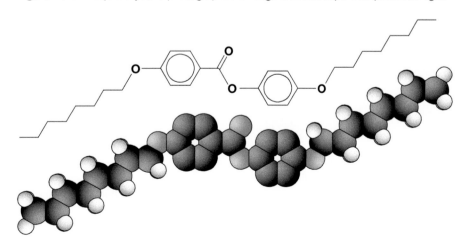

salbutamol

levodopa

chloramphenicol

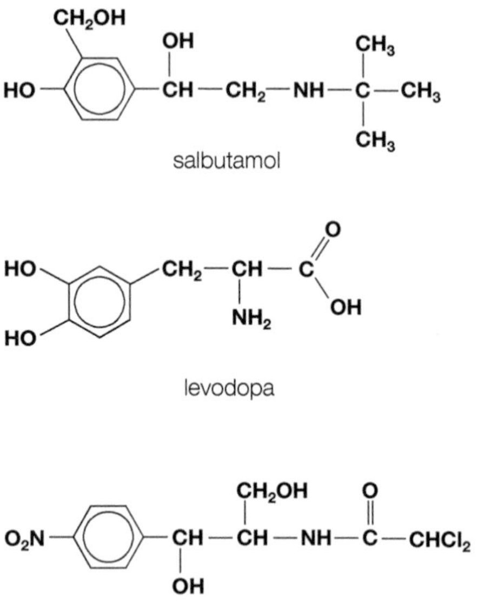

Figure 18.3.3 Three important drugs that have been synthesised by chemists – salbutamol, levodopa and chloramphenicol.

Tip

The synthetic routes discussed in this chapter use reactions of organic compounds studied in earlier chapters of this book and also in Student Book 1. Successful and efficient synthesis depends on a good knowledge and understanding of the reactions of all the functional groups studied. Some of the 'Test yourself' questions are designed to help revise ideas from earlier in the A Level course.

One of the major areas of chemical research today involves the synthesis of drugs and medicines. Every day, large numbers of compounds are synthesised for testing in pharmaceutical laboratories as potential drugs to cure or alleviate a particular disease. Medicines that have been synthesised by chemists include aspirin and paracetamol to relieve pain (Section 18.2.3), salbutamol to prevent asthma, chloramphenicol to treat typhoid and levodopa to alleviate Parkinson's disease (Figure 18.3.3).

Three other important areas of synthetic chemical research involve catalysts (Section 15.11), antiseptics and polymers (Sections 17.3.6 and 18.2.9).

The essential job of synthetic organic chemists is to consider the proposed structure for a target molecule and then devise a way of making it from simpler, readily available starting materials. The scale of work involved and the difficulties encountered in a complex organic synthesis are illustrated by the painstaking and ingenious first synthesis of the anti-cancer drug Taxol® by a team of chemists led by Robert Halton at Florida State University in 1994.

The synthesis of Taxol

In 1962, it was discovered that an extract of the bark of the Pacific yew tree (Figure 18.3.4) was effective as an anti-cancer drug, particularly against ovarian cancer. The structure of the active compound (Figure 18.3.5) was determined in 1971 and its method of action also discovered. But the Pacific yew tree was very slow growing and several trees had to be felled to produce even a small amount of the drug so, in order to supply the demand, chemists had to find a way to make the drug from simpler substances.

Figure 18.3.4 The Pacific yew tree (*Taxus brevifolia*) – about 2000 trees were used to produce 1 kg of Taxol.

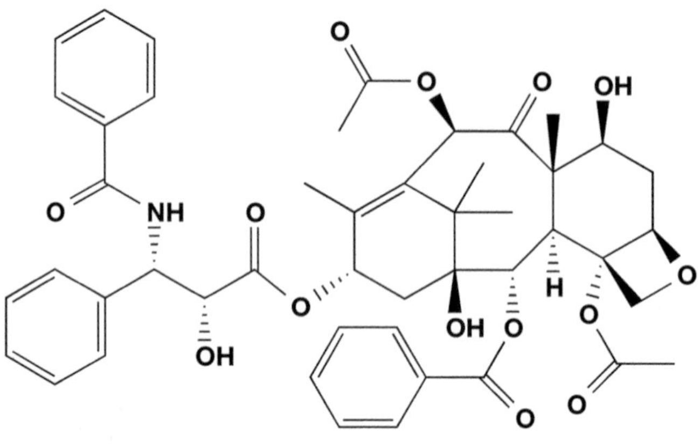

Figure 18.3.5 The structure of paclitaxel, better known under its trademark name Taxol®.

It took many years before a total synthesis was achieved. From the start, the project was planned in great detail. The chemists drew on their understanding of the mechanisms of organic reactions to predict the likely products at each stage and suggest routes to their target molecule.

The synthesis of Taxol would have been impossible without the newer methods of separation, purification and identification that had become available. The variety of spectroscopy techniques was also crucial to success. Halton's team published their paper in 1994 describing the successful 37-step total synthesis and just beat several other teams in the race to discover a synthesis. Halton also used a method starting from a related compound found in the needles of the European yew which involved only four steps and was therefore commercially more viable. Also, as yew needles were used rather than bark, the trees did not need to be felled. Needles of the European yew are also a source in the production of Taxotere®, which is potentially an even better anti-cancer agent than Taxol. In 2002, a further development came with the discovery of a biochemical route using a fermentation process from yew-derived plant cells.

Organic analysis

When complex molecules such as Taxol have been synthesised, chemists must use a variety of methods to analyse them and identify their precise composition and structure.

Traditionally, chemical tests were used to identify functional groups in organic molecules, together with combustion and quantitative analysis. Nowadays, however, modern laboratories rely on a range of highly sensitive, automated and instrumental techniques to identify the products of synthesis. These include chromatography (Section 19.5), mass spectrometry (Section 19.2) and various kinds of spectroscopy (Sections 19.3 and 19.4).

Sensitive methods of analysis are very important in monitoring organic syntheses for several reasons.

- Sensitive methods of analysis determine the degree of purity of a synthetic product.
- Sensitive methods of analysis also identify any impurities, some of which may be toxic and in very small concentration.
- If analysis reveals an impurity in the product, it may be possible to limit its formation by changing the operating conditions for the reaction. Changes in the temperature, the pressure, the solvent used or the choice of catalyst may promote the formation of a desired product while reducing the formation of impurities.
- Many pharmaceutical laboratories that specialise in the development of new drugs produce thousands of compounds every year for further testing. Some of their products are obtained in very small concentrations and particularly sensitive techniques are needed to analyse and identify them. The food and drugs industries operate very high standards of purity in their products. Impurities, depending on their toxicity, may interfere with the health and well-being of consumers. For example, traces of sodium chloride in a medicine would probably not be considered a problem, but the slightest trace of sodium cyanide would be cause for alarm.

18.3.2 The formulae of organic molecules

Empirical formulae

From previous study you should know that the empirical formula of a compound shows the simplest ratio of the number of atoms of each element in it (Section 5.2 in Student Book 1).

Modern analysis usually begins with mass spectrometry to determine the M_r of an organic compound (Section 7.1 in Student Book 1), but finding its empirical formula is an important step in understanding its chemistry. The usual way of doing this is to oxidise a weighed sample of the organic compound completely by burning it in pure dry oxygen, and then determine the masses of carbon dioxide and water produced. This process of combustion analysis then enables you to calculate the masses of carbon and hydrogen in a sample of the compound and hence its empirical formula.

Figure 18.3.6 shows a simplified diagram of the method and apparatus used. Modern methods of combustion analysis include refinements to ensure that the organic compound is completely oxidised and that all the carbon dioxide and water are absorbed and weighed.

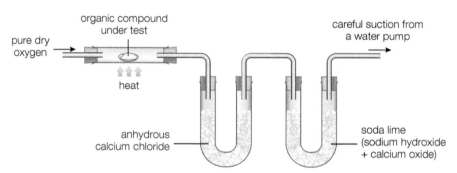

Figure 18.3.6 Using careful suction, draw pure dry oxygen over a heated sample of the solid organic compound. (Liquid organic compounds can be burnt from a wick.) Pass the product gases through anhydrous calcium chloride (or anhydrous copper(II) sulfate) to absorb any water produced, and then through anhydrous soda lime (sodium hydroxide and calcium oxide) to absorb the carbon dioxide produced.

Molecular formulae

Molecular formulae are more helpful than empirical formulae because they show the actual number of atoms of each element in one molecule of a compound. All that is needed to find the molecular formula from the empirical formula is the molar mass of the compound. This can be determined from the mass spectrum of the compound.

A molecular formula is always a simple multiple of the empirical formula. Methane, for example, has the empirical formula CH_4 and the molecular formula CH_4, benzene has the empirical formula CH and the molecular formula C_6H_6, and ethanoic acid has the empirical formula CH_2O and the molecular formula $C_2H_4O_2$.

Structural formulae

An empirical formula can represent many different compounds and a molecular formula can represent many different isomers, but the structural formula of a compound shows how the atoms link together in a molecule of a particular compound. Figure 18.3.7 shows empirical, molecular and structural formulae for cyclohexene.

Given the molecular formula of an organic compound, and knowing something about its characteristic reactions, it is often possible to predict its structure by assuming that:

- carbon atoms form four covalent bonds
- nitrogen atoms form three covalent bonds
- oxygen atoms form two covalent bonds
- hydrogen and halogen atoms form one covalent bond.

The structural formulae of organic compounds can often be determined by combustion analysis to find their percentage composition, followed by a study of their chemical reactions. However, structural formulae can be obtained more definitively, and more precisely, by spectrometry and spectroscopy.

Using mass spectrometry it is possible to identify the fragments of an organic molecule and then piece the whole molecule together.

Spectroscopic methods such as infrared spectroscopy (see Section 7.2 in Student Book 1 and Section 19.3 in this book) and, particularly, nuclear magnetic resonance spectroscopy (Section 19.4), provide information about the various bonds and functional groups in organic molecules in order to confirm their structural formulae.

Sometimes it is enough to show structural formulae in a condensed form, such as $CH_3CH_2CH_2CH{=}CHCH_3$ for hex-2-ene. At other times it is more helpful to write a full structural formula showing all the atoms and all the bonds. This type of formula is called a displayed formula.

Chemists have also devised a useful shorthand for showing the formulae of more complex molecules as skeletal structures. These skeletal formulae need careful study because they represent the hydrocarbon part of the molecule simply as lines for the bonds between carbon atoms, leaving out the symbols for carbon and hydrogen atoms (Figure 18.3.8).

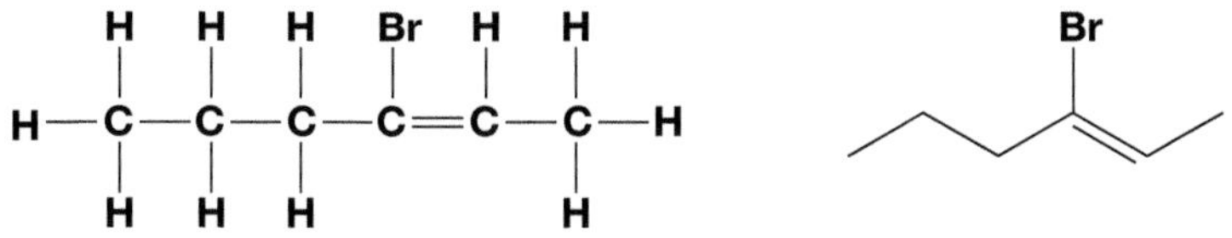

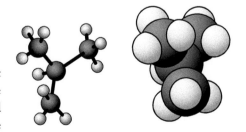

Figure 18.3.8 The displayed and skeletal formulae of 3-bromohex-2-ene.

Although the structural, displayed and skeletal formulae of an organic compound show how its atoms link together, they do not show its true shape in three dimensions. Sometimes, it is important to know and understand what the three-dimensional shape of a molecule is like and chemists use various models to do this. These include ball-and-stick models, space-filling models (Figure 18.3.9) and various types of computer models.

Tip

Don't forget that pairs of stereoisomers have the same structural formula but the atoms are arranged differently in space (see Chapter 17.1).

C_3H_5 empirical formula

C_6H_{10} molecular formula

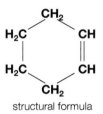

structural formula

Figure 18.3.7 The empirical, molecular and structural formulae of cyclohexene.

Figure 18.3.9 A ball-and-stick model and a space-filling model of 2-methylpropane.

Test yourself

1 A compound of carbon, hydrogen and oxygen with $M_r = 46.0$ contains 52.2% carbon and 13.0% hydrogen. Calculate its molecular formula and draw and name two possible structural formulae for the compound.

2 Write out the empirical, molecular, structural, displayed and skeletal formulae of 2-methylpropane in Figure 18.3.9.

3 Describe briefly how you would determine the percentages of carbon and hydrogen in a solid organic compound.

4 A compound containing only carbon, hydrogen and fluorine was burnt in excess oxygen. 0.32 g of the compound produced 0.44 g of carbon dioxide and 0.09 g of water.

   a) What is the empirical formula of the compound?

   b) The relative molecular mass of the compound is 64.0. What is its molecular formula?

   c) Write all the possible displayed formulae of the compound and give the systematic name for each formula.

5 A compound, containing only carbon, hydrogen and oxygen has prominent peaks in its mass spectrum at *m/z* values of 60, 43, 31, 29 and 17.

   a) What is the relative molecular mass of the compound?

   b) Suggest possible fragment ions for the peaks at *m/z* values of 17, 29 and 31.

   c) Suggest a possible structure for the compound.

18.3.3 Functional groups – the keys to organic molecules

Functional groups provide the key to organic molecules. A knowledge of the properties and reactions of a limited number of functional groups has opened up our understanding of most organic compounds.

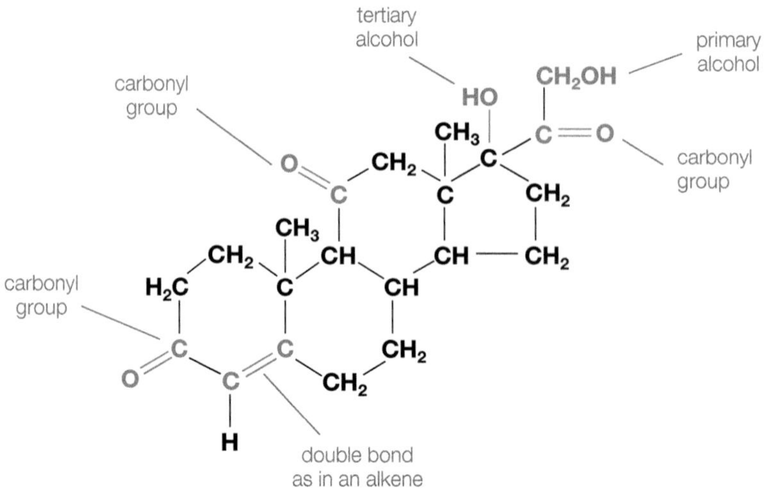

Figure 18.3.10 The structure of the steroid cortisone, labelled to show the reactive functional groups and the hydrocarbon skeleton.

A functional group is the atom or atoms that give a series of organic compounds their characteristic properties and reactions. Chemists often think of an organic molecule as a relatively unreactive hydrocarbon skeleton with one or more functional groups in place of hydrogen atoms. The functional group in a molecule is responsible for most of its reactions. In contrast, the carbon–carbon single bonds and carbon–hydrogen bonds are relatively unreactive, partly because they are both strong and non-polar.

Table 18.3.1 shows the major functional groups that you have met during your A Level studies, together with an example of one compound containing each group.

Key term

A **functional group** is the atom or group of atoms that give an organic compound its characteristic properties.

Table 18.3.1 The major functional groups.

Functional group		Example	Functional group		Example
Alcohol	—OH	propan-1-ol $CH_3CH_2CH_2OH$	Ester	—O—C— ‖ O	methyl ethanoate $CH_3-O-C-CH_3$ ‖ O
Alkene	C=C	propene $CH_3CH=CH_2$	Acyl chloride (acid chloride)	—C—Cl ‖ O	ethanoyl chloride CH_3-C-Cl ‖ O
Halogenoalkane Hal = F, Cl, Br, I	—Hal	1-chloropropane $CH_3CH_2CH_2Cl$	Amine	—NH₂	propylamine $CH_3CH_2CH_2NH_2$
Ether	C—O—C	methoxyethane $CH_3OCH_2CH_3$	Amide	—C—N— ‖ | O H	propanamide $CH_3CH_2-C-NH_2$ ‖ O
Aldehyde	—C O H	propanal CH_3CH_2CHO	Nitrile	—C≡N	ethanenitrile $CH_3—C≡N$
Ketone	C=O	propanone CH_3COCH_3	Phenyl	$(C_6H_5—)$	benzene
Carboxylic acid	—C O OH	propanoic acid CH_3CH_2COOH			

The characteristic properties and tests for most of these functional groups are shown on the data sheets headed 'Tests and observations on organic compounds' and 'Tests for gases', accessed via the QR code for this chapter on page 321. These tests are used in Core practical 15 (part 2): Analysis of some organic unknowns.

Functional groups can also be identified by spectroscopic methods (Chapter 19). An absorption in the infrared spectrum of a compound or a peak in its NMR spectrum can identify a particular functional group in the compound (see the Edexcel Data booklet).

Core practical 15 (part 2)

Analysis of some organic unknowns

A student was provided with three organic liquids, A, B and C, each of which has molecules that contain four carbon atoms.

He carried out a series of tests using (a) 2,4-dinitrophenylhydrazine, (b) iodine and sodium hydroxide and (c) acidified potassium dichromate(vi). Table 18.3.2 shows which reactions gave positive results.

1 Describe what he saw when each test gave a positive result.
2 What deductions about the functional group present can be made using the test with 2,4-dinitrophenylhydrazine?
3 What deductions about the functional group present can be made using the test with iodine and sodium hydroxide?
4 What deductions about the functional group present can be made using the test with acidified potassium dichromate(vi)?

5 Deduce possible structures for compounds A, B and C.

A fourth compound, D, with formula $CH_3CH_2COCH_2OH$, was also tested with the same three reagents.

6 Give the results of the three tests and draw the structure of the organic product, if any, of the reaction of D with (b) iodine and sodium hydroxide and (c) acidified potassium dichromate(vi).
7 Draw a structure for compound E, an isomer of D, which gave a negative test with each of the reagents in Table 18.3.2 but gave an effervescence of a colourless gas when added to aqueous sodium hydrogen carbonate.
8 Draw a structure for compound F, also an isomer of D, which gave a negative test with each of the reagents in Table 18.3.2 and did not react with aqueous sodium hydrogen carbonate.

Table 18.3.2

Liquid	2,4-Dinitrophenylhydrazine	Iodine with sodium hydroxide	Acidified potassium dichromate(vi)
A	✓	✗	✓
B	✓	✓	✗
C	✗	✓	✓

Tip

Analysis of an inorganic unknown is covered in Core practical 15 (part 1) in Chapter 15.

For practical guidance, refer to Practical skills sheet 10, 'Analysing organic unknowns', which you can access via the QR code for Chapter 18.3 on page 321.

Tip

Test-tube reactions carried out to identify functional groups have the major disadvantage that they use up some of the compound. A major advantage of spectroscopic analysis is that these techniques are either non-destructive (infrared and NMR) or use up tiny amounts of compound (mass spectrometry).

Test yourself

6 Anaerobic respiration in muscle cells breaks down glucose to simpler compounds including the following two molecules. Identify the functional groups in these molecules:

a) $HOCH_2–CH(OH)–CHO$

b) $CH_3–CO–COOH$.

7 Pheromones are messenger molecules produced by insects to attract mates or to give an alarm signal. Identify the functional groups in the pheromone below, which is produced by queen bees.

8 Use the data sheets headed 'Tests and observations on organic compounds', accessed via the QR codes for this chapter on page 321, to predict six important properties or reactions of the painkiller dextropropoxyphene and its mirror image, which is an ingredient of cough mixtures.

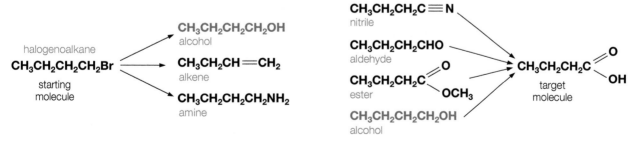

9 For each of parts (a) to (f) below only one of the compounds labelled A to D is correct.

A $CH_3CH_2NH_2$ B $C_6H_5NO_2$ C $C_6H_5NH_2$ D $(CH_3)_4NCl$

a) Which is a strong electrolyte?

b) Which dissolves in dilute hydrochloric acid but not in water?

c) Which is insoluble in water, dilute acid and dilute alkali?

d) Which best forms a blue complex with aqueous copper(II) sulfate?

e) Which has the highest vapour pressure at room temperature?

f) Which combines most readily with H^+ ions?

18.3.4 Organic routes

Organic chemists synthesise new molecules using their knowledge of functional groups, reaction mechanisms and molecular shapes, as well as the factors that control the rate and extent of chemical change.

A synthetic pathway leads from the reactants to the required product in one step or several steps. Organic chemists often start by examining the 'target molecule'. Then, they work backwards through a series of steps to find suitable starting chemicals that are cheap enough and available. This is called 'retrosynthetic analysis' and involves planning a synthesis by transforming a target molecule into simpler precursors without necessarily making any assumptions about starting materials.

Figure 18.3.11 shows an example of the systematic way in which working back can be used in synthesising one 'target molecule' from a 'starting molecule'. In this case, the 'target molecule' is butanoic acid and the 'starting molecule' is 1-bromobutane.

Figure 18.3.11 Working back from the target molecule to find a two-step synthesis of butanoic acid from 1-bromobutane.

1 Begin by writing down the formulae of those compounds that could be readily converted to butanoic acid, the target molecule. These include the nitrile butanenitrile, the aldehyde butanal, the ester methyl butanoate and the alcohol butan-1-ol.

2 Then look at your starting molecule, 1-bromobutane, to see whether it could be converted to one of the compounds that would readily form butanoic acid. If necessary, write down the formulae of compounds that might be produced from 1-bromobutane. These include the alcohol butan-1-ol, the alkene but-1-ene, and the amine butylamine.

3 With any luck, you should now see a possible two-step synthetic route from your starting molecule to the target molecule. In this case, the route can go via butan-1-ol.

4 If a two-step route is not clear at this point, then you might need to consider a three-step route involving the conversion of one of the products from the starting material to one of the reactants that will readily form the target material.

Chemists normally seek a synthetic route that has the least number of steps and produces a high yield of the product. The larger the scale of production, the more important it is to keep the yield high so as to avoid producing large quantities of wasteful by-products.

Changing the functional groups

All the reactions in organic chemistry convert one compound to another, but there are some reactions that are particularly useful for developing synthetic routes. These useful reactions, which do not change the number of carbon atoms, include:

- the addition of hydrogen halides to alkenes
- substitution reactions that replace halogen atoms with other functional groups such as $-OH$ or $-NH_2$
- substitution of a chlorine atom for the $-OH$ group in an alcohol or a carboxylic acid
- elimination of a hydrogen halide from a halogenoalkane to introduce a carbon–carbon double bond
- oxidation of primary alcohols to aldehydes and then carboxylic acids
- reduction of carbonyl compounds to alcohols
- hydrolysis of a nitrile to form a carboxylic acid group.

Test yourself

10 Draw the structural formula of the main organic product in each of the following reactions. Classify each reaction as addition, substitution or elimination and classify the reagent on the arrow as a free radical, nucleophile, electrophile or base.

a) $CH_2=CH_2(g) \xrightarrow{\text{HBr(g)}}$

b) $CH_3CH_2CH_2Br(l) \xrightarrow{\text{KOH(aq)}}$

c) C_6H_6 benzene $\xrightarrow[\text{+ conc. } H_2SO_4]{\text{conc. } HNO_3}$

Changing the carbon chain

Although it is easy to change the reactive part of a molecule, the functional group, chemists sometimes need to lengthen or shorten the carbon skeleton of a molecule. Planning the synthesis of a new drug may involve computer-aided drug design to predict whether replacing methyl groups with ethyl groups, for instance, would increase the effectiveness of the drug.

Increasing the chain length

Use of cyanide ions

The use of cyanide ions to add a single carbon to the chain is described in Section 6.3.4 of Student Book 1. When a halogenoalkane is heated with a solution of potassium cyanide in ethanol, the halogen atom is replaced by the CN group and a nitrile is formed with a longer carbon chain.

For example, the two-carbon chain in bromoethane becomes three in propanenitrile:

$$CH_3CH_2Br \ + \ CN^- \longrightarrow CH_3CH_2CN \ + \ Br^-$$
bromoethane propanenitrile

Use of Grignard reagents

An alternative reaction that can add more than one carbon uses a Grignard reagent.

In most organic compounds, carbon is bonded to a more electronegative atom, such as halogen, oxygen or nitrogen, so that carbon is usually polarised $\delta+$. In a Grignard reagent, carbon is bonded to the much less electronegative element magnesium, so carbon becomes polarised $\delta-$. The $\delta-$ carbon in a Grignard reagent then bonds to the $\delta+$ carbon of a carbonyl compound and a reaction takes place in which a new C–C bond is formed.

Preparation of Grignard reagents

Grignard reagents are unstable and must be prepared immediately before use. Magnesium metal is added to a solution of a halogenoalkane, usually a bromide or an iodide, dissolved in dry ethoxyethane. The magnesium reacts vigorously and dissolves to form a compound that can be represented as RMgX.

Tip

The nitrile group can be reduced to form a primary amine using hydrogen with a nickel catalyst or $LiAlH_4$ in ethoxyethane (Section 18.2.4). The nitrile group can also be hydrolysed to form a carboxylic acid by reacting with hydrochloric acid (Section 17.3.2).

Key term

A Grignard reagent is an organometallic compound with the general formula RMgX formed from a halogenoalkane RX and magnesium.

> **Tip**
>
> New carbon–carbon bonds are formed in Friedel–Crafts reactions of arenes. These involve the replacement of a hydrogen atom by a carbon chain (see Section 18.1.6).

For example, iodoethane reacts with magnesium in ethoxyethane to form ethylmagnesium iodide.

$$CH_3CH_2I + Mg \longrightarrow \underset{\text{ethylmagnesium iodide}}{CH_3CH_2MgI}$$

Reaction of Grignard reagents

Carbonyl compounds or carbon dioxide react with Grignard reagents. The reaction occurs in two steps. An initial product is formed in an addition reaction and this is then hydrolysed in the second step by the addition of dilute acid.

The following examples use the Grignard reagent ethylmagnesium iodide made from iodoethane.

a) Reaction with methanal to form a primary alcohol (Figure 18.3.12).

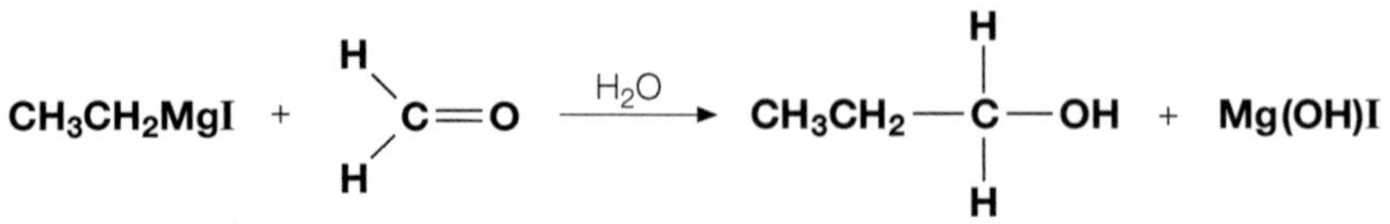

Figure 18.3.12 Reaction of ethylmagnesium iodide with methanal to form propan-1-ol.

b) Reaction with other aldehydes to form a secondary alcohol (Figure 18.3.13).

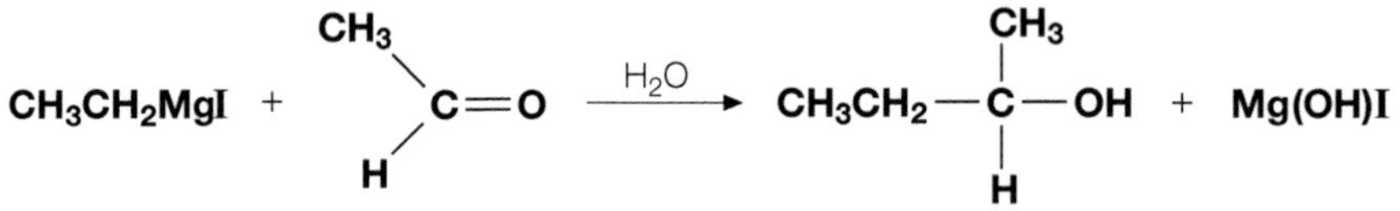

Figure 18.3.13 Reaction of ethylmagnesium iodide with ethanal to form butan-2-ol.

c) Reaction with a ketone to form a tertiary alcohol (Figure 18.3.14).

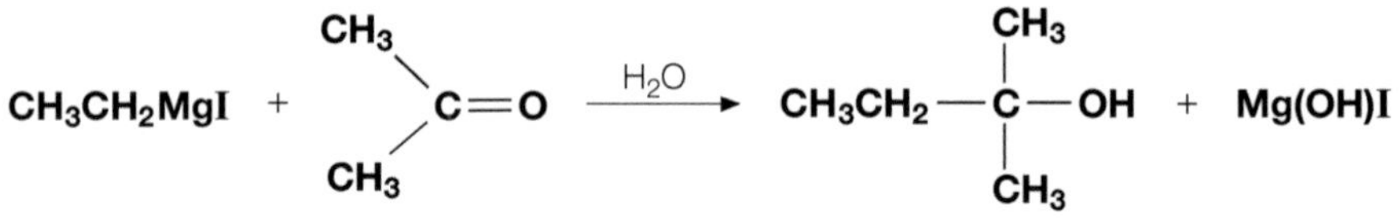

Figure 18.3.14 Reaction of ethylmagnesium iodide with propanone to form 2-methylbutan-2-ol.

d) Reaction with carbon dioxide to form a carboxylic acid (Figure 18.3.15).

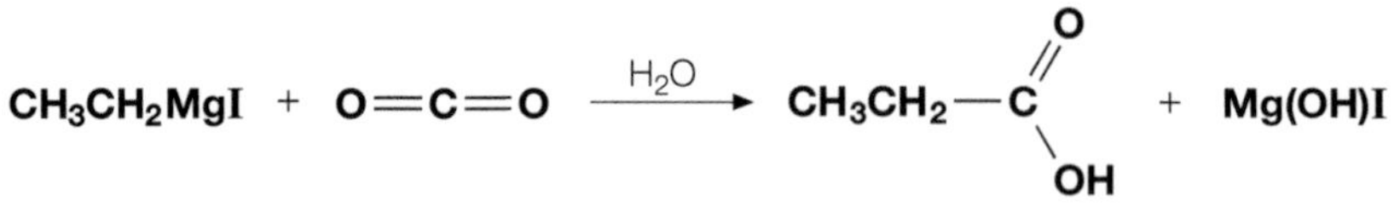

Figure 18.3.15 Reaction of ethylmagnesium iodide with carbon dioxide to form propanoic acid.

Decreasing the chain length

Triiodomethane reaction

The reaction of methyl ketones or methyl secondary alcohols with iodine and sodium hydroxide forms a yellow precipitate of triiodomethane (iodoform) and a carboxylate salt with one fewer carbon (Section 17.2.6).

$$\underset{\text{butanone}}{CH_3CH_2COCH_3} + 3I_2 + 4OH^- \longrightarrow \underset{\text{propanoate ions}}{CH_3CH_2COO^-} + CHI_3 + 3I^- + 3H_2O$$

13 Draw the structural formula of the organic product(s) in each of the following reactions:

a) 1-bromopropane with potassium cyanide in ethanol

b) butan-2-ol with iodine in sodium hydroxide.

14 a) Outline how a Grignard reagent is made from 1-bromopropane.

b) The Grignard reagent formed in part (a) reacts separately with:

i) methanal

ii) carbon dioxide

iii) cyclohexanone

iv) butanal

The intermediates formed are then hydrolysed. Draw the structural formula and name the final product in each case.

15 Give reagents and conditions for the preparation of pentylamine in two-steps starting from 1-bromobutane.

16 Give reagents and conditions for the preparation of 2-phenylbut-2-ene in three steps starting from benzene and show the structure of the intermediates in the synthesis.

Activity

Converting one functional group to another

Make a copy of the flow chart in Figure 18.3.16. For each numbered arrow, write the reagents and conditions needed for the conversion.

1 Using your completed copy of Figure 18.3.16, suggest two-step syntheses, showing the reagents and conditions for each of the following conversions:

a) ethene to ethylamine

b) ethanol to ethyl ethanoate (using ethanol as the only carbon compound)

c) propanoic acid to propanamide.

2 Using your completed copy of Figure 18.3.16, suggest three-step syntheses, showing the reagents and conditions for each of the following conversions:

a) ethene to ethanoic acid

b) propan-2-ol to propane.

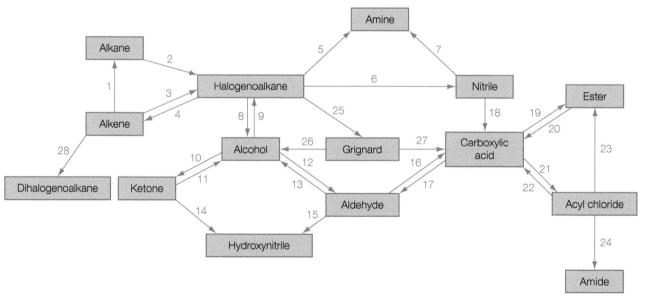

Figure 18.3.16 A flow diagram summarising the methods for converting one functional group to another.

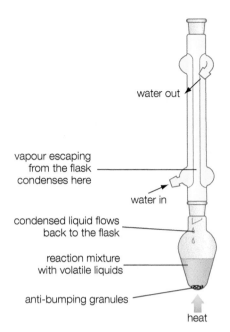

water out

vapour escaping from the flask condenses here

water in

condensed liquid flows back to the flask

reaction mixture with volatile liquids

anti-bumping granules

heat

Figure 18.3.17 Heating in a flask with a reflux condenser prevents vapours escaping while the reaction is happening. Vapours from the reaction mixture condense and flow back (reflux) into the flask.

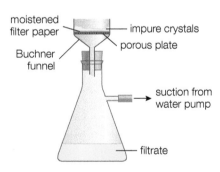

moistened filter paper

impure crystals

porous plate

Buchner funnel

suction from water pump

filtrate

Figure 18.3.18 Filtering a solid using a Buchner funnel by suction from a water pump.

18.3.5 Synthetic techniques

Chemists have developed a range of practical techniques and procedures for the synthesis of solid and liquid organic compounds. These methods allow for the fact that reactions involving molecules with covalent bonds are often slow and that it is difficult to avoid side reactions that produce by-products. There are five key stages in the preparation of an organic compound.

Stage 1: Planning

The starting point of any synthesis is to choose an appropriate reaction or series of reactions as described in the previous section. The next thing to do is to work out suitable reacting quantities from the equation and decide on the conditions for reaction.

An important part of the planning stage is a risk assessment (see Practical skills sheet 3, 'Assessing hazards and risks', which you can access via the QR code for Chapter 18.3 on page 321). This should ensure that hazards have been identified and that appropriate safety precautions and control measures are used in order to reduce the risk during any synthesis.

Stage 2: Carrying out the reaction

During this stage, the reactants are measured out and mixed in suitable apparatus. Most organic reactions are slow at room temperature so it is usually necessary to heat the reactants using a flame, heating mantle or hotplate. One of the commonest techniques is to heat the reaction mixture in a flask fitted with a reflux condenser (Figure 18.3.17).

Organic reagents do not usually mix with aqueous reagents. So another common technique is to shake the immiscible reactants in a stoppered container.

Stage 3: Separating the crude product from the reaction mixture

Filtration
If the crude product is a solid, it can be separated by filtration using a Buchner or Hirsch funnel with suction from a water pump. This is illustrated in Figure 18.3.18.

Distillation
Liquids can often be separated by simple distillation, fractional distillation or **steam distillation**. Distillation with steam at 100 °C allows the separation of compounds that decompose if heated at their boiling temperatures. The technique works only with compounds that do not mix with water. When used to separate the products of organic preparations, steam distillation leaves behind those reagents and products that are soluble in water (Figure 18.3.19).

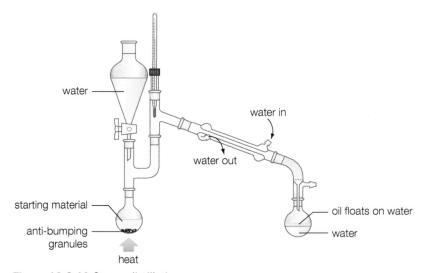

water
water in
water out
starting material
anti-bumping granules
heat
oil floats on water
water

Figure 18.3.19 Steam distillation.

Tip

Some methods of steam distillation generate the steam in a separate flask and pass steam into the impure mixture. The amounts of immiscible liquid and water that distil together depend on the relative vapour pressures of the two liquids at the boiling temperature.

Solvent extraction

Solid or liquid products can be separated from an aqueous reaction mixture using the technique of solvent extraction. The aqueous mixture is shaken in a separating funnel with a solvent which is immiscible with water (Figure 18.3.20). The organic product dissolves preferentially in the organic layer, which in most cases is the upper layer. The lower aqueous layer can then be drained off and the organic product can be obtained from the upper layer by evaporation of the solvent.

Test yourself

17 Pairs of liquids can be separated by (i) simple distillation, (ii) fractional distillation or (iii) steam distillation. For each of the following pairs, suggest the best distillation method to obtain the first liquid from a mixture of it with the second. Boiling temperatures are given in brackets.

 a) methanol (65 °C) from a mixture with ethyl ethanoate (54 °C)

 b) nitrobenzene (211 °C) from a mixture with water (100 °C)

 c) pentan-1-ol (138 °C) from a mixture with pentane (36 °C)

18 A separating funnel is shown in stage C in Figure 17.3.12 on page 200. Explain why the liquid does not drain out if the apparatus is used exactly as shown in the diagram.

Figure 18.3.20 Separating funnels used in pesticide research.

Stage 4: Purifying the product

The 'crude' product separated from the reaction mixture is usually contaminated with by-products and unused reactants. The methods of purifying this 'crude' product depend on whether it is a solid or a liquid.

Purifying organic solids

The usual technique for purifying solids is recrystallisation, which is illustrated in part of Figure 17.2.20 on page 190. The procedure for recrystallisation is based on using a solvent that dissolves the product when hot, but not when cold. The choice of solvent is usually made by trial and error. Use of a Buchner or Hirsch funnel and suction filtration speeds up filtering and facilitates recovery of the purified solid from the filter paper. The procedure is as follows.

1 Dissolve the impure solid in the minimum volume of hot solvent.
2 If the solution is not clear, filter the hot mixture through a heated funnel to remove insoluble impurities.
3 Cool the filtrate so that the product recrystallises, leaving the smaller amounts of soluble impurities in solution.
4 Filter to recover the purified product.
5 Wash the purified solid with small amounts of cold, pure solvent to wash away any solution containing impurities.
6 Allow the solvent to evaporate from the purified solid in the air or place the solid in a **desiccator** (Figure 18.3.21).

> **Tip**
>
> A common drying agent used in a desiccator to remove water is silica gel. This is often used in a form containing a little cobalt (II) chloride. Cobalt chloride is blue when anhydrous, but turns pink in the presence of water. So if the silica gel turns pink, this indicates that it cannot absorb any more water and needs to be heated to return it to the anhydrous state.

> **Key term**
>
> A **desiccator** is a container that is used to dry a solid and also to store materials in a dry atmosphere.

Figure 18.3.21 A desiccator used to store bottles in a dry atmosphere created by the desiccant silica gel.

> **Test yourself**
>
> 19 Explain why connecting a desiccator to a vacuum pump increases the rate of evaporation of the solvent.
> 20 Give the formula of the cobalt-containing complex ion formed when silica gel drying agent containing cobalt(II) chloride turns from blue to pink in a desiccator.

Purifying organic liquids

Washing and drying

Chemists often begin to purify organic liquids that are insoluble in water by shaking with aqueous reagents in a separating funnel to extract impurities. For instance, in the Activity: Preparation of an ester (Section 17.3.3), the impure ester is shaken with aqueous sodium carbonate to remove any acidic impurities.

This first washing is usually followed by a second washing with pure water to remove any inorganic impurities.

The organic liquid is then dried using a solid drying agent such as pieces of anhydrous calcium chloride or anhydrous magnesium sulfate. The organic liquid, left to stand in contact with the solid, goes clear when any water present has been removed. Finally, distillation is used to obtain the pure liquid.

Fractional distillation

Fractional distillation separates mixtures of liquids with different boiling temperatures. On a laboratory scale, the process takes place in the distillation apparatus that has been fitted with a fractionating column between the flask and the still-head (Figure 18.3.22). Separation is improved if the column is packed with inert glass beads or rings to increase the surface area where rising vapour can mix with condensed liquid running back to the flask. The column is hotter at the bottom and cooler at the top. The thermometer reads the boiling temperature of the compound passing over into the condenser.

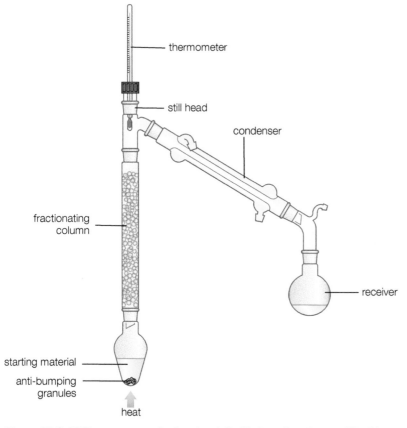

Figure 18.3.22 The apparatus for fractional distillation of a mixture of liquids.

If the flask contains a mixture of liquids, the boiling liquid in the flask produces a **vapour** that is richer in the most **volatile** of the liquids present (the one with the lowest boiling temperature).

Most of the vapour condenses in the column and runs back. As it does so, it meets more of the rising vapour. Some of the vapour condenses. Some of the liquid evaporates. In this way, the mixture evaporates and condenses repeatedly as it rises up the column. But, every time it does so, the vapour becomes richer in the most volatile liquid present. At the top of the column, the vapour contains 100% of the most volatile liquid. So, during fractional distillation, the most volatile liquid with the lowest boiling temperature distils over first, then the liquid with the next lowest boiling temperature, and so on.

Preparation and purification of an organic liquid

A pure sample of the ester methyl benzoate can be prepared by the reaction of methanol with benzoic acid in the presence of concentrated sulfuric acid as follows.

A To a 50 cm³ pear-shaped flask add 8 g of benzoic acid, 15 cm³ of methanol and 2 cm³ of concentrated sulfuric acid. Fit the flask with a reflux condenser and boil the mixture for about 45 minutes.

B Cool the mixture to room temperature and pour it into a separating funnel that contains 30 cm³ of cold water. Add 15 cm³ of hydrocarbon solvent to the pear-shaped flask and then pour this into the separating funnel.

C Mix the contents of the separating funnel by vigorous shaking, releasing the pressure carefully from time to time. Allow the contents of the flask to settle, then run the lower aqueous layer into a conical flask.

D Wash the hydrocarbon solvent layer in the separating funnel first with 15 cm³ of water and secondly with 15 cm³ of 0.5 mol dm⁻³ aqueous sodium carbonate.

E Dry the hydrocarbon solvent extract over anhydrous sodium sulfate, then filter off the solid.

F Remove the hydrocarbon solvent by careful distillation. Collect the distillate boiling above 190 °C and weigh your product.

Questions

1 Why is a reflux condenser necessary in step A?

2 Suggest why the cold water is added in step B.

3 Why is hydrocarbon solvent added to the pear-shaped flask in step B?

4 Describe how the pressure in the separating funnel is released in step C.

5 Describe how the hydrocarbon solvent layer is washed with water in step D and suggest what this washing removes.

6 Suggest what is removed in step D by washing the hydrocarbon solvent layer with sodium carbonate.

7 Describe the appearance of the dry organic layer in step E.

8 Confirm that benzoic acid is the limiting reagent. The density of methanol is 0.79 g cm⁻³.

9 Calculate the percentage yield if the preparation produces 4.8 g of methyl benzoate.

10 What are the hazards posed by the preparation and how might their risk be reduced?

Stage 5: Measuring the yield, identifying the product and checking its purity

Measuring the yield

Comparing the actual yield with the yield expected from the chemical equation is a good measure of the efficiency of a process.

The yield expected from the equation, assuming that the reaction is 100% efficient, is called the theoretical yield.

The efficiency of a synthesis, like that of other reactions, is calculated as a percentage yield. This is given by the relationship:

$$\text{percentage yield} = \frac{\text{actual yield of product}}{\text{theoretical yield of product}} \times 100\%$$

a) What is the theoretical yield of glycine (2-aminoethanoic acid) from 15.5 g of chloroethanoic acid?

b) What is the percentage yield if the actual yield of glycine is 7.9 g?

Notes on the method

Start by writing an equation for the reaction. This need not be a full balanced equation so long as it includes the limiting reactant, the product and the molar amounts of reactant and product in the equation.

In this case we must assume that any other reactants are in excess, the limiting reactant is chloroethanoic acid and the mole ratio is 1:1.

Answer

a) Equation extract: $ClCH_2COOH \rightarrow H_2NCH_2COOH$

The molar mass of chloroethanoic acid, $ClCH_2COOH = 94.5\,g\,mol^{-1}$

The molar mass of glycine, $H_2NCH_2COOH = 75.0\,g\,mol^{-1}$

According to the equation:

1 mol of chloroethanoic acid produces 1 mol of glycine

∴   94.5 g of chloroethanoic acid produce 75.0 g of glycine

So, 15.5 g of chloroethanoic acid produce $\dfrac{15.5}{94.5} \times 75.0\,g$ of glycine

    = 12.3 g of glycine

Theoretical yield of glycine = 12.3 g

b) $percentage\ yield = \dfrac{actual\ yield\ of\ product}{theoretical\ yield\ of\ product} \times 100\%$

$= \dfrac{7.9}{12.3} \times 100\%$

$= 64\%$

Identifying the product and checking its purity

Qualitative tests

Simple chemical tests for functional groups can help to confirm the identity of the product. These tests for functional groups are shown in the data sheets headed 'Tests and observations on organic compounds', which you can access via the QR code for this chapter on page 321.

Measuring melting temperatures and boiling temperatures

Pure solids have sharp melting temperatures, but impure solids soften and melt over a range of temperatures. So watching a solid melt can often show whether or not it is pure. As databases now include the melting temperatures of all known compounds, it is possible to check the identity and purity of a product by checking that it melts sharply at the expected temperature (Figure 18.3.23).

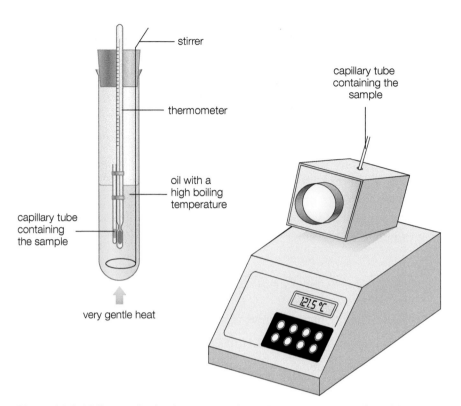

Figure 18.3.23 Two methods of measuring the melting temperature of a solid.

Like melting temperatures for solids, boiling temperatures can be used to check the purity and identity of liquids. If a liquid is pure it will all distil at the expected boiling temperature or in a narrow range including it. The boiling temperature can be measured as the liquid distils over during fractional distillation.

Chromatography and spectroscopy

The use of chromatography and spectroscopy in identifying compounds and checking their purity is covered in Chapter 19. These are the most important modern–day analytical tools.

Test yourself

21 In the diagram for stage E of Figure 17.3.12 (page 200), the thermometer bulb is placed near the top of the apparatus.

   a) Explain why the bulb is placed there.

   b) How and why would the reading differ if atmospheric pressure were higher than normal on the day of the experiment?

   c) How would the reading differ if the thermometer bulb were placed in the liquid in the flask?

22 A possible two-step synthesis of 1,2-diaminoethane first converts an alkene to a dihalogenoalkane and then reacts this with ammonia.

   a) Write out a reaction scheme for the synthesis, giving reagents and conditions.

   b) Calculate the mass of the alkene needed to make 2.0 g of the 1,2-diaminoethane, assuming a 60% yield in step 1 and a 40% yield in step 2.

23 Give three reasons why the actual yield in an organic synthesis is always less than the theoretical yield.

24 A two-step synthesis converts 18 g of benzene first to 22 g of nitrobenzene and then to 12 g of phenylamine.
  a) State the reagents and conditions for each step.
  b) Calculate the theoretical yield and the percentage yield for each step.
  c) What is the overall percentage yield?

25 Read again the sub-section headed 'Purifying organic solids' on page 274 and then answer the following questions.
  a) Why should the impure solid be dissolved in the **minimum** volume of **hot** solvent?
  b) Why is the solution sometimes cooled in ice when the pure product is being recrystallised?
  c) How could you improve the evaporation of excess solvent from the purified solid in the final stage?

Activity

Preparation and purification of N-phenylethanamide

Amines react very rapidly with acyl chlorides to form N-substituted amides. The reaction with acid anhydrides is more easily controlled so the following preparation uses ethanoic anhydride rather than ethanoyl chloride.

A Mix 5.0 cm³ of ethanoic anhydride with 5.0 cm³ of glacial (pure) ethanoic acid in a round-bottomed flask.

B Cool the flask in a beaker of cold water and add 5.0 cm³ of phenylamine, dropwise, with gentle shaking.

C Add anti-bumping granules, fit a reflux condenser and reflux the mixture for 30 minutes.

D Pour the liquid from the flask into a beaker containing 100 cm³ of cold water. Stir, then allow the mixture to stand until no more crystals are formed. Filter off the crystals under reduced pressure.

E Wash the crystals with cold water and recrystallise from the minimum volume of boiling water.

F Dry the crystals between filter papers and then by storing in a desiccator.

G Weigh the dry crystals and measure their melting temperature.

> ### Tip
> For practical guidance, refer to Practical skills sheet 3, 'Assessing hazards and risks', and also to Practical skills sheet 11, 'Synthesis of an organic solid', both of which you can access via the QR code for Chapter 18.3 on page 321.

Questions

1 Write an equation for the reaction between phenylamine and ethanoic anhydride to produce N-phenylethanamide.

2 Apart from difficulty in controlling the reaction rate, give another disadvantage of using ethanoyl chloride rather than ethanoic anhydride in this preparation.

3 Suggest why phenylamine is added dropwise to cold ethanoic anhydride.

4 Explain the function of the anti-bumping granules in step C.

5 Give three practical details in step D which help to keep the loss of product to a minimum.

6 Describe how the crystals are washed with cold water in step E.

7 Name a possible drying agent to use in the desiccator.

8 The yield of N-phenylethanamide is 70.0% of the theoretical yield. Calculate the actual mass obtained given that ethanoic anhydride is in excess and the density of phenylamine is 1.02 g cm⁻³.

9 What are the hazards posed by this preparation and how is their risk reduced?

Exam practice questions

1 A series of tests was carried out on three organic compounds, A, B and C. The results of the tests are described below.

State the deductions which you could make from the tests on each of A, B and C. You are not expected to identify compounds A, B and C.

a) i) A is a colourless liquid which does not mix with water. *(1)*
  ii) After warming a few drops of A with aqueous sodium hydroxide, the resulting solution was acidified with nitric acid. Silver nitrate solution was then added and a cream-coloured precipitate formed. *(2)*
b) i) B is a white solid that chars on heating and gives off a vapour which condenses to a liquid that turns cobalt(II) chloride paper from blue to pink. *(2)*
  ii) A solution of B turns universal indicator red. *(1)*
  iii) A solution of B reacts with aqueous sodium carbonate to produce a colourless gas that turns limewater milky. *(2)*
  iv) When a little of B is warmed with ethanol and one drop of concentrated sulfuric acid, a sweet-smelling product can be detected on pouring the reaction mixture into cold water. *(2)*
c) i) C is a liquid that burns with a very smoky, yellow flame. *(1)*
  ii) C does not react with sodium carbonate solution. *(1)*
  iii) C fizzes with sodium and gives off a gas that produces a 'pop' with a burning splint. *(2)*

2 The following steps were used in one method of synthesising ethyl ethanoate (boiling temperature 77 °C).
  A Heat ethanol and ethanoic acid under reflux for about 45 minutes with a little concentrated sulfuric acid.
  B Then, distil the reaction mixture, collecting all the liquid that distils below 84 °C.
  C Shake the distillate with aqueous sodium carbonate solution.

  D Add two spatula measures of anhydrous sodium sulfate or anhydrous calcium chloride to the organic product.
  E Finally, redistill the organic product, collecting the liquid that boils between 75 and 79 °C.

  a) Draw a diagram of the apparatus for heating under reflux. *(3)*
  b) State the reasons for each of the procedures in steps A to E. *(8)*

3 Salicylic acid has been used as a painkiller. Its displayed formula is shown in the diagram.

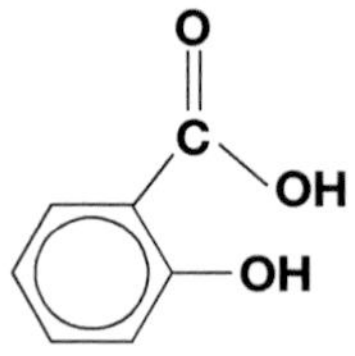

  a) Identify the functional groups in salicylic acid. *(2)*
  b) Write the molecular formula of salicylic acid. *(1)*
  c) Draw the displayed formula of the organic product that forms when salicylic acid:
    i) is heated under reflux with ethanol and concentrated sulfuric acid *(1)*
    ii) reacts with bromine water *(1)*
    iii) is warmed with aqueous sodium hydroxide. *(2)*

4 A sample of the hydrocarbon limonene was obtained from the peel of oranges. Small pieces of chopped orange peel were placed in a 250 cm³ round-bottomed flask with 100 cm³ of water. The flask was then heated on a wire gauze as shown in the diagram. About 50 cm³ of liquid was collected.

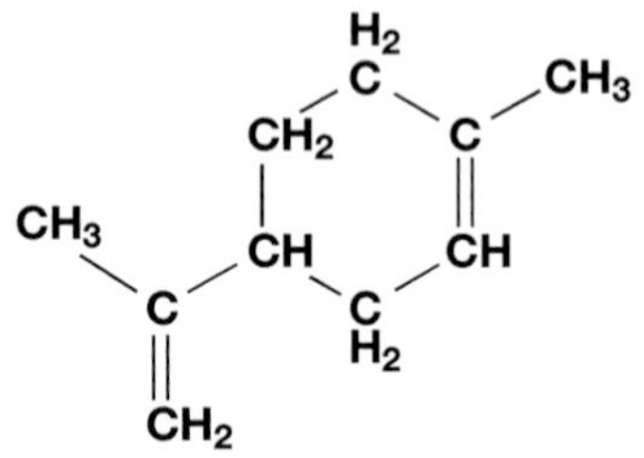

limonene

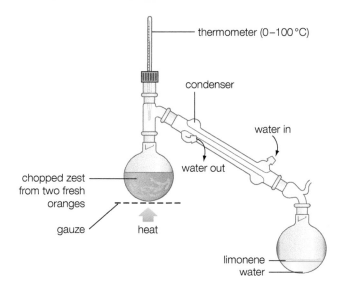

thermometer (0–100 °C)

condenser

water in

water out

chopped zest from two fresh oranges

gauze   heat

limonene water

a) State why the round-bottomed flask was heated on a wire gauze. *(1)*

b) Why was it necessary to chop the orange rind into small pieces? *(1)*

c) Name the process used to obtain the limonene and outline how limonene is obtained from the liquid collected. *(2)*

d) The sample of limonene that you have collected is probably contaminated with a little water. State how could you dry your sample of limonene. *(1)*

e) Why is limonene not obtained by heating the orange rind alone and without adding water? *(1)*

f) Name the functional group present in limonene and suggest two simple tests to show the presence of this functional group in limonene. *(3)*

5 The skeletal formula of compound D is shown below. D is a constituent of jasmine oil and it is partly responsible for the taste and smell of black tea.

Compound D

a) What is the molecular formula of D? *(1)*

b) Name the functional groups in D. *(3)*

c) Compound D is a stereoisomer.
   i) Draw the structure of D and identify each stereochemical component with an asterisk. *(2)*
   ii) Label each of these components with the type of stereoisomerism involved. *(2)*
   iii) How many stereoisomers are there with the skeletal formula shown? Explain your answer. *(3)*

6 a) Two isomeric compounds, F and G, with the molecular formula C_3H_8O can be oxidised to H and J respectively. H reacts with Fehling's solution to produce a red-brown precipitate of copper(I) oxide. J has no reaction with Fehling's solution, but gives a yellow crystalline product with 2,4-dinitrophenylhydrazine. Give the structural formulae and names of F, G, H and J. *(8)*

b) i) Describe what you would observe and name all the products formed when H is warmed with sodium dichromate(VI) solution acidified with dilute sulfuric acid. *(4)*
   ii) Write an equation or equations for the reactions taking place. *(3)*

7 An organic compound, X, containing carbon, hydrogen and oxygen only, was found to have a relative molecular mass of about 70. When 0.36 g of the compound was burned in excess oxygen, 0.88 g of carbon dioxide and 0.36 g of water were formed.

a) Calculate the empirical formula and the molecular formula of X. *(5)*

b) X reacted with 2,4-dinitrophenylhydrazine to produce an orange solid. What can you conclude from this? *(1)*

c) Write all the possible non-cyclic structural formulae for X and give the systematic name for each formula. *(3)*

d) The ^{13}C NMR spectrum of compound X shows that there are carbon atoms in three different environments in the molecule of X. Identify which is the correct structure for X and describe how you could confirm this using the product from the reaction with 2,4-dinitrophenylhydrazine. *(3)*

8 Outline clearly how you would distinguish between the members in each of the following pairs of compounds using one simple chemical test for each pair.

In each case, state the reagents and conditions used, describe what happens with each compound during the test.

a) $CH_3CH_2CH_2Cl$ and $CH_3CH_2CH_2I$ *(4)*

b) $CH_3COCH_2CH_2CH_3$ and $CH_3CH_2COCH_2CH_3$ *(3)*

c) *(3)*

and

d) *(3)*

and

e) $CH_3CH_2CH_2CONH_2$ and $CH_3CH_2CH_2CH_2NH_2$ *(3)*

9 Substance Y is an organic compound. The mass spectrum of Y showed that its molecular ion has a mass-to-charge ratio of 132. When 2.64 g of Y were completely combusted in pure dry oxygen, the only products were 7.92 g of carbon dioxide and 1.44 g of water.

Tests were carried out on Y and the following observations recorded.

Observations
Burns with a smoky flame
The yellow/orange colour of bromine water is decolourised
A yellow/orange precipitate is produced with 2,4-dinitrophenylhydrazine
A silver mirror is formed with Tollens' reagent.

a) Calculate the mass of carbon in 7.92 g of carbon dioxide and the mass of hydrogen in 1.44 g of water. Hence, calculate the molecular formula of Y. *(6)*

b) State what you can deduce from the observations about the functional groups present in Y. *(4)*

c) Draw a possible structural formula for Y. *(1)*

10 The solid halogenoalkane D is known to contain bromine, but no other halogen.
a) Describe how you would determine the percentage by mass of bromine in D. *(6)*
b) Explain the chemistry of your method and show how you would calculate the percentage of bromine in D from your measurements. *(5)*

11 Outline synthetic routes to prepare:
a) N-butyl ethanamide in three steps starting from 1-bromopropane *(6)*
b) but-2-ene using ethene as the only organic reagent. *(13)*

Your answers should use as few steps as possible and should give the reagents, conditions and product of each step. Balanced equations are not required.

12 Gaseous chloromethane reacts with an alloy of aluminium and sodium to form the liquid A.

The composition by mass of compound A is 50.0% carbon, 12.5% hydrogen and 37.5% aluminium.

0.24 g of A reacts with excess water to produce $0.24 \, dm^3$ of a gas, B, and a white gelatinous precipitate, C. The $0.24 \, dm^3$ of B has a mass of 0.16 g. C dissolves in hydrochloric acid and in sodium hydroxide solution.

(The volume of gas B was measured at room temperature and pressure, at which the molar volume of a gas $= 24 \, dm^3 \, mol^{-1}$. H = 1.0, C = 12.0, Al = 27.0)

a) Calculate the empirical formula of A. *(2)*
b) Suggest a structural formula for A. *(1)*
c) Calculate the molecular mass of B. *(1)*
d) Identify the compounds B and C. *(2)*
e) Write equations for:
  i) the reaction of chloromethane with the Al/Na alloy *(1)*
  ii) the reaction of A with excess water *(3)*
  iii) the reaction of C with hydrochloric acid *(1)*
  iv) the reaction of C with sodium hydroxide solution. *(1)*

19.1 Analytical techniques

Analytical chemists use a combination of techniques using **spectroscopes** and **spectrometers** to identify organic compounds and determine their structures.

● Mass spectrometry gives the relative molecular mass of a compound and can suggest a likely structure for a compound from fragmentation peaks.
● Infrared spectroscopy shows the presence of particular functional groups by detecting their characteristic vibration frequencies.
● Nuclear magnetic resonance (NMR) techniques help to detect groups with carbon atoms and hydrogen atoms in particular environments in molecules and is the most useful tool for determining structure.

19.2 Mass spectrometry

Mass spectrometry is used to determine the relative molecular masses and molecular structures of organic compounds. In this way it can be used to identify unknown compounds.

The combination of gas chromatography (GC) with mass spectrometry is of great importance in modern chemical analysis (Section 19.5). First, gas chromatography separates the chemicals in an unknown mixture, such as a sample of urine; then mass spectrometry detects and identifies the components (Figure 19.1).

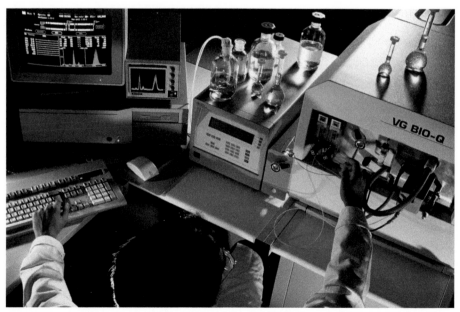

Figure 19.1 Using a mass spectrometer in a forensic laboratory to detect drugs in a urine sample.

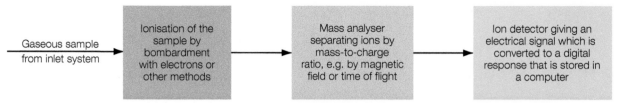

| Gaseous sample from inlet system | → | Ionisation of the sample by bombardment with electrons or other methods | → | Mass analyser separating ions by mass-to-charge ratio, e.g. by magnetic field or time of flight | → | Ion detector giving an electrical signal which is converted to a digital response that is stored in a computer |

Figure 19.2 A schematic diagram to show the key features of a mass spectrometer.

All mass spectrometers have the components shown in Figure 19.2.

In a mass spectrometer, a beam of high-energy electrons bombards the molecules of the sample. This turns them into ions by knocking out one or more electrons.

Bombarding molecules with high-energy electrons not only ionises them but usually splits them into fragments. As a result, the mass spectrum consists of a 'fragmentation pattern'.

Molecules break up more readily at weak bonds or at bonds which give rise to stable fragments. The highest peaks correspond to positive ions which are relatively more stable, such as tertiary carbocations or ions such as RCO^+ (the acylium ion) or the fragment $C_6H_5^+$ from aromatic compounds related to benzene, C_6H_6.

After ionisation and fragmentation, the charged species are separated to produce the mass spectrum that distinguishes the fragments on the basis of their ratio of mass to charge (m/z).

Chemists study mass spectra with these ideas in mind and, as a result, can gain insight into the structure of new molecules. They identify the fragments from their masses and then piece together likely structures with the help of evidence from other methods of analysis, such as infrared spectroscopy and NMR spectroscopy.

Chemists have also built up a very large database of mass spectra of known compounds for use in analysis. They regard the spectra in databases as 'fingerprints' for identifying chemicals during analysis. The computer of a mass spectrometer is programmed to search its database to find a good match between the spectrum of a compound being analysed and a spectrum in the database (Figure 19.3).

Figure 19.3 The mass spectrum of methyl benzoate, $C_6H_5COOCH_3$. The pattern of fragments is characteristic of this compound.

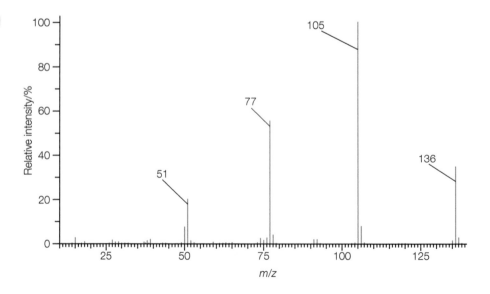

Analysing mass spectra

When analysing molecular compounds, the peak of the ion with the largest m/z value is usually the whole molecule ionised. So the mass of this 'parent ion', M^+, gives the relative molecular mass of the compound.

The presence of isotopes shows up in spectra of organic compounds containing chlorine or bromine atoms (Figure 19.4). Chlorine has two isotopes, ^{35}Cl and ^{37}Cl. Chlorine-35 is three times more abundant than chlorine-37. If a molecule contains one chlorine atom, its two molecular ions appear as two peaks separated by two mass units. The peak with the lower value of m/z is three times higher than the peak with the higher value of m/z.

Bromine consists largely of two isotopes, ^{79}Br and ^{81}Br, in roughly equal proportions. If a molecule contains one bromine atom, the molecular ion shows up as two peaks of roughly equal intensity separated by two mass units.

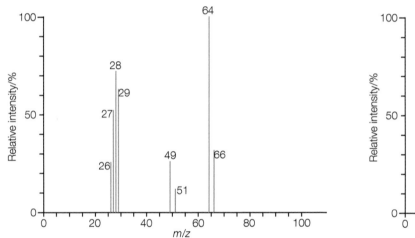 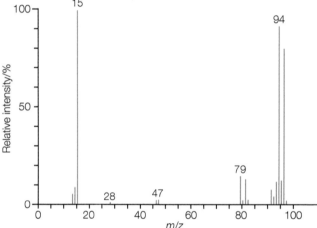

Figure 19.4 The mass spectra of two compounds containing halogen atoms.

High-resolution mass spectrometry

Modern mass spectrometers can measure a **relative isotopic mass** to four or five decimal places. Using these values, the **relative atomic mass** of an element can be found to four decimal places and therefore the accurate relative molecular mass of a compound can be calculated, given these accurate relative atomic masses. This makes it possible to identify one compound from several with the same integral mass.

For example, both C_4H_{10} and C_3H_6O have relative molecular mass equal to 58 to the nearest whole number.

Table 19.1 shows relative atomic masses to four decimal places.

Table 19.1 Relative atomic masses to four decimal places.

Element	Relative atomic mass
H	1.0079
C	12.0107
N	14.0067
O	15.9994

Key terms

The **relative isotopic mass** of an isotope is the mass of the isotope on a scale on which a ^{12}C atom has a mass of exactly 12.000 units.

The **relative atomic mass** of an element is the weighted average of the masses of its isotopes on a scale on which a ^{12}C atom has a mass of exactly 12.000 units.

Tip

Note that the distinction using accurate M_r values is between different molecular formulae. Since a molecular formula may represent several isomers, the structure of the particular isomer cannot be determined using the M_r value. However, an isomer may be identified using the fragmentation pattern of its molecular ion or, alternatively, by studying its reactions.

Tip

Section 19.3 revisits the content of Section 7.2 in Student Book 1 and is included here as revision.

Tip

Infrared absorptions are measured in wavenumbers with the unit cm^{-1}. The wavenumber is the number of waves in 1 cm.

Key term

An **absorption spectrum** is a plot showing how strongly a sample absorbs radiation over a range of frequencies. Absorption spectra from infrared spectroscopy give chemists valuable information about the bonding and structure of chemicals.

Tip

Most organic molecules contain C—H bonds. As a result most organic compounds have a peak at around $3000\,cm^{-1}$ in their IR spectrum.

Using the figures from Table 19.1, the relative molecular mass of C_4H_{10} is 58.1218 and that of C_3H_6O is 58.0789, both to four decimal places. It is therefore possible to distinguish between the two compounds.

Test yourself

1 An organic molecule M can be represented as a combination of two parts: $m_1.m_2$. Draw a diagram to represent the ionisation and then fragmentation of the molecule and explain why only one of the two fragments shows up in the mass spectrum.

2 Suggest the identity of the peaks labelled in the mass spectrum shown in Figure 19.3.

3 Account for these facts about the mass spectrum of dichloroethene:

   a) It includes three peaks at m/z values of 96, 98 and 100 with intensities in the ratio 9:6:1.

   b) It includes two peaks at m/z values 61 and 63 with intensities in the ratio of 3:1.

4 One of the mass spectra in Figure 19.4 is bromomethane and the other is chloroethane. Match the spectra to the compounds and identify as many fragments in the spectra as you can.

5 An organic compound is found to have the accurate $M_r = 46.0682$. Use the data in Table 19.1 to decide which of the following molecular formulae is correct: CH_2O_2, CH_6N_2 or C_2H_6O.

19.3 Infrared spectroscopy

Spectroscopists have found that it is possible to correlate absorptions in the region $4000\text{–}1500\,cm^{-1}$ of an **absorption spectrum** with the stretching or bending vibrations of particular bonds. As a result, the infrared spectrum gives valuable clues to the presence of functional groups in organic molecules.

Figure 19.5 The essential features of a modern single-beam IR spectrometer.

The important correlations between different bonds and observed absorptions are shown in Figure 19.6. Hydrogen bonding broadens the absorption peaks of −OH groups in alcohols and even more so in carboxylic acids.

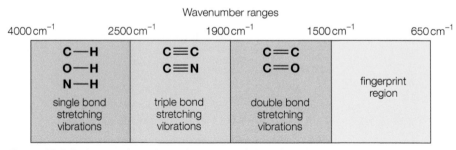

Figure 19.6 A chart to show the main regions of the infrared spectrum and important correlations between bonds and observed absorptions.

Molecules with several atoms can vibrate in many ways because the vibrations of one bond affect others close to it. The complex pattern of vibrations, particularly in the region 1500–650 cm⁻¹, can be used as a 'fingerprint' to be matched against the recorded IR spectrum in a database.

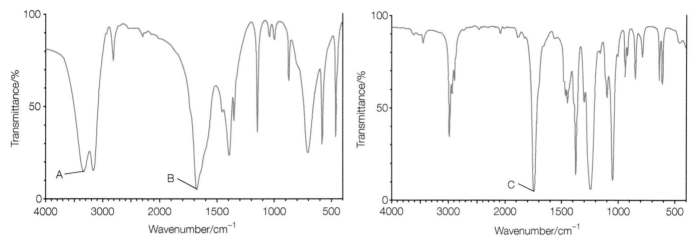

Figure 19.7 Two IR spectra.

19.4 Nuclear magnetic resonance spectroscopy (NMR)

Nuclear magnetic resonance spectroscopy (NMR) is a powerful analytical technique for finding the structures of carbon compounds. The technique is used to identify unknown compounds, to check for impurities and to study the shapes of molecules.

This type of spectroscopy studies the behaviour of the nuclei of atoms in magnetic fields. It is limited to those nuclei which behave like tiny magnets because they have a property called spin. In common organic compounds the only nuclei to do so are those of carbon-13 atoms, ¹³C, and of hydrogen atoms, ¹H. The nuclei of the much more common carbon-12, oxygen-16 and nitrogen-14 atoms do not show up in NMR spectra.

When placed in a very strong magnetic field, magnetic nuclei line up either in the same direction as the field or in the opposite direction to it. Those aligned in the same direction are slightly lower in energy and the difference between the two energy levels is known as the energy gap (Figure 19.8).

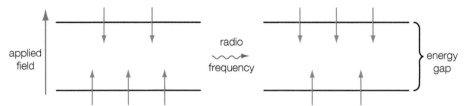

Figure 19.8 Energy levels and the energy gap for protons in an applied magnetic field.

This energy gap corresponds to a particular frequency in the radio-frequency range. When exactly the right frequency is supplied, a proton in the lower energy level can flip into the higher level and the protons are said to be in resonance. The radio-frequency needed for this is recorded.

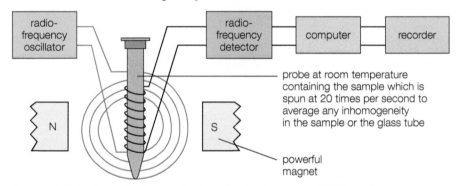

Figure 19.9 A schematic diagram to show the key features of NMR spectroscopy.

The tube with the sample is supported in a strong magnetic field in the spectrometer (Figure 19.9). The operator turns on a source of radiation at radio-frequencies. The radio-frequency detector records the intensity of the signal from the sample as the oscillator emits pulses of radiation across a range of wavelengths.

The sample is dissolved in a solvent. Also in the solution is some tetramethylsilane (TMS), which is a **standard reference compound** that produces a single, sharp absorption peak well away from the peaks produced by samples for analysis.

Each peak corresponds to one or more magnetic atoms in a particular chemical environment. Nuclei in different parts of a molecule experience slightly different magnetic fields in an NMR machine. This is because they are shielded to a greater or lesser extent from the field applied by the spectrometer by the tiny magnetic fields associated with the electrons of neighbouring bonds and atoms.

The recorder prints out a spectrum that has been analysed by computer to show peaks wherever the sample absorbs radiation strongly. The zero on the scale is fixed by the absorption of magnetic atoms in the reference chemical.

The distances of the sample peaks from this zero are called their 'chemical shifts'. The symbol used for chemical shift is δ. Tables of chemical shifts are included in the Edexcel Data booklet.

Carbon-13 NMR

Carbon-13 NMR relies on the magnetic properties of the ^{13}C isotope. This isotope makes up only about 1% of all naturally occurring carbon atoms, but this is enough for a signal to be detected in an NMR machine.

Figure 19.10 shows the ^{13}C NMR spectrum for ethanol. Spectra of this kind are available from the Spectral Data Base System (SDBS) for Organic Compounds at the National Institute of Advanced Industrial Science and Technology (AIST), Japan.

Key term

The horizontal scale of an NMR spectrum shows the chemical shifts of the peaks measured in parts per million (ppm). The symbol δ stands for the 'chemical shift' relative to the zero on the scale that is given by the signal from tetramethylsilane.

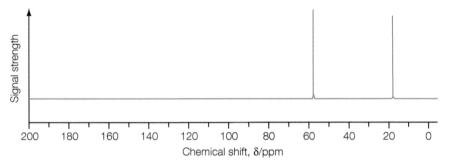

Figure 19.10 ^{13}C NMR spectrum for ethanol dissolved in CDCl$_3$.

There are two peaks in the ^{13}C NMR spectrum for ethanol. This reflects the fact that there are two carbon atoms in an ethanol molecule and they are in different environments. The carbon in the CH$_3$ group is attached to three hydrogen atoms and a carbon atom. The carbon in the CH$_2$ group is attached to two hydrogen atoms, a carbon atom and an oxygen atom.

Spectra of the type shown in Figure 19.10 are usually recorded with the sample in solution. The chosen solvent is commonly CDCl$_3$. The molecules of CDCl$_3$ contain one carbon atom and so produce a single line in ^{13}C spectra that is easy to recognise. This line is usually removed from the spectra in databases such as SDBS to avoid any confusion. The line produced by the solvent is not shown in any of the ^{13}C spectra in this book.

Also omitted from Figure 19.10 is the peak at zero produced by the reference chemical tetramethylsilane (TMS). The chemical shifts are measured relative to the TMS peak at the zero mark. This peak is usually removed from the spectra for clarity.

Test yourself

8 a) What is the difference in the structure of the nuclei of ^{12}C and ^{13}C atoms?

 b) What is the difference between the formula of CDCl$_3$ and the formula of trichloromethane?

9 Other than providing a suitable peak, what other requirements must there be for a standard added to a solution in an NMR test?

10 TMS is related to silane, SiH$_4$, but has the four hydrogen atoms replaced by methyl groups. Why does this reference chemical produce just one peak in ^{13}C NMR spectra?

Figure 19.11 illustrates the principles of ^{13}C NMR in a more complex example. Here, too, every carbon atom, or group of carbon atoms, in a chemically distinct environment gives a separate peak in the NMR spectrum.

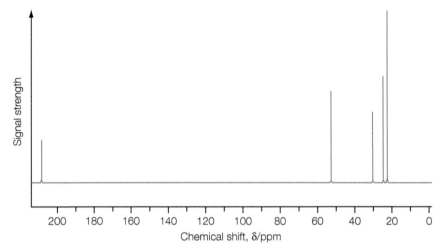

Figure 19.11 ^{13}C NMR spectrum for 4-methylpentan-2-one dissolved in CDCl$_3$.

There are six carbon atoms in 4–methylpentan–2–one but only five peaks in the ^{13}C NMR spectrum. The reason for this is shown in Figure 19.12. The carbon atoms of two of the methyl groups in the molecule are in exactly the same environment so they give rise to only one peak. The other four carbon atoms, including the carbon atom in the third methyl group, are in slightly different environments and give rise to separate peaks.

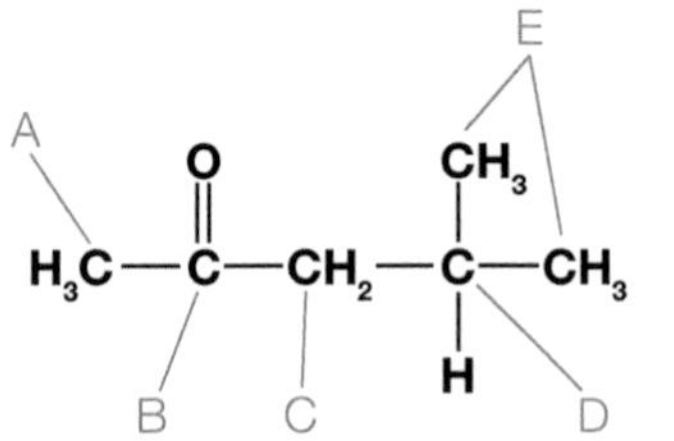

Figure 19.12 The structure of 4-methylpentan-2-one labelled to show the five different environments for the six carbon atoms. The two carbon atoms labelled E are in exactly the same chemical environment.

Test yourself

11 Predict the number of peaks in the ^{13}C NMR spectrum of:

a) pentane

b) propyl ethanoate

c) 2-methylbutanal

d) benzene

e) methyl benzene.

12 Explain how you could distinguish between propanal and propanone by inspection of the ^{13}C NMR spectra of the two compounds.

13 Deduce the number of peaks in the ^{13}C spectrum of ibuprofen (Figure 19.13).

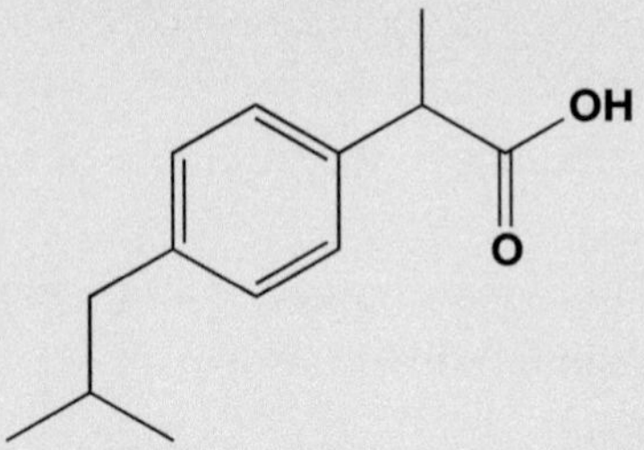

Figure 19.13 A skeletal formula of ibuprofen.

Generally carbon atoms attached to electronegative atoms such as oxygen show the largest chemical shifts. Chemists have found that it is possible to draw up tables to show the likely range of values for the chemical shifts of ^{13}C atoms in different environments. For example, ^{13}C atoms in alkanes typically give chemical shifts in the range 5–50 ppm, while carbon atoms in the carbonyl groups of aldehydes and ketones give chemical shifts in the range 190–220 ppm (see the Edexcel Data booklet).

Tip

You are only expected to interpret ^{13}C NMR spectra in which each chemical environment for carbon atoms is represented by a single peak. Also note that you cannot draw any conclusions by looking at the size of the peaks in ^{13}C NMR. In these two ways ^{13}C NMR differs from proton NMR, where the peaks may be split and the area of each peak is significant.

Test yourself

14 Refer to the data sheet from the Edexcel Data booklet showing chemical shifts for ^{13}C NMR. To what extent do the data show that the presence of electronegative atoms increases the chemical shift values?

15 Use the ^{13}C NMR chemical shift values from the data sheet to suggest which carbon atoms give rise to which peaks in the spectrum of:

a) ethanol (Figure 19.10)

b) 4-methylpentan-2-one (Figures 19.11 and 19.12).

16 Sketch the ^{13}C NMR spectrum you would expect to observe for:

a) ethyl ethanoate

b) cyclohexene.

Proton NMR spectroscopy

Information obtained from proton NMR spectra

Proton NMR relies on the magnetic properties of the ^1H isotope. As with ^{13}C NMR, the number of main peaks in the spectrum shows how many different chemical environments there are for hydrogen atoms. The values of the chemical shifts are a useful indication of the types of chemical environment for the hydrogen atoms corresponding to each peak.

Even more information can be deduced from a proton NMR spectrum than from a ^{13}C spectrum because it is possible to work out the number of hydrogen atoms in each environment. In a proton NMR spectrum the area under a peak is proportional to the number of nuclei.

Tip

The word 'proton' is used in NMR spectroscopy, but there are no H^+ ions present in these molecules. The term strictly refers to the proton in the nucleus of the ^1H atoms which are affected by the magnetic field.

Figure 19.14 Researcher adding a sample to an NMR spectrometer.

An NMR instrument (Figure 19.14) can work out the ratios of the areas below the peaks. Sometimes the results of the calculation are shown by an **integration** trace, such as the blue line shown in the spectrum for methyl ethanoate in Figure 19.15. Alternatively, the instrument's computer prints a number below each peak that is a measure of the relative area under the curve, as shown for the spectrum in Figure 19.16.

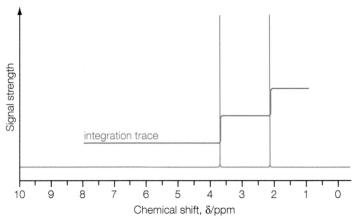

Figure 19.15 Proton NMR spectrum of methyl ethanoate in $CDCl_3$.

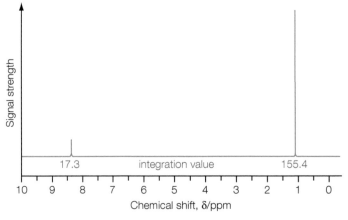

Figure 19.16 Proton NMR spectrum of a compound $C_5H_{10}O$ in $CDCl_3$.

Low-resolution proton NMR spectra, such as those in Figures 19.15 and 19.16, show the main peaks but no fine detail.

Proton NMR spectra, like carbon-13 NMR spectra, are usually recorded with the sample in solution. It is important that the solvent does not contain hydrogen atoms that would give peaks with chemical shifts similar to those in the sample. One possibility is to use a solvent that contains no hydrogen atoms, such a tetrachloromethane, CCl_4. The other is to use a solvent in which atoms of the ^1H isotope have been replaced by deuterium atoms. A compound that is often used is $CDCl_3$. Deuterium atoms produce peaks in regions of the spectrum well away from the chemical shifts for proton NMR.

The reference chemical for proton NMR is again tetramethylsilane (TMS). There are 12 hydrogen atoms in a molecule of TMS and they all have the same chemical environment. TMS provides a single strong peak that marks the zero on the scale of chemical shifts.

17 Refer to the proton NMR spectrum in Figure 19.15 and the Edexcel Data booklet for chemical shifts.

   a) Explain why there are two peaks in the spectrum.

   b) Use the chemical shift values to decide which hydrogen atoms give rise to each peak.

   c) Show that the integration trace is as expected for the molecule.

18 Deduce the number of peaks in the low-resolution proton NMR spectrum of each compound.

   a) 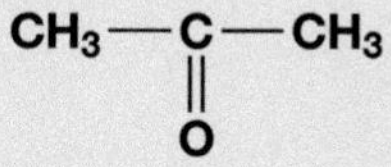 CH_3-C-CH_3 with =O below C

   b) $Cl-CH_2-C-CH_3$ with =O below C

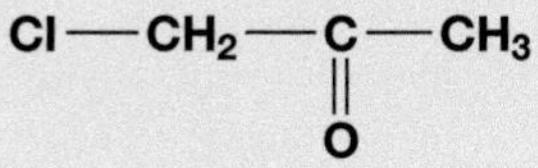

   c) H_3C, H_3C $C=C$ CH_2-CH_3, CH_2-CH_3

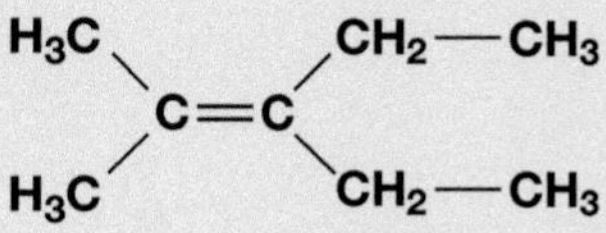

   d) 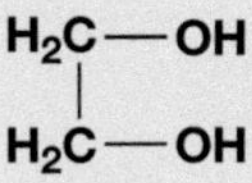 H_2C-OH H_2C-OH

   e)

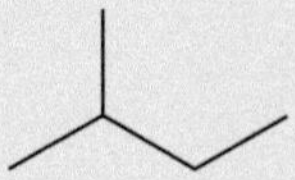

   f)

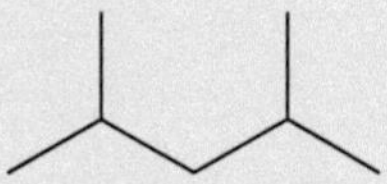

19 Refer to the proton NMR spectrum in Figure 19.16 and the chemical shift data from the Edexcel Data booklet.

   a) How many different chemical environments are there for hydrogen atoms in the molecule?

   b) Use the integration values under each peak to work out the ratios of hydrogen atoms in each environment.

   c) Use your answers to (a) and (b) and chemical shift values to suggest a structure for the compound.

   d) Describe two chemical tests that could be used to confirm the presence of the main functional group in the molecule. State what you would do and what you would expect to observe.

20 With the help of the chemical shift data from the Edexcel Data booklet, sketch the low-resolution proton NMR spectrum you would expect for:

   a) butanone

   b) 2-methylpropan-2-ol.

Key term

Equivalent protons are those in the same chemical environment with the same chemical shift. They do not couple with each other.

Tip

Do not confuse the solvent with the standard. The solvent, such as $CDCl_3$, does not contain protons so does not give rise to any peaks in the ^1H NMR spectrum. The standard, TMS, has 12 equivalent protons and a peak for it may be seen in spectra at $\delta = 0$.

Key term

Non-equivalent protons are those in different chemical environments with different chemical shifts. Protons on adjacent, that is neighbouring, carbon atoms couple with each other.

Coupling

At high resolution it is possible to produce proton NMR spectra with more detail that provide even more information about molecular structures (Figure 19.17). The spins of **non-equivalent protons** connected to neighbouring carbon atoms interact with each other. Chemists call this interaction 'spin–spin coupling' and they find that the effect is to split the peaks into a number of lines.

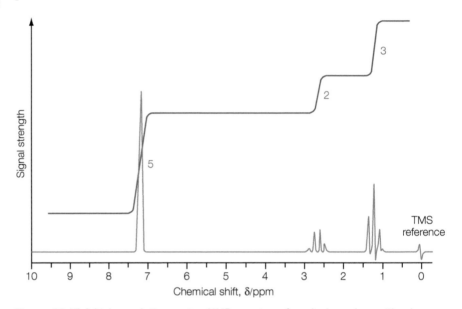

Figure 19.17 A high-resolution proton NMR spectrum for a hydrocarbon with a benzene ring. Note the extra peaks compared with a low-resolution spectrum.

A peak from protons bonded to an atom that is next to an atom with two protons splits into three lines, with the central line being twice as large as the other two. This happens because there are three energy states available to the two protons, depending on whether each proton is aligned with or against the magnetic field:

- both aligned with the field
- one aligned with the field and one against the field (with two possible combinations)
- both aligned against the field.

For similar reasons, a peak from protons bonded to an atom that is next to an atom with three protons splits into four lines. In general the '$n + 1$' rule makes it possible to work out the splitting, where n is the number of protons on the adjacent atom. The number of lines and their intensities can also be worked out using Pascal's triangle (Figure 19.18).

Figure 19.18 Pascal's triangle predicts the pattern of peaks and the relative peak heights.

number of equivalent protons causing splitting	splitting pattern and relative intensity of the peaks
1	1 1
2	1 2 1
3	1 3 3 1
4	1 4 6 4 1

Explain the splitting pattern in the spectrum of ethanol, shown in Figure 19.19.

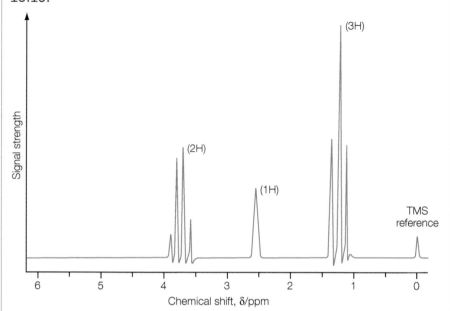

Figure 19.19 The high-resolution spectrum of ethanol.

Answer

Consider the peak at $\delta = 1.2$ (integration value 3); this corresponds to the CH_3 group.

The group adjacent to the CH_3 in the molecule is a CH_2 group, so the number of adjacent protons is 2, therefore $n = 2$.

Using the $n + 1$ rule gives $n + 1 = 3$ and so the peak is a triplet.

Therefore the CH_3 peak is a triplet because it is split by the two protons in CH_2.

Consider the peak at $\delta = 3.8$ (integration value 2); this corresponds to the CH_2 group.

The group adjacent to the CH_2 in the molecule is a CH_3 group, so the number of adjacent protons is 3, therefore $n = 3$.

Using the $n + 1$ rule gives $n + 1 = 4$ and so the peak is a quartet.

Therefore the CH_2 peak is a quartet because it is split by the three protons in CH_3.

The peak at $\delta = 2.5$ due to the proton in the OH group is not split. This is usual for hydrogens in alcohol groups and is explained in the section on labile protons on page 296.

Ethyl groups, CH_3CH_2-, are often present in organic molecules. If a proton NMR spectrum contains a triplet (relative integration 3) and a quartet (relative integration 2) this means that the molecule contains an ethyl group. The ethyl group is not adjacent to other protons.

Labile protons

Hydrogen bonding affects the properties of compounds with hydrogen atoms attached to highly electronegative atoms such as oxygen or nitrogen (Figure 19.20). These molecules can rapidly exchange protons as they move from one electronegative atom to another. Chemists describe these protons as labile.

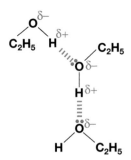

Figure 19.20 Hydrogen bonding in ethanol.

Labile protons do not couple with the protons linked to neighbouring atoms. This means that the NMR peak for a proton in an –OH group appears as a single peak in a high-resolution spectrum.

Note that in the high-resolution spectrum of ethanol (Figure 19.19) there is no coupling between the proton in the –OH group and the protons in the next-door –CH$_2$ group.

A useful technique for detecting labile protons is to measure the NMR spectrum in the presence of deuterium oxide (heavy water), D$_2$O. Deuterium nuclei can exchange rapidly with labile protons. Deuterium nuclei do not show up in the proton NMR region of the spectrum and so the peaks of any labile protons disappear.

25 Write an equation to show the reversible exchange of deuterium and hydrogen nuclei between ethanol and deuterium oxide.

26 a) Account for the splitting pattern shown for the peaks in the proton NMR spectrum of 3-chloropropanoic acid, $Cl-CH_2-CH_2-COOH$, in Figure 19.21.

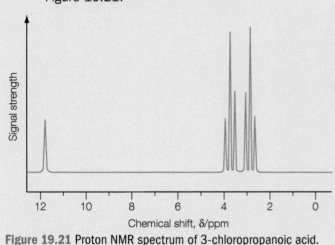

Figure 19.21 Proton NMR spectrum of 3-chloropropanoic acid.

b) How would the spectrum in Figure 19.21 change if 3-chloropropanoic acid were used in the presence of D_2O?

c) Describe the high-resolution proton NMR spectrum you would expect to observe with 2-chloropropanoic acid.

27 Table 19.2 shows the main features of the high-resolution NMR spectrum of a compound containing carbon, hydrogen and oxygen. The peak with a chemical shift of 11.7 disappears in the presence of D_2O. Deduce the structure of the compound.

Table 19.2

Chemical shift/ppm	Number of lines	Integration ratio
1.2	Triplet	3
2.4	Quartet	2
11.7	Singlet	1

Medical benefits from NMR

In medicine, magnetic resonance imaging uses NMR to detect the hydrogen nuclei in the human body, especially in water and lipids. A computer translates the information from a body scan into images of the soft tissue and internal organs that are normally transparent to X-rays (Figure 19.22).

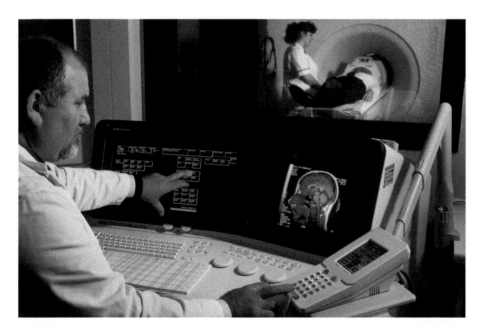

Figure 19.22 A brain scan in progress in an advanced magnetic resonance imaging (MRI) machine.

28 Suggest a reason why doctors and radiographers refer to MRI scanning rather than to NMR imaging, even though the technologies used in medicine and chemical research are essentially the same.

Analysing a perfume chemical

Jasmine blossom is a source of chemicals used in perfumes (Figure 19.23).
Analysis of an extract from the blossom by gas chromatography shows that it can
contain over 200 compounds. Two of the compounds in the mixture are mainly
responsible for the smell of the blossom. One of these two chemicals is jasmone,
which has the empirical formula $C_{11}H_{16}O$.

Figure 19.23 Harvesters gathering jasmine flowers for the French perfume industry.

The structure of jasmone has been studied by mass spectrometry and by NMR and
IR spectroscopy, with the results shown in Figures 19.24–19.26.

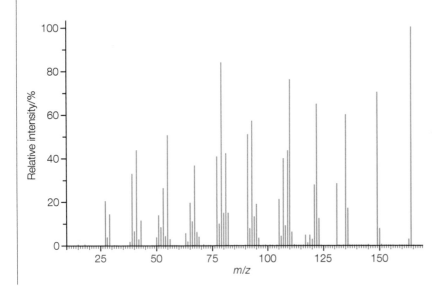

Figure 19.24 The mass spectrum of the perfume chemical jasmone.

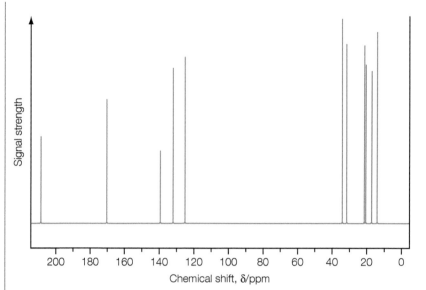

Figure 19.25 The ^{13}C NMR spectrum of the perfume chemical jasmone.

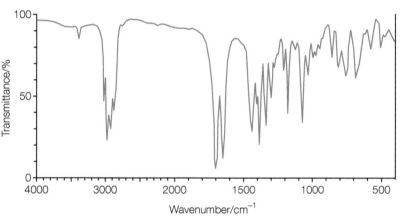

Figure 19.26 The IR spectrum of the perfume chemical jasmone.

1 a) Use the mass spectrum of jasmone to determine its relative molecular mass.
  b) What is the molecular formula of jasmone?
  c) How many double bonds and/or rings are there in the molecule?
2 Refer to the ^{13}C NMR spectrum in Figure 19.25.
  a) How many different environments for carbon are there in the molecule?
  b) Use the table of ^{13}C NMR chemical shifts in the Edexcel Data booklet to suggest which chemical environments for carbon are in the molecule.
  c) What can you conclude about the structure of the molecule from the spectrum?
3 Refer to the IR spectrum in Figure 19.26 and the IR correlation tables in the Edexcel Data booklet. What functional groups are present in the molecule?
4 Suggest a possible structure for jasmone that is consistent both with the information from the spectra and with the fact that the full name of the compound is *cis*-jasmone.
5 Describe what you would expect to observe if you tested a sample of *cis*-jasmone with:
  a) a solution of bromine in an organic solvent
  b) Tollens' reagent
  c) 2,4-dinitrophenylhydrazine.

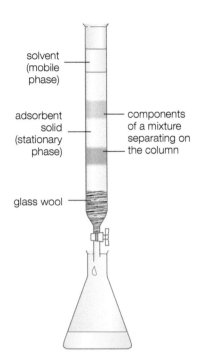

solvent
(mobile
phase)

adsorbent
solid
(stationary
phase)

components
of a mixture
separating on
the column

glass wool

Figure 19.27 Column chromatography. A solution of the mixture to be analysed is added to the top of the column. Then a solvent is added slowly and continuously to run through the column. The substances in the mixture separate and emerge at different times from the column.

Figure 19.28 In column chromatography there is an equilibrium between molecules adsorbed onto the surface of tiny particles of the silica or alumina stationary phase and molecules dissolved in the solvent.

19.5 Chromatography

Principles of chromatography

In 1903, the Russian botanist Michel Tswett developed the technique of column chromatography to study plant pigments. 'Chroma' means colour. The name chromatography was chosen because the technique was first used to separate coloured chemicals from mixtures. The general set up for column chromatography is shown in Figure 19.27.

There are now a range of chromatography techniques that can be used to:

- separate and identify the components of a mixture of chemicals
- check the purity of a chemical
- identify the impurities in a chemical preparation
- purify a chemical product.

Every type of chromatography has a **stationary phase** and a **mobile phase** that flows through it. Chemicals in a mixture separate because they differ in the extent to which they mix with the mobile phase or stick to the stationary phase.

Powdered solids now used in column chromatography include silicon oxide (silica) and aluminium oxide (alumina). Both these solids can **adsorb** chemicals onto their surfaces (Figure 19.28). The greater the tendency for molecules to be adsorbed by the stationary phase, the slower they move during chromatography.

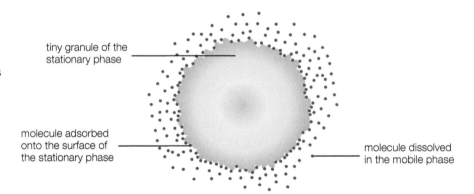

tiny granule of the
stationary phase

molecule adsorbed
onto the surface of
the stationary phase

molecule dissolved
in the mobile phase

Key terms

The **stationary phase** in chromatography may be a solid or a liquid held by a solid support. The **mobile phase** moves through the stationary phase and may be a liquid or a gas.

In column chromatography the liquid flowing through the column is the eluent. It washes the components of the mixture through the column. This is the process of elution.

Solids can **adsorb** very thin films of liquids or gases onto their external surfaces.

By contrast, a sponge absorbs water internally into its pores as it soaks up the liquid. Paper is absorbent and soaks up the moving solvent during paper chromatography as does silica or alumina on a TLC plate.

Liquid chromatography

The column chromatography developed by Michel Tswett was an example of liquid chromatography. Modern versions of the column technique continue to be widely used. Other variants include thin-layer chromatography and high-performance liquid chromatography.

Thin-layer chromatography, TLC

Thin-layer chromatography is another type of liquid chromatography in which the stationary phase is a thin layer of a solid, such as silica or alumina, supported on a glass or plastic plate. As in column chromatography, the rate at which a sample moves up a TLC plate also depends on the equilibrium between adsorption on the solid and solution in the solvent. The position of equilibrium varies from one compound to another so the components of a mixture separate.

TLC is quick, cheap and only needs a very small sample for analysis. The technique is widely used both in research laboratories and in industry.

The amino acids in peptides and proteins can be investigated by hydrolysing the peptides and proteins with concentrated hydrochloric acid, then separating and identifying the amino acids produced using either thin-layer or paper chromatography (Figure 19.29).

> **Tip**
>
> TLC can be used quickly to check that a chemical reaction is going as expected and making the required product. After attempts to purify a chemical, TLC can show whether or not all the impurities have been removed.

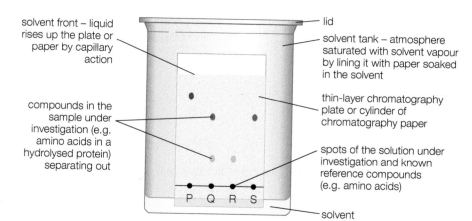

solvent front – liquid rises up the plate or paper by capillary action

lid

solvent tank – atmosphere saturated with solvent vapour by lining it with paper soaked in the solvent

thin-layer chromatography plate or cylinder of chromatography paper

compounds in the sample under investigation (e.g. amino acids in a hydrolysed protein) separating out

spots of the solution under investigation and known reference compounds (e.g. amino acids)

P Q R S

solvent

Figure 19.29 Thin-layer chromatography or paper chromatography can be used to separate and identify amino acids after a protein has been hydrolysed. The sample under investigation is spotted at Q and known amino acids are spotted at P, R and S.

Identifying amino acids – ninhydrin and R_f values

After chromatography, it is easy to see coloured compounds on the TLC plate or chromatography paper. With colourless compounds such as amino acids, it is necessary to use a locating agent. Locating agents can be sprayed onto the plate or paper, where they react with the separated colourless compounds to form coloured products. Alternatively, the plate can be placed in a covered beaker with a locating agent such as iodine crystals, and the iodine vapour stains the spots.

The locating agent for amino acids is ninhydrin (Figure 19.30).

Figure 19.30 The structure of ninhydrin.

When ninhydrin is sprayed on the chromatography plate or paper and then heated in an oven at about 100°C, it reacts with any amino acids to form purple spots which fade and turn brown with time.

R_f values are used to record the distances moved by chemicals in a mixture relative to the distance moved by the solvent. R_f stands for 'relative to the solvent front'.

The values are ratios calculated using this formula, where x and y are as shown in Figure 19.31:

$$R_f = \frac{\text{distance moved by chemical}}{\text{distance moved by solvent front}} = \frac{x}{y}$$

Using R_f values it is possible to identify the different separated spots from the sample under investigation by comparison with known reference compounds.

R_f values can help to identify components of mixtures so long as the conditions are carefully controlled. The values vary with the type of TLC plate (or paper) and the nature of the solvent.

Tip

An alternative method is to use a TLC plate impregnated with a fluorescent chemical. Under a UV lamp the whole plate glows except in the areas where organic compounds absorb radiation, so that they show up as dark spots.

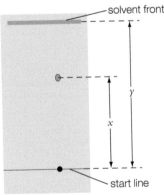

Figure 19.31 The distances on a TLC plate used to calculate R_f values.

32 Why is it important to handle the TLC plate only on the edges and not to touch the surface with fingers?

33 Why is the air inside the container of the TLC solvent saturated with vapour of the solvent before adding the TLC plate?

34 Which of the four samples P, Q, R and S in Figure 19.29:

   a) is the mixture being analysed

   b) are reference compounds also present in the mixture

   c) is a reference compound not present in the mixture?

35 Estimate the R_f value for the yellow component in Figure 19.29.

36 Suggest two factors that will change the R_f value of a particular amino acid.

Activity

Using TLC to investigate the aspirin produced in Core practical 16

Core practical 16 (Section 17.3.5) describes the preparation of aspirin and how the purity of the product can be assessed by measuring its melting temperature.

A student followed the instructions below to investigate the purity of samples of the crude product, the recrystallised aspirin and a commercial sample of aspirin by using thin-layer chromatography (TLC).

A Make sure that that you handle the TLC plate only by the edges and do not touch the surface. Using a pencil, draw a line across the plate about 1 cm from the bottom and mark three evenly spaced points on this line. Place a small amount (about one-third of a spatula measure) of the crude product, the recrystallised aspirin and the commercial sample of aspirin in three separate test tubes and label the tubes.

B In a fume cupboard or a well-ventilated room, place 2 cm³ of ethanol and 2 cm³ of dichloromethane in a test tube. Ensure the liquids are mixed, then add 1 cm³ of this solvent mixture to each of the labelled test tubes to dissolve the samples.

C Using separate capillary tubes, place a small spot of each of the three sample solutions onto the TLC plate. Allow the spots to dry and then add more sample to each spot, three times in all. Do not let the spots become larger than about 2 mm across.

D After the spots are dry, place the TLC plate in a developing tank containing 0.5 cm depth of ethyl ethanoate. Make sure that the original pencil line and spots are above the level of the developing solvent.

E Place a lid on the tank and allow it to stand in a fume cupboard until the solvent front has risen to about 1 cm from the top of the plate. Remove the plate from the tank, mark the position of the solvent front and allow the plate to dry.

F Observe the plate under a UV lamp and mark any spots observed carefully with a pencil. In a fume cupboard, place the plate in a beaker containing two or three iodine crystals. Cover the beaker and warm gently on a steam bath until spots begin to appear.

Questions

1 Why is the start line for spotting TLC samples on a plate drawn in pencil and not with ink?

2 The recrystallised aspirin and the commercial sample both produced only one spot on the TLC plate at an R_f value of 0.65. The crude product also produced this same spot but in addition gave spots at R_f values 0.22 and 0.48. Draw a diagram of the plate, similar to that in Figure 19.31, to show these results.

3 What conclusions can you draw about the nature of the three samples tested?

4 In a different experiment, the developing solvent used was a mixture of ethyl ethanoate and hexane. The R_f value for aspirin in this experiment was 0.15. Suggest a reason why the R_f value with this solvent mixture was lower that the R_f value with pure ethyl ethanoate.

5 Why should the spot size not be larger than 0.2 mm in step C?

High-performance liquid chromatography

High-performance liquid chromatography (HPLC) is a sophisticated version of column chromatography. The technique can be used to separate components in a mixture which are very similar to each other (Figure 19.32).

High performance is achieved by packing very small particles of a solid such a silica into a steel column. A typical column is about 10–30 cm long and has an internal diameter of about 4 mm.

The use of fine particles increases the surface area of the stationary phase. This makes the separation efficient but it means that a high-pressure pump is necessary to force the solvent through the tightly packed column. As a result, the technique is sometimes called high-pressure liquid chromatography

An advantage of HPLC is that it is carried out at room temperature and so can analyse mixtures that would decompose on heating, such as many biological molecules. Organic molecules that break down on heating cannot be studied using gas chromatography. HPLC is also suitable for separating biological molecules such as proteins. Another important application of HPLC is to study urine or blood samples to investigate what happens to drugs as they are metabolised in the body.

Figure 19.32 A scientist checking a sample tube in front of a set of high-performance liquid chromatography (HPLC) columns.

Gas chromatography

Gas chromatography (GC) is a sensitive technique for analysing complex mixtures. The technique is used for compounds that vaporise on heating without decomposing. This type of chromatography not only separates the chemicals in a sample, but also gives a measure of how much of each is present.

In gas chromatography the mobile phase is a gas, commonly helium, argon or nitrogen, which carries the mixture of volatile chemicals through a long tube containing the stationary phase. The column is coiled inside an oven. Heating the column makes it possible to analyse any chemicals that turn to vapour at the temperature of the oven (Figure 19.33).

Figure 19.33 The main features of gas chromatography. The carrier gas takes the mixture of chemicals through the column, where they separate. As the compounds leave the column they are detected and measured.

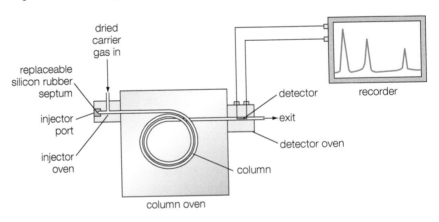

The analyst injects a small sample into the column where it enters the oven. Volatile solids are dissolved in a solvent before injection. The chemicals in the sample turn to gases and mix with the carrier gas. The gases then pass through the column.

The components in the mixture separate as they pass through the column. After a time the chemicals emerge one by one. They pass into a detector which sends a signal to a recorder as each compound appears. A series of peaks, one for each compound in the mixture, make up the chromatogram (Figure 19.34).

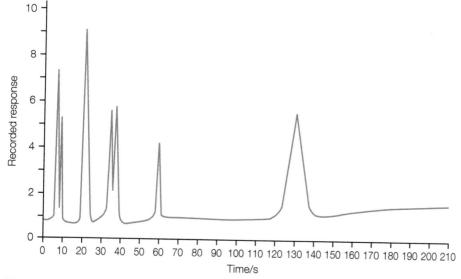

Figure 19.34 The printout from a gas chromatography instrument.

The position of a peak on a GC printout is a record of how long it takes for a compound to pass through the column. This is called the compound's **retention time**.

The area under each peak gives an indication of the relative amounts of the compounds in the mixture. If the peaks are narrow it is sufficient to measure the peak heights to get an indication of the relative amounts.

Some GC instruments have capillary columns. These are 20–60 metres long with a very small internal diameter. Capillary columns are often made of silica with an outer polymer coating. The stationary phase is the inner surface of the column, which adsorbs chemicals to a greater or lesser extent. The inner surface may be coated with a solid adsorbent or a thin film of a liquid.

Other GC columns are steel or glass tubes packed with a powder. The powder is an inert solid coated with a thin film of a liquid that has a high boiling temperature. In these columns the stationary phase is the liquid coating. Chemicals in the carrier gas separate in these columns because they differ in their solubility in the liquid of the stationary phase. When this type of column is used the technique is sometimes called gas–liquid chromatography.

Applications of gas chromatography include:

- tracking down the source of oil pollution from the pattern of peaks, which acts like a fingerprint for any batch of oil
- monitoring the presence of chemicals in industrial processes
- measuring the level of alcohol in blood samples from drivers
- detecting pesticides in river water.

Key term

The **retention time** in gas chromatography is the time it takes for a compound in a mixture to pass through a chromatographic column and reach the detector.

37 Suggest a reason for choosing helium, argon or nitrogen as the carrier gas in gas chromatography.

38 Refer to Figure 19.34.
   a) How many chemicals were there in the mixture?
   b) What was the retention time of the least abundant chemical in the mixture?

39 Why must the liquid for the stationary phase in gas–liquid chromatography have a high boiling temperature?

40 What are the implications of the fact that similar compounds may have very similar retention times in gas chromatography?

41 Explain why it is not possible to identify a previously unknown chemical by gas chromatography.

Activity

Forensic investigations of arson

Arsonists sometimes use flammable liquids such as petrol or paraffin to accelerate fires. Firefighters collect samples from the burnt remains which forensic scientists can analyse in the search for clues as to how the fire started (Figure 19.35).

Suitable samples for analysis come from areas where furniture and fittings have not been completely destroyed. Useful samples include carpet underlays, soil from pot plants, bedding, clothing and material collected from underneath floorboards.

1 Suggest a reason why firefighters collect only partially burned materials for analysis.
2 Suggest a reason why soil from pot plants can provide good evidence that there have been flammable liquids present.

Figure 19.35 Firefighters searching through the wreckage of a burnt-out house. They are looking for evidence of how the fire started to determine whether it was an accident or arson.

Analysts use solvents to extract chemicals from the samples and then investigate the solutions by gas chromatography. They compare the chromatograms with those from standard samples of common flammable substances (Figure 19.36).

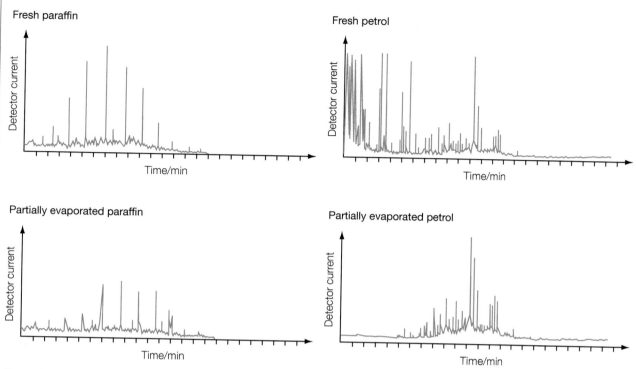

Figure 19.36 Gas chromatograms of standard samples of fuels.

3 The gas chromatogram for fresh petrol is very different from the chromatogram for fresh paraffin. Describe and explain the differences.

4 Suggest why the chromatogram for partially evaporated petrol differs from the chromatogram for fresh petrol.

The two chromatograms in Figure 19.37 show the results of analysing the chemicals from samples collected from a burnt-out house. Sample A was collected from the charred floorboards just inside the front door. Sample B came from the partly burned carpet under a table near the window of the front room.

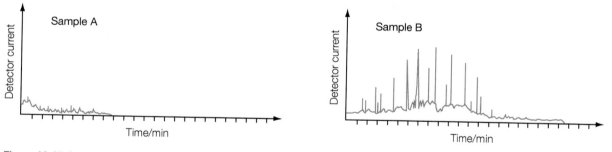

Figure 19.37 Gas chromatograms of chemical extracts from two samples collected from a burnt-out house.

5 What can you conclude from the chromatogram for sample A?

6 What can you conclude from the chromatogram for sample B?

7 Suggest the further evidence that the analysts would need to collect before deciding whether the house fire was an accident or the result of arson.

Chromatography combined with mass spectrometry

Coupling gas chromatography (GC) with mass spectrometry (MS) gives a very powerful system for separating, identifying and measuring complex mixtures of chemicals. The combined technique (GC–MS) is widely used in drug detection in sport and elsewhere, in the forensics investigation of fires, in environmental monitoring and in airport security (Figure 19.38). Some space probes carry tiny GC–MS systems for analysing samples collected in space or on the surface of planets (Figure 19.39).

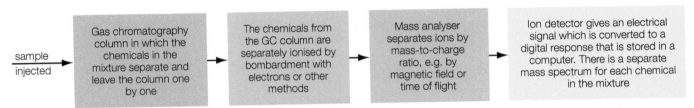

sample injected → Gas chromatography column in which the chemicals in the mixture separate and leave the column one by one → The chemicals from the GC column are separately ionised by bombardment with electrons or other methods → Mass analyser separates ions by mass-to-charge ratio, e.g. by magnetic field or time of flight → Ion detector gives an electrical signal which is converted to a digital response that is stored in a computer. There is a separate mass spectrum for each chemical in the mixture

Figure 19.38 A schematic diagram showing the key features of a GC-MS system.

GC–MS overcomes some of the limitations of gas chromatography. Similar compounds often have similar retention times in GC, which means that they cannot be identified by chromatography alone, even if the conditions are carefully standardised. Also, GC alone cannot identify any new chemicals because there are no standards that can be used to determine retention times under given conditions.

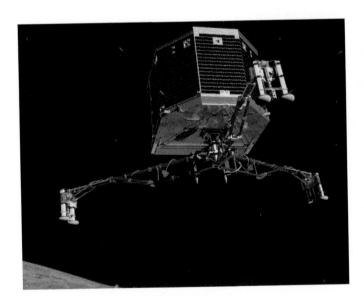

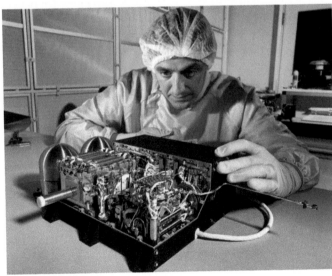

Figure 19.39 The Philae lander (left) from the Rosetta mission in November 2014 analysed samples of the comet 67P/Churyumov-Gerasimenko using a GC–MS system the size of a shoe box (right) and detected organic molecules just above the comet's surface.

GC–MS produces a mass spectrum for each of the chemicals separated on the GC column. These spectra can be used like fingerprints to identify the compounds because every chemical has a unique mass spectrum.

A computer receives the data from the GC–MS system. This computer can be linked to a library of spectra of known compounds. The computer compares the mass spectrum of each chemical in a mixture to mass spectra in the library. It automatically reports a list of likely identifications along with the probability that the matches are correct.

Solving a pollution problem with GC–MS

With complex mixtures, computers can help an analyst to look for a 'needle in a haystack'. Figure 19.40 shows the chromatogram from an investigation of the air in a home where the family was feeling very sick. During the analysis by GC-MS the computer stored 700 mass spectra as the mixture of chemicals emerged from the chromatography column.

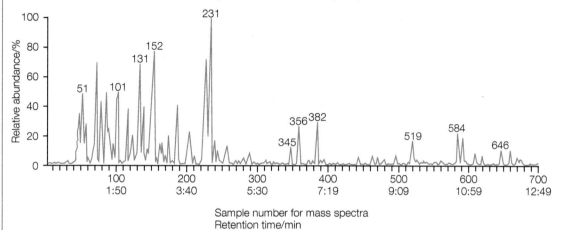

Figure 19.40 The gas chromatogram of chemicals sampled from the air in a house. The numbers 1–700 on the time axis indicate the points at which mass spectra were recorded and stored. Underneath these numbers are the retention times.

The analysts suspected that the chemicals causing the family's sickness might have come from petrol, so they asked the computer to plot a chromatogram showing only those chemicals producing a peak with mass-to-charge ratio of 91 in their spectra. The result is shown in Figure 19.41, which indicates that methylbenzene, ethylbenzene and three dimethylbenzenes were present in the mixture. These are all chemicals that are distinctive for the mixture of hydrocarbons found in petrol. With this evidence the investigators carried out further searches and tracked down the source of the petrol vapour.

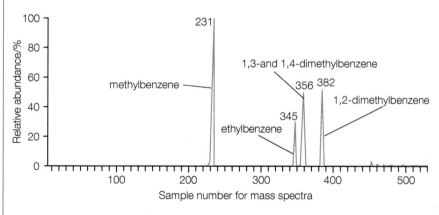

Figure 19.41 The GC-MS printout for the same sample as in Figure 19.40 but showing only the chemicals with a prominent peak with a mass-to-charge ratio of 91 in their mass spectra.

1 Why, in a mass spectrometer, does each chemical:
   a) have to be ionised
   b) pass through a region with electric and/or magnetic fields
   c) produce a spectrum with several peaks?

2 Suggest the identity of the ion fragment with a mass-to-charge ratio of 91 in the mass spectra of methylbenzene, $C_6H_5-CH_3$, and related compounds.

3 How does the computer identify a chemical with a mass spectrum recorded at a particular retention time?

4 Suggest two reasons why forensic scientists find GC-MS particularly valuable.

5 HPLC can also be combined with MS. Give an example of a sample that could be analysed by HPLC-MS but not by GC-MS and explain your choice.

Exam practice questions

1 a) The relative molecular mass of a compound is 74.1212. Use the data in Table 19.1 to confirm that the molecular formula of this compound is $C_4H_{10}O$ rather than $C_3H_6O_2$ or $C_2H_6N_2O$. *(2)*

b) Draw the four isomeric alcohols with molecular formula $C_4H_{10}O$ and predict the number of peaks in the ^{13}C NMR spectra of each compound. *(4)*

c) The compound in part (a) is unaffected by acidified potassium dichromate(VI). Identify this compound and justify your choice. *(2)*

2 Benzene was heated with a mixture of concentrated nitric and sulfuric acids and the major product obtained was found to be 1,3-dinitrobenzene.

a) Write an equation for the formation of 1,3-dinitrobenzene from benzene. *(1)*

b) Draw the structures of the three possible isomers of dinitrobenzene. *(3)*

c) Explain how ^{13}C NMR could be used to confirm that the major product was 1,3-dinitrobenzene. *(3)*

3 The diagram shows the mass spectra of two isomers: benzoic acid (benzenecarboxylic acid, C_6H_5COOH) and 3-hydroxybenzaldehyde (3-hydroxybenzenecarbaldehyde, HOC_6H_4CHO).

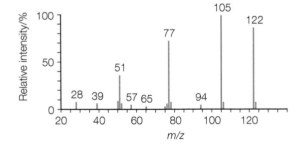

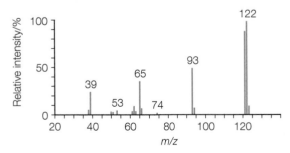

a) Identify the peaks at $m/z = 77$ and 105 in the top spectrum and $m/z = 93$ in the bottom spectrum. *(3)*

b) Hence match the compounds to the spectra and give your reasons. *(2)*

c) Describe two test-tube reactions that will confirm your choice. *(6)*

4 The proton NMR spectrum of a compound of carbon, oxygen and hydrogen is shown below.

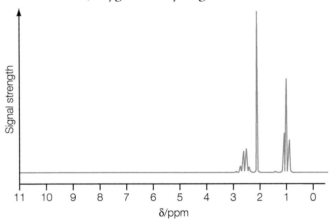

An integration trace gives the ratios for the peaks, as shown in the table.

Chemical shift/ppm	Relative values from the integration trace
1.0	0.9
2.1	0.9
2.5	0.6

a) i) How many chemical environments for hydrogen atoms are there in the molecule? *(1)*

ii) What is the ratio of the numbers of each type of hydrogen atom? *(1)*

b) Suggest likely chemical environments of the protons in the molecule with the help of the chart of chemical shift values in the Edexcel Data booklet. *(2)*

c) What can you deduce from the splitting patterns of the peaks at chemical shifts 1.0 and 2.5? *(2)*

d) The compound gives an orange precipitate with 2,4-dinitrophenylhydrazine but does not react with Fehling's solution or phosphorus(V) chloride. What does this tell you about the functional groups in the molecule? *(3)*

e) Suggest a structure for the compound *(1)*

5 Gas chromatography can be used to analyse the chemicals formed during the production of beer. Below is the GC chromatogram for a sample taken during beer making. The instrument was fitted with a capillary column with solid stationary phase.

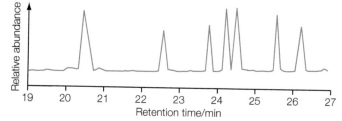

The table gives the retention times for GC under the conditions used to analyse the beer sample.

Compound	Retention time/min
Methanol	19.5
Ethanol	20.5
Propan-1-ol	20.9
Propan-2-ol	22.5
2-Methylpropan-1-ol	22.7
Butan-2-ol	24.6
Ethanal	20.2
Propanal	24.5
Butanal	25.5
Propanone	23.8
Ethanoic acid	24.2
Butanoic acid	26.2

a) Suggest a reason for the trend in the retention times of the four primary alcohols. *(2)*

b) Suggest a reason why 2-methylpropan-1-ol has a shorter retention time than butan-2-ol. *(2)*

c) i) Use the gas chromatogram and the table to identify the chemicals in the beer sample. *(3)*

ii) Which peaks were hard to identify and why? *(2)*

d) The mixture was analysed by GC–MS. Below is the mass spectrum of the fifth peak to emerge from the GC column. Use this mass spectrum to decide whether this peak is butan-2-ol or propanal and explain how you decided on your answer. *(4)*

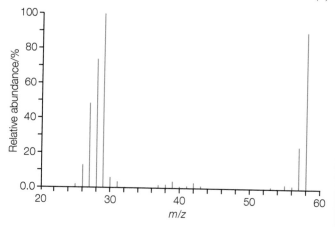

6 A naturally occurring dipeptide, A, has the molecular formula $C_7H_{14}O_3N_2$. The dipeptide is hydrolysed forming two amino acids, B and C, on heating with concentrated hydrochloric acid. The two amino acids can be separated by paper chromatography using a solvent in which B has an R_f value of 0.60 and C has an R_f value of 0.26.

a) Draw a labelled diagram, to scale, showing the original and final spots and the solvent front on the chromatogram. *(4)*

b) Amino acid B is chiral, but C is non-chiral.

i) Draw the displayed formula of C. *(1)*

ii) What is the molecular formula of amino acid B? *(1)*

iii) Draw a possible structural formula for B. *(2)*

c) Draw a possible structural formula for the dipeptide A. *(2)*

d) What procedure would you use to make the 'spots' of amino acids visible on the chromatogram? *(2)*

e) How would you show that a sample of amino acid B was chiral? *(2)*

7 Two isomeric ketones, X and Y, with the formula $C_5H_{10}O$ and unbranched carbon chains have the mass spectra shown below.

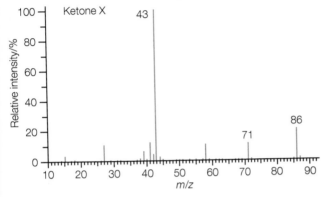

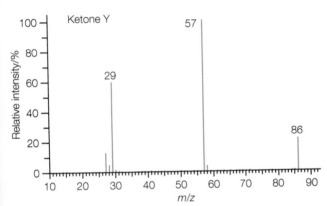

a) Draw the skeletal formulae of the two ketones and name them. *(2)*

b) Why do both spectra have peaks at *m/z* values of 86? *(2)*

c) i) Suggest possible identities for the four fragments in the two spectra with *m/z* values of 29, 43, 57 and 71. *(4)*

ii) Hence show which spectrum belongs to which compound. *(2)*

d) Use the chart of chemical shifts in the Edexcel Data booklet to predict:

i) the carbon-13 NMR spectrum of ketone X *(3)*

ii) the proton NMR spectrum of ketone Y. *(6)*

8 Identify the compound containing carbon, hydrogen and oxygen only that gives rise to the spectra below. Draw the displayed formula of the compound and name it. Give your reasoning and show how you can account for the key features in the three spectra. *(10)*

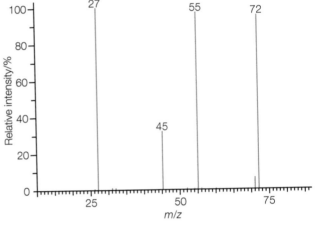

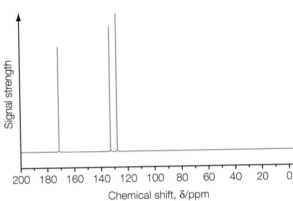

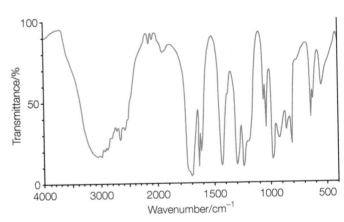

9 An organic compound Z contains carbon, hydrogen, oxygen and one other element. The carbon-13 NMR spectrum of Z has four peaks, showing that there are carbon atoms in a Z molecule in four distinct chemical environments.

The mass spectrum and the proton NMR spectrum of the compound are shown below.

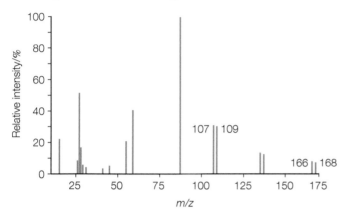

Mass spectrum of Z.

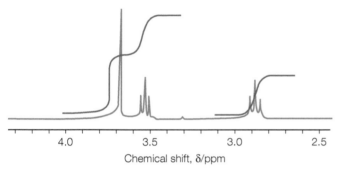

Proton NMR spectrum of Z.

a) What can you conclude from the pattern of the mass spectrum and, in particular, from the two peaks at 166 and 168 in the mass spectrum? *(2)*

b) The carbon-13 NMR spectrum suggests that there are four carbon atoms in the molecule. Use this information, your answer to (a) and the proton NMR spectrum to suggest a possible molecular formula for the compound. *(2)*

c) Measure the step heights on the integration trace of the proton NMR spectrum and explain what the values show. *(2)*

d) What can you conclude from the chemical shift values and splitting patterns in the proton NMR spectrum? *(4)*

e) Suggest a structure for the compound. *(1)*

f) Account for the peaks at $m/z = 107$ and 109 in the mass spectrum. *(1)*

10 The proton NMR spectrum of a compound with the formula $C_6H_{12}O_3$ is shown below.

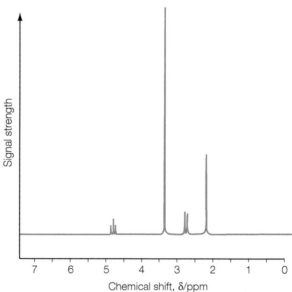

The integration trace gives the ratios for the peaks as shown in the table.

Chemical shift	Ratios of values from integration trace
2.2	1.8
2.7	1.2
3.4	3.6
4.8	0.6

a) i) How many chemical environments for hydrogen atoms are there in the molecule? *(1)*
   ii) What is the ratio of the numbers of each type of hydrogen atom? *(1)*

b) The compound gives an orange precipitate with 2,4-dinitrophenylhydrazine but does not react with Fehling's solution or phosphorus(v) chloride. What does this tell you about the functional groups in the molecule? *(2)*

c) Suggest likely chemical environments of the protons in the molecule with the help of a table of chemical shift values. *(3)*

d) i) What can you deduce from the splitting patterns of the peaks at chemical shifts 2.7 and 4.8? *(2)*
   ii) What can you deduce from the absence of splitting of the peaks at chemical shifts at 2.2 and 3.4? *(1)*

e) Suggest a structure for the compound. *(1)*

11 a) Compound A has the molecular formula $C_4H_8O_2$. The infrared spectrum of compound A is shown below.

Compound A

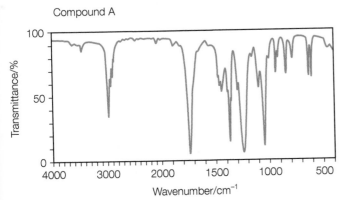

Infrared spectrum of compound A.

The proton NMR spectrum of A contains only three peaks. The table contains data about these peaks.

δ/ppm	Integration	Splitting
4.12	2	Quartet
2.04	3	Singlet
1.26	3	Triplet

Peak data for compound A.

Identify compound A and justify your answer. (6)

b) Compound B has the molecular formula $C_4H_8O_2$. The infrared spectrum of compound B is shown below.

Compound B

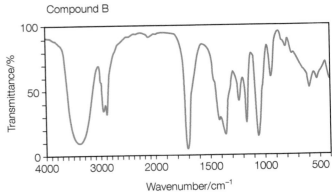

Infrared spectrum of compound B.

The proton NMR spectrum of B contains only four peaks. The table contains data about these peaks.

δ/ppm	Integration	Splitting
3.84	1.2	Triplet
3.40	0.6	Singlet
2.70	1.2	Triplet
2.20	1.8	Singlet

Peak data for compound B.

Identify compound B and justify your answer. (9)

Index

Note: page numbers in **bold** refer to illustrations.

Free online resources

Answers for the following features found in this book are available online:

- Test yourself questions
- Activities

You'll also find Practical skills sheets and Data sheets. Additionally there is an Extended glossary to help you learn the key terms and formulae you'll need in your exam.

You can also access extra chapters that will help you with the mathematical requirements of A Level chemistry and to prepare for written examinations.

- Mathematics in A Level chemistry
- Preparing for the exam

Scan the QR codes below for each chapter.

Alternatively, you can browse through all chapters at **www.hoddereducation.co.uk/EdexcelAChemistry2**

How to use the QR codes

To use the QR codes you will need a QR code reader for your smartphone/tablet. There are many free readers available, depending on the smartphone/tablet you are using. We have supplied some suggestions below, but this is not an exhaustive list and you should only download software compatible with your device and operating system. We do not endorse any of the third-party products listed below and downloading them is at your own risk.

- for iPhone/iPad, search the App store for Qrafter
- for Android, search the Play store for QR Droid
- for Blackberry, search Blackberry World for QR Scanner Pro
- for Windows/Symbian, search the Store for Upcode

Once you have downloaded a QR code reader, simply open the reader app and use it to take a photo of the code. You will then see a menu of the free resources available for that topic.

11 Equilibrium II

12 Acid–base equilibria

13.1 Lattice energy

13.2 Entropy

14 Redox II

18.1 Arenes – benzene compounds

15 Transition metals

18.2 Amines, amides, amino acids and proteins

16 Kinetics II

18.3 Organic synthesis

17.1 Chirality

19 Modern analytical techniques II

17.2 Carbonyl compounds

Mathematics in A Level chemistry

17.3 Carboxylic acids and their derivatives

Preparing for the exam

The periodic table of elements

Key

relative atomic mass
atomic symbol
name
atomic (proton) number

1.0
H
hydrogen
1

(1) 1	(2) 2	(3)	(4)	(5)	(6)	(7)	(8)	(9)	(10)	(11)	(12)	(13) 3	(14) 4	(15) 5	(16) 6	(17) 7	0(8) (18)
																	4.0 **He** helium 2
6.9 **Li** lithium 3	9.0 **Be** beryllium 4											10.8 **B** boron 5	12.0 **C** carbon 6	14.0 **N** nitrogen 7	16.0 **O** oxygen 8	19.0 **F** fluorine 9	20.2 **Ne** neon 10
23.0 **Na** sodium 11	24.3 **Mg** magnesium 12											27.0 **Al** aluminium 13	28.1 **Si** silicon 14	31.0 **P** phosphorus 15	32.1 **S** sulfur 16	35.5 **Cl** chlorine 17	39.9 **Ar** argon 18
39.1 **K** potassium 19	40.1 **Ca** calcium 20	45.0 **Sc** scandium 21	47.9 **Ti** titanium 22	50.9 **V** vanadium 23	52.0 **Cr** chromium 24	54.9 **Mn** manganese 25	55.8 **Fe** iron 26	58.9 **Co** cobalt 27	58.7 **Ni** nickel 28	63.5 **Cu** copper 29	65.4 **Zn** zinc 30	69.7 **Ga** gallium 31	72.6 **Ge** germanium 32	74.9 **As** arsenic 33	79.0 **Se** selenium 34	79.9 **Br** bromine 35	83.8 **Kr** krypton 36
85.5 **Rb** rubidium 37	87.6 **Sr** strontium 38	88.9 **Y** yttrium 39	91.2 **Zr** zirconium 40	92.9 **Nb** niobium 41	95.9 **Mo** molybdenum 42	[98] **Tc** technetium 43	101.1 **Ru** ruthenium 44	102.9 **Rh** rhodium 45	106.4 **Pd** palladium 46	107.9 **Ag** silver 47	112.4 **Cd** cadmium 48	114.8 **In** indium 49	118.7 **Sn** tin 50	121.8 **Sb** antimony 51	127.6 **Te** tellurium 52	126.9 **I** iodine 53	131.3 **Xe** xenon 54
132.9 **Cs** caesium 55	137.3 **Ba** barium 56	138.9 **La*** lanthanum 57	178.5 **Hf** hafnium 72	180.9 **Ta** tantalum 73	183.8 **W** tungsten 74	186.2 **Re** rhenium 75	190.2 **Os** osmium 76	192.2 **Ir** iridium 77	195.1 **Pt** platinum 78	197.0 **Au** gold 79	200.6 **Hg** mercury 80	204.4 **Tl** thallium 81	207.2 **Pb** lead 82	209.0 **Bi** bismuth 83	[209] **Po** polonium 84	[210] **At** astatine 85	[222] **Rn** radon 86
[223] **Fr** francium 87	[226] **Ra** radium 88	[227] **Ac**** actinium 89	[261] **Rf** rutherfordium 104	[262] **Db** dubnium 105	[266] **Sg** seaborgium 106	[264] **Bh** bohrium 107	[277] **Hs** hassium 108	[268] **Mt** meitnerium 109	[271] **Ds** damstadtium 110	[272] **Rg** roentgenium 111							

Elements with atomic numbers 112–116 have been reported but not fully authenticated

*Lanthanide series

140 **Ce** cerium 58	141 **Pr** praseodymium 59	144 **Nd** neodymium 60	[147] **Pm** promethium 61	150 **Sm** samarium 62	152 **Eu** europium 63	157 **Gd** gadolinium 64	159 **Tb** terbium 65	163 **Dy** dysprosium 66	165 **Ho** holmium 67	167 **Er** erbium 68	169 **Tm** thulium 69	173 **Yb** ytterbium 70	175 **Lu** lutetium 71

**Actinide series

232 **Th** thorium 90	[231] **Pa** protactinium 91	238 **U** uranium 92	[237] **Np** neptunium 93	[242] **Pu** plutonium 94	[243] **Am** americium 95	[247] **Cm** curium 96	[245] **Bk** berkelium 97	[251] **Cf** californium 98	[254] **Es** einsteinium 99	[253] **Fm** fermium 100	[256] **Md** mendelevium 101	[254] **No** nobelium 102	[257] **Lr** lawrencium 103